Intermediate
Heat
Transfer

MECHANICAL ENGINEERING
A Series of Textbooks and Reference Books

Founding Editor

L. L. Faulkner

Columbus Division, Battelle Memorial Institute
and Department of Mechanical Engineering
The Ohio State University
Columbus, Ohio

1. *Spring Designer's Handbook*, Harold Carlson
2. *Computer-Aided Graphics and Design*, Daniel L. Ryan
3. *Lubrication Fundamentals*, J. George Wills
4. *Solar Engineering for Domestic Buildings*, William A. Himmelman
5. *Applied Engineering Mechanics: Statics and Dynamics*, G. Boothroyd and C. Poli
6. *Centrifugal Pump Clinic*, Igor J. Karassik
7. *Computer-Aided Kinetics for Machine Design*, Daniel L. Ryan
8. *Plastics Products Design Handbook, Part A: Materials and Components; Part B: Processes and Design for Processes*, edited by Edward Miller
9. *Turbomachinery: Basic Theory and Applications*, Earl Logan, Jr.
10. *Vibrations of Shells and Plates*, Werner Soedel
11. *Flat and Corrugated Diaphragm Design Handbook*, Mario Di Giovanni
12. *Practical Stress Analysis in Engineering Design*, Alexander Blake
13. *An Introduction to the Design and Behavior of Bolted Joints*, John H. Bickford
14. *Optimal Engineering Design: Principles and Applications*, James N. Siddall
15. *Spring Manufacturing Handbook*, Harold Carlson
16. *Industrial Noise Control: Fundamentals and Applications*, edited by Lewis H. Bell
17. *Gears and Their Vibration: A Basic Approach to Understanding Gear Noise*, J. Derek Smith
18. *Chains for Power Transmission and Material Handling: Design and Applications Handbook*, American Chain Association
19. *Corrosion and Corrosion Protection Handbook*, edited by Philip A. Schweitzer
20. *Gear Drive Systems: Design and Application*, Peter Lynwander
21. *Controlling In-Plant Airborne Contaminants: Systems Design and Calculations*, John D. Constance
22. *CAD/CAM Systems Planning and Implementation*, Charles S. Knox
23. *Probabilistic Engineering Design: Principles and Applications*, James N. Siddall
24. *Traction Drives: Selection and Application*, Frederick W. Heilich III and Eugene E. Shube
25. *Finite Element Methods: An Introduction*, Ronald L. Huston and Chris E. Passerello

26. *Mechanical Fastening of Plastics: An Engineering Handbook*, Brayton Lincoln, Kenneth J. Gomes, and James F. Braden
27. *Lubrication in Practice: Second Edition*, edited by W. S. Robertson
28. *Principles of Automated Drafting*, Daniel L. Ryan
29. *Practical Seal Design*, edited by Leonard J. Martini
30. *Engineering Documentation for CAD/CAM Applications*, Charles S. Knox
31. *Design Dimensioning with Computer Graphics Applications*, Jerome C. Lange
32. *Mechanism Analysis: Simplified Graphical and Analytical Techniques*, Lyndon O. Barton
33. *CAD/CAM Systems: Justification, Implementation, Productivity Measurement*, Edward J. Preston, George W. Crawford, and Mark E. Coticchia
34. *Steam Plant Calculations Manual*, V. Ganapathy
35. *Design Assurance for Engineers and Managers*, John A. Burgess
36. *Heat Transfer Fluids and Systems for Process and Energy Applications*, Jasbir Singh
37. *Potential Flows: Computer Graphic Solutions*, Robert H. Kirchhoff
38. *Computer-Aided Graphics and Design: Second Edition*, Daniel L. Ryan
39. *Electronically Controlled Proportional Valves: Selection and Application*, Michael J. Tonyan, edited by Tobi Goldoftas
40. *Pressure Gauge Handbook*, AMETEK, U.S. Gauge Division, edited by Philip W. Harland
41. *Fabric Filtration for Combustion Sources: Fundamentals and Basic Technology*, R. P. Donovan
42. *Design of Mechanical Joints*, Alexander Blake
43. *CAD/CAM Dictionary*, Edward J. Preston, George W. Crawford, and Mark E. Coticchia
44. *Machinery Adhesives for Locking, Retaining, and Sealing*, Girard S. Haviland
45. *Couplings and Joints: Design, Selection, and Application*, Jon R. Mancuso
46. *Shaft Alignment Handbook*, John Piotrowski
47. *BASIC Programs for Steam Plant Engineers: Boilers, Combustion, Fluid Flow, and Heat Transfer*, V. Ganapathy
48. *Solving Mechanical Design Problems with Computer Graphics*, Jerome C. Lange
49. *Plastics Gearing: Selection and Application*, Clifford E. Adams
50. *Clutches and Brakes: Design and Selection*, William C. Orthwein
51. *Transducers in Mechanical and Electronic Design*, Harry L. Trietley
52. *Metallurgical Applications of Shock-Wave and High-Strain-Rate Phenomena*, edited by Lawrence E. Murr, Karl P. Staudhammer, and Marc A. Meyers
53. *Magnesium Products Design*, Robert S. Busk
54. *How to Integrate CAD/CAM Systems: Management and Technology*, William D. Engelke
55. *Cam Design and Manufacture: Second Edition*; with cam design software for the IBM PC and compatibles, disk included, Preben W. Jensen
56. *Solid-State AC Motor Controls: Selection and Application*, Sylvester Campbell
57. *Fundamentals of Robotics*, David D. Ardayfio
58. *Belt Selection and Application for Engineers*, edited by Wallace D. Erickson
59. *Developing Three-Dimensional CAD Software with the IBM PC*, C. Stan Wei
60. *Organizing Data for CIM Applications*, Charles S. Knox, with contributions by Thomas C. Boos, Ross S. Culverhouse, and Paul F. Muchnicki

61. *Computer-Aided Simulation in Railway Dynamics*, by Rao V. Dukkipati and Joseph R. Amyot
62. *Fiber-Reinforced Composites: Materials, Manufacturing, and Design*, P. K. Mallick
63. *Photoelectric Sensors and Controls: Selection and Application*, Scott M. Juds
64. *Finite Element Analysis with Personal Computers*, Edward R. Champion, Jr., and J. Michael Ensminger
65. *Ultrasonics: Fundamentals, Technology, Applications: Second Edition, Revised and Expanded*, Dale Ensminger
66. *Applied Finite Element Modeling: Practical Problem Solving for Engineers*, Jeffrey M. Steele
67. *Measurement and Instrumentation in Engineering: Principles and Basic Laboratory Experiments*, Francis S. Tse and Ivan E. Morse
68. *Centrifugal Pump Clinic: Second Edition, Revised and Expanded*, Igor J. Karassik
69. *Practical Stress Analysis in Engineering Design: Second Edition, Revised and Expanded*, Alexander Blake
70. *An Introduction to the Design and Behavior of Bolted Joints: Second Edition, Revised and Expanded*, John H. Bickford
71. *High Vacuum Technology: A Practical Guide*, Marsbed H. Hablanian
72. *Pressure Sensors: Selection and Application*, Duane Tandeske
73. *Zinc Handbook: Properties, Processing, and Use in Design*, Frank Porter
74. *Thermal Fatigue of Metals*, Andrzej Weronski and Tadeusz Hejwowski
75. *Classical and Modern Mechanisms for Engineers and Inventors*, Preben W. Jensen
76. *Handbook of Electronic Package Design*, edited by Michael Pecht
77. *Shock-Wave and High-Strain-Rate Phenomena in Materials*, edited by Marc A. Meyers, Lawrence E. Murr, and Karl P. Staudhammer
78. *Industrial Refrigeration: Principles, Design and Applications*, P. C. Koelet
79. *Applied Combustion*, Eugene L. Keating
80. *Engine Oils and Automotive Lubrication*, edited by Wilfried J. Bartz
81. *Mechanism Analysis: Simplified and Graphical Techniques, Second Edition, Revised and Expanded*, Lyndon O. Barton
82. *Fundamental Fluid Mechanics for the Practicing Engineer*, James W. Murdock
83. *Fiber-Reinforced Composites: Materials, Manufacturing, and Design, Second Edition, Revised and Expanded*, P. K. Mallick
84. *Numerical Methods for Engineering Applications*, Edward R. Champion, Jr.
85. *Turbomachinery: Basic Theory and Applications, Second Edition, Revised and Expanded*, Earl Logan, Jr.
86. *Vibrations of Shells and Plates: Second Edition, Revised and Expanded*, Werner Soedel
87. *Steam Plant Calculations Manual: Second Edition, Revised and Ex panded*, V. Ganapathy
88. *Industrial Noise Control: Fundamentals and Applications, Second Edition, Revised and Expanded*, Lewis H. Bell and Douglas H. Bell
89. *Finite Elements: Their Design and Performance*, Richard H. MacNeal
90. *Mechanical Properties of Polymers and Composites: Second Edition, Revised and Expanded*, Lawrence E. Nielsen and Robert F. Landel
91. *Mechanical Wear Prediction and Prevention*, Raymond G. Bayer

92. *Mechanical Power Transmission Components*, edited by David W. South and Jon R. Mancuso
93. *Handbook of Turbomachinery*, edited by Earl Logan, Jr.
94. *Engineering Documentation Control Practices and Procedures*, Ray E. Monahan
95. *Refractory Linings Thermomechanical Design and Applications*, Charles A. Schacht
96. *Geometric Dimensioning and Tolerancing: Applications and Techniques for Use in Design, Manufacturing, and Inspection*, James D. Meadows
97. *An Introduction to the Design and Behavior of Bolted Joints: Third Edition, Revised and Expanded*, John H. Bickford
98. *Shaft Alignment Handbook: Second Edition, Revised and Expanded*, John Piotrowski
99. *Computer-Aided Design of Polymer-Matrix Composite Structures*, edited by Suong Van Hoa
100. *Friction Science and Technology*, Peter J. Blau
101. *Introduction to Plastics and Composites: Mechanical Properties and Engineering Applications*, Edward Miller
102. *Practical Fracture Mechanics in Design*, Alexander Blake
103. *Pump Characteristics and Applications*, Michael W. Volk
104. *Optical Principles and Technology for Engineers*, James E. Stewart
105. *Optimizing the Shape of Mechanical Elements and Structures*, A. A. Seireg and Jorge Rodriguez
106. *Kinematics and Dynamics of Machinery*, Vladimír Stejskal and Michael Valášek
107. *Shaft Seals for Dynamic Applications*, Les Horve
108. *Reliability-Based Mechanical Design*, edited by Thomas A. Cruse
109. *Mechanical Fastening, Joining, and Assembly*, James A. Speck
110. *Turbomachinery Fluid Dynamics and Heat Transfer*, edited by Chunill Hah
111. *High-Vacuum Technology: A Practical Guide, Second Edition, Revised and Expanded*, Marsbed H. Hablanian
112. *Geometric Dimensioning and Tolerancing: Workbook and Answerbook*, James D. Meadows
113. *Handbook of Materials Selection for Engineering Applications*, edited by G. T. Murray
114. *Handbook of Thermoplastic Piping System Design*, Thomas Sixsmith and Reinhard Hanselka
115. *Practical Guide to Finite Elements: A Solid Mechanics Approach*, Steven M. Lepi
116. *Applied Computational Fluid Dynamics*, edited by Vijay K. Garg
117. *Fluid Sealing Technology*, Heinz K. Muller and Bernard S. Nau
118. *Friction and Lubrication in Mechanical Design*, A. A. Seireg
119. *Influence Functions and Matrices*, Yuri A. Melnikov
120. *Mechanical Analysis of Electronic Packaging Systems*, Stephen A. McKeown
121. *Couplings and Joints: Design, Selection, and Application, Second Edition, Revised and Expanded*, Jon R. Mancuso
122. *Thermodynamics: Processes and Applications*, Earl Logan, Jr.
123. *Gear Noise and Vibration*, J. Derek Smith
124. *Practical Fluid Mechanics for Engineering Applications*, John J. Bloomer
125. *Handbook of Hydraulic Fluid Technology*, edited by George E. Totten
126. *Heat Exchanger Design Handbook*, T. Kuppan

127. *Designing for Product Sound Quality*, Richard H. Lyon
128. *Probability Applications in Mechanical Design*, Franklin E. Fisher and Joy R. Fisher
129. *Nickel Alloys,* edited by Ulrich Heubner
130. *Rotating Machinery Vibration: Problem Analysis and Troubleshooting*, Maurice L. Adams, Jr.
131. *Formulas for Dynamic Analysis,* Ronald L. Huston and C. Q. Liu
132. *Handbook of Machinery Dynamics*, Lynn L. Faulkner and Earl Logan, Jr.
133. *Rapid Prototyping Technology: Selection and Application*, Kenneth G. Cooper
134. *Reciprocating Machinery Dynamics: Design and Analysis*, Abdulla S. Rangwala
135. *Maintenance Excellence: Optimizing Equipment Life-Cycle Decisions*, edited by John D. Campbell and Andrew K. S. Jardine
136. *Practical Guide to Industrial Boiler Systems*, Ralph L. Vandagriff
137. *Lubrication Fundamentals: Second Edition, Revised and Expanded*, D. M. Pirro and A. A. Wessol
138. *Mechanical Life Cycle Handbook: Good Environmental Design and Manufacturing*, edited by Mahendra S. Hundal
139. *Micromachining of Engineering Materials*, edited by Joseph McGeough
140. *Control Strategies for Dynamic Systems: Design and Implementation*, John H. Lumkes, Jr.
141. *Practical Guide to Pressure Vessel Manufacturing*, Sunil Pullarcot
142. *Nondestructive Evaluation: Theory, Techniques, and Applications*, edited by Peter J. Shull
143. *Diesel Engine Engineering: Thermodynamics, Dynamics, Design, and Control*, Andrei Makartchouk
144. *Handbook of Machine Tool Analysis*, Ioan D. Marinescu, Constantin Ispas, and Dan Boboc
145. *Implementing Concurrent Engineering in Small Companies*, Susan Carlson Skalak
146. *Practical Guide to the Packaging of Electronics: Thermal and Mechanical Design and Analysis*, Ali Jamnia
147. *Bearing Design in Machinery: Engineering Tribology and Lubrication*, Avraham Harnoy
148. *Mechanical Reliability Improvement: Probability and Statistics for Experimental Testing*, R. E. Little
149. *Industrial Boilers and Heat Recovery Steam Generators: Design, Applications, and Calculations*, V. Ganapathy
150. *The CAD Guidebook: A Basic Manual for Understanding and Improving Computer-Aided Design*, Stephen J. Schoonmaker
151. *Industrial Noise Control and Acoustics*, Randall F. Barron
152. *Mechanical Properties of Engineered Materials*, Wolé Soboyejo
153. *Reliability Verification, Testing, and Analysis in Engineering Design*, Gary S. Wasserman
154. *Fundamental Mechanics of Fluids: Third Edition*, I. G. Currie
155. *Intermediate Heat Transfer*, Kau-Fui Vincent Wong

Additional Volumes in Preparation

HVAC Water Chillers and Cooling Towers: Fundamentals, Application, and Operations, Herbert W. Stanford III

Handbook of Turbomachinery: Second Edition, Revised and Expanded, Earl Logan, Jr., and Ramendra Roy

Progressing Cavity Pumps, Downhole Pumps, and Mudmotors, Lev Nelik

Gear Noise and Vibration: Second Edition, Revised and Expanded, J. Derek Smith

Piping and Pipeline Engineering: Design, Construction, Maintenance, Integrity, and Repair, George A. Antaki

Turbomachinery: Design and Theory: Rama S. Gorla and Aijaz Ahmed Khan

Mechanical Engineering Software

Spring Design with an IBM PC, Al Dietrich

Mechanical Design Failure Analysis: With Failure Analysis System Software for the IBM PC, David G. Ullman

Intermediate Heat Transfer

Kau-Fui Vincent Wong

University of Miami
Coral Gables, Florida, U.S.A.

CRC Press
Taylor & Francis Group
Boca Raton London New York

CRC Press is an imprint of the
Taylor & Francis Group, an **informa** business

CRC Press
Taylor & Francis Group
6000 Broken Sound Parkway NW, Suite 300
Boca Raton, FL 33487-2742

First issued in paperback 2019

© 2003 by Taylor & Francis Group, LLC
CRC Press is an imprint of Taylor & Francis Group, an Informa business

ISBN-13: 978-0-8247-4236-2 (hbk)
ISBN-13: 978-0-367-39556-8 (pbk)

Library of Congress Cataloging-in-Publication Data
A catalog record for this book is available from the Library of Congress.

Visit the Taylor & Francis Web site at
http://www.taylorandfrancis.com

and the CRC Press Web site at
http://www.crcpress.com

This book is dedicated to all my students.

Preface

Heat transfer is a required course for mechanical, aerospace, nuclear, and chemical engineering undergraduates. Advanced courses in heat transfer are also required for most graduate students in the same four fields. Generally, these advanced courses are named "Conduction," "Convection," or "Radiation". In many universities, however, there is an Intermediate Heat Transfer course for seniors and first year graduate engineering students. This is their textbook. For this second course in heat transfer; this volume evolved from a series of lecture notes developed by the author in almost twenty-five years of teaching a graduate-level course of this type at the Mechanical Engineering Department, University of Miami, Coral Gables, Florida, U.S.A.

There are several distinguishing features that set *Intermediate Heat Transfer* apart from existing texts on the subject. A discussion of these features follows.

A major difficulty of engineering graduates in studying heat transfer at the advanced level is the big jump of knowledge required. It is difficult for them to comprehend the advanced material because the introductory course did not adequately prepare them. This book bridges this gap in knowledge about heat transfer.

Intermediate Heat Transfer provides the necessary background for seniors and first-year graduate students so that they can independently read and understand research papers in heat transfer. The one standard course in heat transfer, usually at the junior undergraduate level, does not cover enough material for the student to be cognizant of most of the archival material on heat transfer. This book fills the knowledge gap of heat transfer for most of our engineering graduates who have taken only one course in heat transfer.

The special features of the book are as follows:

- Confusing and unnecessary details have been eliminated. Only essential facts and methods have been provided to make the work of the student easier.
- Wherever a two-dimensional treatment is effective in imparting the knowledge, it is used instead of a three-dimensional treatment, to ensure that the material is more understandable.

- Examples and problems are used to drive the concepts home. Many advanced books on heat transfer are devoid of examples and problems. Many introductory books on heat transfer have too many problems that are not classified or grouped together, so such books do not build on the information embodied in each of their problems. Students see such problems as many dissociated problems; even though they illustrate certain points, they do not systematically increase the students' own bodies of knowledge.

- A chapter on numerical analysis in conduction and one on numerical analysis in convection are considered important features of the book. In this modern age of computers, the typical student uses software to help in solving heat transfer problems. For many, the software is a "black box", a clever one, but nonetheless a black box. These chapters are written to enlighten the students about the methods and techniques used and programmed into the black boxes.

These special features of the book are geared towards making heat transfer a less difficult field for graduate engineers who are returning to undertake graduate studies, when their own undergraduate experience only included one course in heat transfer.

Kau-Fui Vincent Wong

Contents

PREFACE.. v

SECTION I   FUNDAMENTALS OF HEAT TRANSFER

1   Fundamentals of Heat Transfer   1
     1.1 Conduction   1
     1.2 Convection   4
     1.3 Radiation   6
     1.4 Combined Convection and Radiation   9

SECTION II   CONDUCTION

2   The General Heat Conduction Equation   13
     2.1 Introduction   13
     2.2 Governing Differential Equation of
           Heat Conduction   13
     2.3 Laplace Equation   15
     2.4 Poisson's Equation   15
     2.5 Fourier's Equation   15
     2.6 Initial and Boundary Conditions   16
     2.7 First Kind (Dirichlet) Boundary Conditions   17
     2.8 Second Kind (Neumann) Boundary Conditions   17
     2.9 Third Kind (Robin or Mixed) Boundary
           Conditions   18
     2.10 Temperature-Dependent Thermal
           Conductivity   19
     2.11 Dimensionless Heat Conduction Numbers   21

3   One-Dimensional Steady-State Heat Conduction   27
     3.1 The Slab (One-Dimensional Cartesian Coordinates) 27
     3.2 The Cylinder (One-Dimensional Cylindrical
           Coordinates)   28
     3.3 The Sphere (One-Dimensional Spherical

Coordinates)   29
3.4 Thermal Resistance   30
3.5 Conduction Through a Slab from One Fluid to Another Fluid   32
3.6 Composite Medium   33
3.7 Thermal Contact Resistance   35
3.8 Standard Method of Determining the Thermal Conductivity in a Solid   36
3.9 Temperature-Dependent Thermal Conductivity   37
3.10 Critical Radius of Insulation   43
3.11 Effects of Radiation   44
3.12 One-Dimensional Extended Surfaces   45

4 Two-Dimensional Steady and One-Dimensional Unsteady Heat Conduction   61
4.1 Method of Separation of Variables   61
4.2 Steady-State Heat Conduction in a Rectangular Region   64
4.3 Heat Flow   67
4.4 Separation into Simpler Problems   70
4.5 Summary of Steps Used in the Method of Separation of Variables   71
4.6 Steady-State Heat Conduction in a Two-Dimensional Fin   72
4.7 Transient Heat Conduction in a Slab   76

5 Numerical Analysis in Conduction   83
5.1 Introduction   83
5.2 Finite Difference of Derivatives   83
5.3 Finite Difference Equations for 2-D Rectangular Steady-State Conduction   86
5.4 Finite Difference Representation of Boundary Conditions   88
5.5 Solution of Finite Difference Equations   92
5.6 Finite Difference Equations for 1-D Unsteady-State Conduction in Rectangular Coordinates   101
5.7 Finite Difference of 2-D Unsteady-State Problems in Rectangular Coordinates   109

5.8 Finite Difference Method Applied in Cylindrical
Coordinates 113
5.9 Truncation Errors and Round-Off Errors in Finite
Difference Method 115
5.10 Stability and Convergence 117

SECTION III   CONVECTION

6 Equations for Convection 125
6.1 Continuity 125
6.2 Momentum Equations 126
6.3 Energy Equation 131
6.4 Summary of Governing Equations 137
6.5 Dimensionless Numbers 140
6.6 The Boundary Layer Equations 143

7 External Forced Convection 149
7.1 Momentum Integral Method of Analysis 149
7.2 Integral Method of Analysis for Energy
Equation 154
7.3 Hydrodynamic and Thermal Boundary Layers on a
Flat Plate, Where Heating Starts at $x = x_o$ 157
7.4 Similarity Solution 179

8 Internal Forced Convection 191
8.1 Couette Flow 191
8.2 Heat Transfer and Velocity Distribution in
Hydrodynamically and Thermally Developed
Laminar Flow in Conduits 197
8.3 The Circular Tube Thermal-Entry-Length, with
Hydrodynamically Fully Developed Laminar Flow 205
8.4 The Rectangular Duct Thermal-Entry Length, with
Hydrodynamically Fully Developed Laminar Flow 211

9 Natural Convection 221
9.1 Boundary Layer Concept for Free Convection 221
9.2 Similarity Solution: Boundary Layer with Uniform
Temperature 225

9.3   Similarity Solution: Boundary Layer with
      Uniform Heat Flux   230
9.4   Integral Method of Solution   234

10   Numerical Analysis in Convection   243
10.1 Introduction   243
10.2 Burgers Equation   244
10.3 Convection Equations   245
10.4 Numerical Algorithms   247
10.5 Boundary Layer Equations   253
10.6 Convection with Incompressible Flow   259
10.7 Two-Dimensional Convection with
      Incompressible Flow   261
10.8 Convection in a Two-Dimensional Porous
      Medium   262

SECTION IV   RADIATION

11   Basic Relations of Radiation   275
11.1 Thermal Radiation   275
11.2 Radiation Intensity and Blackbody   276
11.3 Reflectivity, Absorptivity, Emissivity and
      Transmissivity   281
11.4 Kirchhoff's Law of Radiation   284

12   Radiative Exchange in a Non-Participating Medium   287
12.1 Radiative Exchange Between Two Differential
      Area Elements   287
12.2 Concept of View Factor   291
12.3 Properties of Diffuse View Factors   293
12.4 Determination of Diffuse View Factor by Contour
      Integration   295
12.5 Relations Between View Factors   304
12.6 Diffuse View Factor Between an Elemental
      Surface and an Infinitely Long Strip   307
12.7 Diffuse View Factor Algebra   308

13   Radiation Exchange in Long Enclosures   319
13.1 Diffuse View Factor in Long Enclosures   319

Contents

14   Radiation with Other Modes of Heat Transfer   357
     14.1   Introduction   357
     14.2   Radiation with Conduction   357
     14.3   Radiation with Convection   368
     14.4   Radiation with Conduction and Convection   377

Appendix A   Bessel Functions   389

Appendix B   Physical Constants and Thermophysical Properties   397

Index   *401*

1

Fundamentals of Heat Transfer

Heat transfer takes place only when there is a temperature difference. Heat energy moves from a higher potential (measured by temperature) to a lower potential.

Usually experimental methods are used to measure heat transfer. Basic transfer mechanisms commonly recognized are conduction and radiation. Convection is often used as a third classification. The convection classification is also used in the current work.

1.1 Conduction

Conduction is the heat transfer mechanism that takes place when the media is stationary. It can take place in solids, gases and liquids. It may be thought of as the transfer of energy from the more energetic particles of a medium to nearby particles that are less energetic owing to particle interactions. Conduction heat transfer is described macroscopically by Fourier's law, which is

$$Q = -kA\nabla T \tag{1.1}$$

where k is a property of the medium (substance) called the thermal conductivity, and A is the area through which the heat is flowing. In conduction, Fourier's law states that the driving potential is the temperature gradient. The energy flows in the direction of decreasing temperature; hence, the negative sign. Gold and iron are two substances with large values of thermal conductivity, and they are called good thermal conductors. Others with small conductivities like air and wood are good thermal insulators.

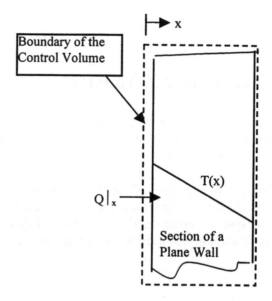

Figure 1.1   Conduction in a plane wall.

In one dimension, Fourier's law becomes

$$Q\big|_x = -kA \frac{dT}{dx}\bigg|_x .$$

(1.2)

Fourier's law is applied to conduction in a plane wall, as shown in Fig. 1.1.   The heat flow $Q\big|_x$ is the heat energy transfer in the x direction. The area is the cross-area of the control volume normal to the heat flow, i.e. the x direction.   In this case where the temperature gradient is constant, the equation becomes

$$Q\big|_x = -kA \frac{\Delta T}{\Delta x}\bigg|_x .$$

(1.3)

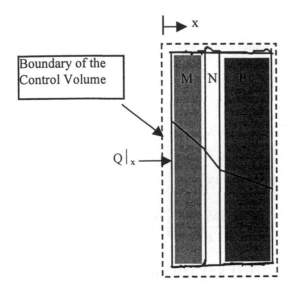

Figure 1.2 Application of Fourier's law to a plane wall with three layers.

A control volume drawn around a plane wall with three layers is shown in Fig. 1.2. Three different materials, M, N and P, of different thicknesses, Δx_M, Δx_N and Δx_P, make up the three layers. The thermal conductivities of the three substances are k_M, k_N and k_P respectively. By the conservation of energy, the heat conducted through each of the three layers have to be equal. Fourier's law for this control volume gives

$$Q|_x = -k_M A \frac{\Delta T_M}{\Delta x_M} = -k_N A \frac{\Delta T_N}{\Delta x_N} = -k_P A \frac{\Delta T_P}{\Delta x_P} \qquad (1.4)$$

In Fig. 1.2, for the particular illustration shown, the temperature gradient in N is observed to be larger than that in M, which is in turn larger than that in P. It can be deduced from Eq. (1.4) that $k_N < k_M < k_P$. In other words, material P is the best conductor and material N is the worst conductor of the three.

Example 1.1

Problem: The thickness of a silver plate is 6 cm. One face is at 300°C and the other is at 0°C. The thermal conductivity for silver is 369 W/(m.°C) at 150°C. Find the heat conducted through the plate.

Solution
    From Fourier's law,

$$q\big|_x = \frac{Q\big|_x}{A} = -k\frac{\Delta T}{\Delta x}$$

$$= \frac{-(369)(0-300)}{6\times10^{-2}}\ \frac{W}{m.°C}.°C.\frac{1}{m}$$

$$= 1.845 \text{ MW/m}^2.$$

1.2 Convection

When heat transfer occurs in a moving medium, it is usually called convection. As an example, heat energy can be transferred from a solid plane surface at one temperature to an adjacent moving fluid at another temperature.   Consider the case shown in Fig. 1.3. Heat energy is conducted from the solid to the moving fluid, where energy is carried away by the combined effects of conduction within the fluid and the bulk motion of the fluid. The heat transfer from the solid system to the fluid can be expressed by the empirical equation

$$Q = h_{conv}A(\ T_s - T_f\ ) \tag{1.5}$$

known as Newton's law of cooling. In this equation, A is the surface area, T_f is the fluid temperature away from the surface (bulk or mean temperature of the fluid), and T_s is the temperature of the surface. For $T_f < T_s$, heat energy flows from the solid to the fluid. The proportionality factor h_{conv} is referred to as the heat transfer coefficient. This coefficient is not a thermodynamic property. It is an empirical parameter that may be found experimentally. This coefficient incorporates into the heat transfer relationship the geometry of the system, the fluid flow pattern near the surface and the fluid properties. If fans, pumps or turbines make the fluid flow, the value of the heat transfer coefficient is normally greater than when relatively slow bouyancy-driven motion takes place. These two classifications are called forced and free (or natural) convection, respectively. In Fig. 1.3, it is observed that the heat transfer between the solid plane and the fluid may be described by Fourier's conduction law according to

$$Q\big|_y = -k_f A \frac{dT}{dy}\bigg|_{wall} \cong -k_f A \frac{\Delta T}{\Delta y}\bigg|_{wall} = \frac{k_f}{\Delta y} A(T_s - T_f). \qquad (1.6)$$

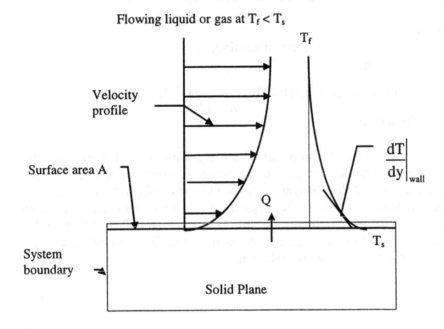

Figure 1.3 Relationship between Fourier's law and Newton's law of cooling.

Examining Eqs. (1.5) and (1.6), it can be deduced that the heat transfer coefficient h_{conv} is an approximation of the quantity $k_f/\Delta y$. The thermal conductivity of the fluid k_f is a thermodynamic property, however Δy is a function of the fluid flow pattern near the surface, the geometry of the system and the fluid properties. As stated previously, the heat transfer coefficient is not a thermodynamic property, but an empirical parameter.

It is also clear that Newton's law of cooling is a special case of Fourier's law. The foregoing provides the reason for only two commonly recognized basic heat transfer mechanisms. But owing to the complexity of fluid motion, convection is often treated as a separate heat transfer mode.

Example 1.2

Problem: Air at 18°C blows over a hot plate at 210°C. The convection heat transfer coefficient is 32 W/(m².°C). The dimensions of the plate are 10 by 40 cm. Determine the heat transfer.

Solution

From Newton's law of cooling,

$Q = h_{conv}A(T_s-T_f)$

$$Q = 32(0.04)(210-18)\frac{W}{m^2.^oC}.m^2.^oC = 246\,W.$$

1.3 Radiation

Thermal radiation can take place without a medium. Thermal radiation may be understood as being emitted by matter that is a consequence of the changes in the electronic configurations of its atoms or molecules. Solid surfaces, gases, and liquids all emit, absorb, and transmit thermal radiation to different extents. The radiation heat transfer phenomenon is described macroscopically by a modified form of the Stefan-Boltzmann law, which is

$$Q = \varepsilon\sigma A\, T_s^4 \tag{1.7}$$

where σ is the Stefan-Boltzmann constant and ε is a property of the surface that characterizes how effectively the surface radiates ($0 \le \varepsilon \le 1$). This property is called the emissivity of the surface. The Stefan-Boltzmann constant σ is 5.669 x 10^{-8} W/(m².K⁴) or 0.1714 x 10^{-8} Btu/(ft².°R⁴.h). Thermal radiation takes place according to the fourth power of the absolute temperature of the surface, T_s. The net thermal radiation heat transfer between two surfaces, in general, involves complex relationships among the properties of the surfaces, their orientations with respect to each other, and the extent to which the medium in between scatters, emits, and absorbs thermal radiation, and other factors.

Consider a simple two-body radiation problem with a non-participating intervening medium. The radiation equation is

$$Q = \varepsilon_1 \sigma A_1 (T_1^4 - T_2^4) \qquad (1.8)$$

where Q is the net radiation heat transfer from body 1 with the higher temperature T_1 to body 2 of the lower temperature T_2. The parameters ε_1 and A_1 are the emissivity of body 1 and the effective area of radiation for body 1, respectively. Likewise, the net radiation can be expressed as

$$Q = \varepsilon_2 \sigma A_2 (T_2^4 - T_1^4) . \qquad (1.9)$$

In this equation, ε_2 and A_2 are the emissivity of body 2 and the effective area of radiation for body 2, respectively. Since radiation heat transfer has to obey the principle of the conservation of energy, the magnitude of the net heat radiated in Eq. (1.8) has to equal the magnitude of the heat radiated in Eq. (1.9).

Consider the situation where there are more than two bodies. The radiation equation may be modified further as

$$Q_{1\text{-}2} = \varepsilon_1 \sigma F_{1-2} A_1 (T_1^4 - T_2^4) \qquad (1.10)$$

where $F_{1\text{-}2}$ is the view factor of body 2 from body 1. This view factor $F_{1\text{-}2}$ is the percentage of the thermal radiation from body 1 that arrives at body 2. In addition, if body 1 also sees body 3, then $F_{1\text{-}3}$ is the percentage of the thermal radiation from body 1 that arrives at body 3. Consequently the net radiation from body 1 to body 3 is given by

$$Q_{1\text{-}3} = \varepsilon_1 \sigma F_{1-3} A_1 (T_1^4 - T_3^4) . \qquad (1.11)$$

It follows that the sum of the view factors from body 1 is equal to unity, i.e., $F_{1\text{-}2} + F_{1\text{-}3} = 1$.

Describing the net radiation from body 2 to the other bodies, the following equations apply:

$$Q_{2\text{-}1} = \varepsilon_2 \sigma F_{2-1} A_2 (T_2^4 - T_1^4) \qquad\qquad (1.12)$$

and $\qquad Q_{2\text{-}3} = \varepsilon_2 \sigma F_{2-3} A_2 (T_2^4 - T_3^4) \qquad\qquad (1.13)$

For body 3,

$$Q_{3\text{-}1} = \varepsilon_3 \sigma F_{3-1} A_3 (T_3^4 - T_1^4) \qquad\qquad (1.14)$$

and $\qquad Q_{3\text{-}2} = \varepsilon_3 \sigma F_{3-2} A_3 (T_3^4 - T_2^4). \qquad\qquad (1.15)$

From the definition of view factors, it is clear that the sum of the view factors from body 2 is equal to unity, i.e., $F_{2\text{-}1} + F_{2\text{-}3} = 1$, and also that the sum of the view factors from body 3 is equal to unity, i.e., $F_{3\text{-}1} + F_{3\text{-}2} = 1$. In the three-body radiation problem discussed, it has been assumed that none of the bodies can radiate to itself. Expressed technically, the view factor of body 1 to itself is zero, and this is the case for bodies 2 and 3 also, that is, $F_{1\text{-}1} = F_{2\text{-}2} = F_{3\text{-}3} = 0$.

Example 1.3

Problem: Two extremely large parallel plates at 800°C and 500°C exchange heat via radiation. Determine the heat transfer per unit area. Assume that $\varepsilon = 1$ for the plates.

Solution
Assumptions:
(1) The medium in between does not participate in the heat transfer.
Analysis:
    From the Stefan-Boltzmann law,

$$\frac{q}{A} = \sigma\varepsilon\left(T_1^4 - T_2^4\right)$$

$$\frac{q}{A} = \left(5.669x10^{-8}\right)\left(1073.15^4 - 773.15^4\right)\frac{W}{m^2 K^4}.K^4$$

$$q/A = 54.93 \text{ kW/m}^2.$$

Example 1.4

Problem: The view factor of body 1 to 2 is 0.5 and that from body 1 to itself is 0.1. If the temperatures of bodies 1,2 and 3 are 450, 325 and

225°C respectively, calculate the heat radiated from body 1 to body 3. It is known that all the emissivities are 0.75. The surface area of body 1 is 1.5 m².

Solution
Assumptions:
(1) The medium in between does not participate in the heat transfer.
Analysis:
From the Stefan-Boltzmann's law,

$$Q_{1\text{-}3} = \varepsilon_1 \sigma F_{1\text{-}3} A_1 (T_1^4 - T_3^4)$$
$$F_{1\text{-}1} + F_{1\text{-}2} + F_{1\text{-}3} = 1.$$

Thus, $F_{1\text{-}3} = 1 - 0.1 - 0.5 = 0.4$

$$Q_{1\text{-}3} = 0.75\left(5.669 \times 10^{-8}\right)(0.4)(1.5)\left(723.15^4 - 498.15^4\right) \frac{W}{m^2 K^4} . m^2 . K^4 = 5405\, W.$$

1.4   Combined Convection and Radiation

Heat transfer by convection may be added to the heat transfer by radiation. So the total heat transfer from a body 1 to the surrounding fluid f is the sum of the convective heat transfer and the radiative heat transfer between body 1 and body 2, say. The convective heat transfer from body 1 to the fluid f is

$$Q_{conv} = h_{conv} A_{conv} (T_1 - T_f). \tag{1.16}$$

Similarly, the radiative heat transfer from body 1 to body 2 is

$$Q_{rad} = \varepsilon \sigma F_{1\text{-}2} A_{rad} (T_1^4 - T_2^4). \tag{1.17}$$

The total heat transfer from body 1 by convection and radiation is thus

$$Q_{total} = h_{conv} A_{conv} (T_1 - T_f) + \varepsilon \sigma F_{1\text{-}2} A_{rad} (T_1^4 - T_2^4). \tag{1.18}$$

Please observe that the area available for convective heat transfer is not necessarily the same as that available for radiative heat transfer between bodies 1 and 2. Additionally, the temperature of the surrounding fluid T_f is in general not the same as the fluid of the body 2, T_2.

Consider the situation where T_1 is near T_2, Eq. (1.16) may be simplified according to

$$
\begin{aligned}
Q_{rad} &= \varepsilon \sigma F_{1-2} A_{rad} (T_1^2 + T_2^2)(T_1^2 - T_2^2) \\
&= \varepsilon \sigma F_{1-2} A_{rad} (T_1^2 + T_2^2)(T_1 - T_2)(T_1 + T_2) \\
&\approx \varepsilon \sigma F_{1-2} A_{rad} 4T_1^3 (T_1 - T_2) \\
&= h_{rad} A_{rad} (T_1 - T_2)
\end{aligned}
\tag{1.19}
$$

The radiative heat transfer has been approximated and expressed to be similar to Newton's law of cooling, with a heat transfer coefficient due to radiation. The radiative heat transfer coefficient, like the convective heat transfer coefficient, is not a property of either bodies. The total heat transfer from body 1 is then expressed as

$$
Q_{total} = h_{conv} A_{conv} (T_1 - T_f) + h_{rad} A_{rad} (T_1 - T_2).
\tag{1.20}
$$

Example 1.5

Problem: For a body that is being considered, the convection heat transfer coefficient to the adjacent air is 33 W/(m^2.°C), and the radiative heat transfer coefficient from this body to another body is approximately 36 W/(m^2.°C). If the temperature of the first body is 188° C , that of the adjacent air is 22°C, determine the temperature of the second body so that the heat transferred by convection is equal in magnitude to the heat transferred by radiation. The area ratio A_{conv}:A_{rad} is 1:1.2.

Solution

Assume T_2 is the temperature of the second body. It is given that $Q_{conv} = Q_{rad}$

$$
h_{conv} A_{conv} (T_1 - T_f) = h_{rad} A_{rad} (T_1 - T_2)
$$

$$
\frac{(T_1 - T_2)}{(T_1 - T_f)} = \frac{h_{conv} A_{conv}}{h_{rad} A_{rad}}
$$

$$
T_2 = T_1 - (T_1 - T_f) \frac{h_{conv} A_{conv}}{h_{rad} A_{rad}}
$$

$$T_2 = 188°C - 166°C \frac{33(1)}{36(1.2)} = 61.2°C.$$

The temperature of the second body is 61.2°C.

PROBLEMS

Conduction

1.1. The thickness of a copper plate is 8 cm. One surface is at 450°C and the other is at 150°C. The thermal conductivity for copper is 369 W/(m.°C) at 300°C. Find the heat conducted through the plate in MW/m².

1.2. The thickness of a silver plate is 7 cm. The higher temperature surface is at 440°C. The thermal conductivity for silver can be taken to be 362 W/(m.°C). The heat conducted through the plate is 2 MW/m², determine the temperature of the other surface.

1.3. The wall is a third as thick as the building insulation. The thermal conductivities of the wall and the building insulation are in the ratio 3:1. If the temperature drop across the wall is 3°C, find the temperature drop across the insulation.

Convection

1.4. Air at 25°C flows over a plate at 350°C. The convection heat transfer coefficient is 30 W/(m².°C). The plate is 50 by 90 cm. Determine the heat transfer in kW.

1.5. Air at 24°C moves over a plate at 100°C. The dimensions of the plate are 25 cm by 50 cm. If the heat transfer is 300 W, compute the convection heat transfer coefficient between the air and the plate.

1.6. Carbon dioxide at 15°C moves over a hot plate at 250°C, such that the convection heat transfer coefficient is 30 W/(m².°C). If the heat transfer is 3 kW, calculate the area of the plate.

Radiation

1.7. Two extremely big parallel plates at 900°C and 300°C exchange heat by radiation. Calculate the heat transfer per unit area. Assume that $\varepsilon = 1$ for the plates.

1.8. Two extremely big parallel plates exchange heat via radiation at the rate of 148 kW/m². The hotter plate is at 1000°C. Determine the temperature of the cooler plate. Assume that $\varepsilon = 1$ for the plates.

1.9. Body 1 sees bodies 2 and 3, besides itself. The view factor of body 1 to 2 is 0.2, and that to itself is 0.45. The temperature of body 1 is 550°C, and the heat radiated from body 1 to body 3 is 1000 W. Compute the temperature of body 3. All the emissivities are 0.6. The surface area of body 1 is 0.5 m^2.

1.10. The view factor of body 1 to 2 is 0.4. When the temperatures of bodies 1, 2 and 3 are 700, 400 and 250°C respectively, the heat radiated from body 1 to body 3 is 5000 W. All the emissivities are 0.5. Body 1 has a surface area of 0.6 m^2. Calculate the view factor of body 1 to itself.

Combined Convection and Radiation

1.11. Consider an oven in which the convection heat transfer coefficient to the adjacent air is 32 W/(m^2.°C), and the radiative heat transfer coefficient from this oven to another oven is 40 W/(m^2.°C). If the temperature of the oven under discussion is 220°C, and that of the second oven is 80°C, find the adjacent air temperature when the heat transfer by convection is equal in magnitude to the heat transfer by radiation. Assume that the area ratio A_{conv}:A_{rad} is 1:1.1.

1.12. In this chapter, only heat radiation with a nonparticipating medium has been discussed. Write an essay or have a discussion session regarding the phenomena when the medium in between participates in the heat transfer.

Heat Transfer Fundamentals

It is energy transfer by conduction
It is energy transfer by convection
It is energy transfer by radiation
It includes conduction, convection with radiation.

Conduction heat transfer follows Fourier's law
Convection heat transfer follows Newton's law
Radiation follows Stefan-Boltzmann's law
It is a fact that heat transfer follows laws.

K.V. Wong

2

The General Heat Conduction Equation

2.1 Introduction

For isotropic and homogeneous media, the conductive heat flux is given by Fourier's heat conduction law as

$$\vec{q} = -k\,\vec{\nabla}T \tag{2.1}$$

where k is the thermal conductivity of the medium and T is the temperature.

2.2 Governing Differential Equation of Heat Conduction

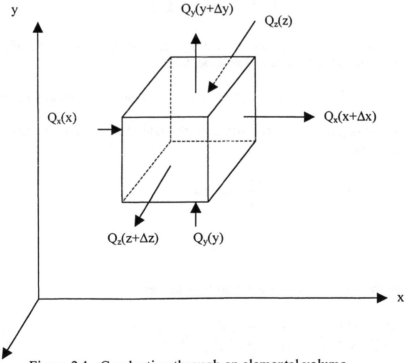

Figure 2.1   Conduction through an elemental volume.

The conservation of energy for conduction through an elemental volume is (I) Net rate of heat entering by conduction + (II) Rate of energy generated internally = (III) Rate of increase of internal energy.

Consider conduction in the x direction:

$$Q_x(x) = q_x \Delta y \Delta z$$

and $\quad Q_x(x + \Delta x) = (q_x + \dfrac{\partial q_x}{\partial x} \Delta x + ...) \Delta y \Delta z.$

Hence, net rate of heat entering in the x direction is $-\dfrac{\partial q_x}{\partial x} \Delta x \Delta y \Delta z.$

Similarly, the net rate of heat entering the y direction is $-\dfrac{\partial q_y}{\partial y} \Delta x \Delta y \Delta z,$

and that entering the z direction is $-\dfrac{\partial q_z}{\partial z} \Delta x \Delta y \Delta z.$ The net rate of heat entering by conduction is thus

$$(I) \qquad = -\left(\frac{\partial q_x}{\partial x} + \frac{\partial q_y}{\partial y} + \frac{\partial q_z}{\partial z} \right) \Delta x \Delta y \Delta z .$$

If g(x,y,z,t) is the rate of energy generation (within the elemental volume) per unit time and volume, then

$$(II) \qquad = g(x,y,z,t) \Delta x \Delta y \Delta z \qquad \text{is the rate of energy generation.}$$

Assuming a constant specific heat, the rate of increase of internal energy is given by

$$(III) \qquad = \rho C_P \frac{\partial T}{\partial t} \Delta x \Delta y \Delta z.$$

Therefore, $\qquad -\vec{\nabla} \cdot \vec{q} + g = \rho C_P \dfrac{\partial T}{\partial t}.$ $\qquad\qquad$ (2.2)

In three-dimensional Cartesian coordinates,

$$\frac{\partial}{\partial x}\left(k \frac{\partial T}{\partial x} \right) + \frac{\partial}{\partial y}\left(k \frac{\partial T}{\partial y} \right) + \frac{\partial}{\partial z}\left(k \frac{\partial T}{\partial z} \right) + g = \rho C_P \frac{\partial T}{\partial t}. \qquad (2.3)$$

The full conduction equation is Eq. (2.2), that is, conduction with heat generation. The general heat conduction equations with variable thermal conductivity, in the three principal coordinate systems are listed in Table 2.1. When the thermal conductivity is constant, the first term of Eq. (2.3) becomes the Laplacian of the temperature, T. The Laplacians of the temperature in the three principal coordinate systems are listed in Table 2.2.   There are three other special forms of the conduction equation with constant thermal conductivity, as listed below.

2.3 Laplace Equation

This is for constant k, steady state heat transfer so that the term in $\dfrac{\partial}{\partial t}$ is zero, and no heat generation or g = 0.

$$\nabla^2 T = 0 \qquad\qquad (2.4)$$

where $\nabla^2 T$ is the Laplacian of the temperature.

2.4 Poisson's Equation

This is for constant k and steady state heat transfer so that the term in $\dfrac{\partial}{\partial t}$ is zero.

$$\nabla^2 T + \frac{g}{k} = 0. \qquad\qquad (2.5)$$

2.5 Fourier's Equation

This is for constant k and no heat generation or g is zero.

$$\nabla^2 T = \frac{1}{\alpha}\frac{\partial T}{\partial t} \qquad\qquad (2.6)$$

The parameter α is the thermal diffusivity, $\alpha = k/\rho C_p$.

Table 2.1 Heat conduction equation with variable thermal conductivity in the three principal coordinate systems.

Coordinate system	$\nabla.(k\nabla T) + g = \rho C_p \dfrac{\partial T}{\partial t}$
Rectangular	$\dfrac{\partial}{\partial x}\left(k\dfrac{\partial T}{\partial x}\right) + \dfrac{\partial}{\partial y}\left(k\dfrac{\partial T}{\partial y}\right) + \dfrac{\partial}{\partial z}\left(k\dfrac{\partial T}{\partial z}\right) + g = \rho C_p \dfrac{\partial T}{\partial t}$
Cylin-drical	$\dfrac{1}{r}\dfrac{\partial}{\partial r}\left(kr\dfrac{\partial T}{\partial r}\right) + \dfrac{1}{r^2}\dfrac{\partial}{\partial \phi}\left(k\dfrac{\partial T}{\partial \phi}\right) + \dfrac{\partial}{\partial z}\left(k\dfrac{\partial T}{\partial z}\right) + g = \rho C_p \dfrac{\partial T}{\partial t}$
Spherical	$\dfrac{1}{r^2}\dfrac{\partial}{\partial r}\left(kr^2\dfrac{\partial T}{\partial r}\right) + \dfrac{1}{r^2 \sin \theta}\dfrac{\partial}{\partial \theta}\left(k \sin \theta \dfrac{\partial T}{\partial \theta}\right) + \dfrac{1}{r^2 \sin^2 \theta}\dfrac{\partial}{\partial \phi}\left(k\dfrac{\partial T}{\partial \phi}\right) + g$ $= \rho C_p \dfrac{\partial T}{\partial t}$

Table 2.2 The Laplacian of temperature in the three principal coordinate systems.

Coordinate System	$\nabla^2 T$
Rectangular	$\dfrac{\partial^2 T}{\partial x^2} + \dfrac{\partial^2 T}{\partial y^2} + \dfrac{\partial^2 T}{\partial z^2}$
Cylindrical	$\dfrac{\partial^2 T}{\partial r^2} + \dfrac{1}{r}\dfrac{\partial T}{\partial r} + \dfrac{1}{r^2}\dfrac{\partial^2 T}{\partial \phi^2} + \dfrac{\partial^2 T}{\partial z^2}$
Spherical	$\dfrac{1}{r^2}\dfrac{\partial}{\partial r}\left(r^2\dfrac{\partial T}{\partial r}\right) + \dfrac{1}{r^2 \sin \theta}\dfrac{\partial}{\partial \theta}\left(\sin \theta \dfrac{\partial T}{\partial \theta}\right) + \dfrac{1}{r^2 \sin^2 \theta}\dfrac{\partial^2 T}{\partial \phi^2}$

2.6 Initial and Boundary Conditions

To find the solutions to various conduction problems, we need boundary conditions in space and time since both the temperature T and the heat generation term g are functions of x, y, z and time t. In general, there are seven constants of integration. There is the first-order derivative with respect to the time variable and second-order derivatives with respect to each space variable. The number of conditions for each independent variable is equal to the order of the highest derivative of that variable in the equation. Hence, one initial condition is required for all time dependent problems; two boundary conditions are needed for each coordinate.

The spatial boundary conditions may be classified into three principal classes: the first kind or Dirichlet boundary conditions, the second kind or Neumann boundary conditions, and the third kind or Robin boundary conditions.

2.7 First Kind (Dirichlet) Boundary Conditions

Here, the temperatures are known at the boundaries.

$$T\left(\vec{x},t\right)\Big|_{surface} = T_s \tag{2.7}$$

An example of the first kind of boundary conditions for one-dimensional heat conduction is

$$T(x,t)\big|_{x=0} = T_o \quad \text{and} \quad T(x,t)\big|_{x=L} = T_L.$$

An example of the first kind of boundary conditions for two-dimensional heat conduction is

$$T(x,y,t)\big|_{x=0} = T_o(y) \quad \text{and} \quad T(x,y,t)\big|_{x=L} = T_L(y) \qquad \text{where } T_o \text{ and } T_L \text{ are}$$
prescribed functions of y.   If these functions are zero, these boundary conditions are called first kind homogeneous boundary conditions.

2.8 Second Kind (Neumann) Boundary Conditions

Here, the heat fluxes are known at the boundaries.

$$q_s = -k\frac{\partial T}{\partial x}\Big|_{surface} \quad \text{is known.} \tag{2.8}$$

An example of the second kind of boundary conditions for one-dimensional heat conduction is

$$\frac{\partial T}{\partial x}\bigg|_{x=0} = \frac{-q_1(y)}{k} = f_1(y) \text{ where } f_1 \text{ is a prescribed function of y.}$$

If this function is zero, the boundary condition is called the second kind homogeneous boundary condition.

2.9 Third Kind (Robin or Mixed) Boundary Conditions

Here, the convection heat transfer coefficients are known at the boundaries.

$$q = h\Delta T = -k\frac{\partial T}{\partial \eta} \quad \text{is known.} \tag{2.9}$$

An example of the third kind of boundary conditions for one-dimensional heat conduction is

$$h_1(T_\infty - T_{x=0}) = -k\frac{\partial T}{\partial x}\bigg|_{x=0} \quad \text{or} \quad \left[-k\frac{\partial T}{\partial x}\bigg|_{x=0} + h_1 T_{x=0}\right] = h_1 T_\infty = f_1$$

where f_1 is a prescribed function of y.

Other boundary conditions include nonlinear type boundary conditions. When there is radiation, phase change or a transient heat transfer at the boundary, the boundary conditions are nonlinear in nature.

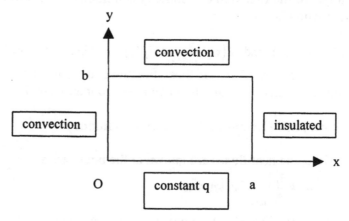

Figure 2.2   Sketch for Example 2.1.

Example 2.1

Problem: For a steady-state heat conduction problem with heat generation in a rectangular medium, write the governing equation and the mathematical representation of the boundary conditions. For x = 0, there is convection with heat transfer coefficient h_1. For x = a, the boundary is insulated. For y = 0, there is constant heat flux q. For y = b, there is convection with heat transfer coefficient h_2.

Solution

The governing energy conservation equation is

$$\frac{\partial^2 T}{\partial x^2} + \frac{\partial^2 T}{\partial y^2} + \frac{g}{k} = 0 \quad \text{for } 0 \le x \le a, 0 \le y \le b.$$

The boundary conditions are

$$-k\frac{\partial T}{\partial x} + h_1 T = h_1 T_\infty \quad \text{at x = 0} \tag{i}$$

$$\frac{\partial T}{\partial x} = 0 \quad \text{at } x = a \tag{ii}$$

$$-k\frac{\partial T}{\partial y} = q \quad \text{at y = 0} \tag{iii}$$

$$-k\frac{\partial T}{\partial x} + h_2 T = h_2 T_\infty \quad \text{at y = b.} \tag{iv}$$

2.10 Temperature-Dependent Thermal Conductivity

When the thermal conductivity is dependent on temperature, the general heat conduction equation is

$$\nabla \cdot \{k(T)\nabla T\} + g(\vec{x},t) = \rho C_p \frac{\partial T}{\partial t}. \tag{2.10}$$

Equation (2.10) is a nonlinear equation and difficult to solve. Equation (2.10) may be reduced to a linear differential equation by introducing a new temperature function θ by means of the Kirchhoff transformation as

$$\theta = \frac{1}{k_r} \int_{T_r}^{T} k(\hat{T}) d\hat{T}$$

(2.11)

where T_r is a convenient reference temperature and $k_r = k(T_r)$. It follows from Eq. (2.11) that

$$\nabla \theta = \frac{k(T)}{k_r} \nabla T$$

(2.12)

and

$$\frac{\partial \theta}{\partial t} = \frac{k(T)}{k_r} \frac{\partial T}{\partial t}.$$

(2.13)

Thus, Eq. (2.10) can be written as

$$\nabla^2 \theta + \frac{g\left(\vec{x}, t\right)}{k_r} = \frac{1}{\alpha} \frac{\partial \theta}{\partial t}.$$

(2.14)

If the thermal diffusivity is constant, Eq. (2.14) is linear. If the thermal diffusivity is not constant, then Eq. (2.14) is not nonlinear. The dependence of the thermal diffusivity on temperature can generally be neglected compared to that of the thermal conductivity, for many solids. If the thermal diffusivity is assumed to be independent of temperature, and thus a constant, Eq. (2.14) is not dissimilar to the heat conduction equation with constant k. The transformed equation may be solved with the usual techniques, as long as the boundary conditions can also be transformed. Boundary conditions of the first and second kind can be transformed; boundary conditions of the third kind usually cannot be transformed. Equations with boundary conditions of the third kind are generally solved using numerical methods.

For steady-state problems, Eq. (2.14) is a linear differential equation regardless of the behavior of the thermal diffusivity. Hence, the equation may be solved with the methods for linear equations.

2.11   Dimensionless Heat Conduction Numbers

By transforming the heat conduction equations to nondimensional form, the number of variables may be reduced.   Consider a slab in the region $0 \leq x \leq L$ with constant thermal properties, which is initially at a uniform temperature T_i.   For times t greater than zero, the boundary at x = 0 is kept at a uniform temperature T_1 and the boundary at x = L loses heat by convection to a fluid at temperature T_2 with a heat transfer coefficient h. Heat is generated within the slab at a rate of g W/m^3.   The governing equation of this problem is

$$\frac{\partial^2 T}{\partial x^2} + \frac{g}{k} = \frac{1}{\alpha}\frac{\partial T}{\partial t} \qquad \text{for } t > 0, \text{ in region } 0 \leq x \leq L. \qquad (2.15)$$

The initial condition is

$$T(x, t = 0) = T_i \qquad \text{in region } 0 \leq x \leq L. \qquad (2.16)$$

The boundary conditions are

$$T(x = 0, t) = T_1 \qquad \text{for } t > 0 \qquad (2.17)$$

$$k\frac{\partial T}{\partial x} + hT = T_2 \quad \text{at } x = L, \text{ for } t > 0. \qquad (2.18)$$

The following dimensionless variables are defined, using given quantities as reference values:

$$X = x/L = \text{dimensionless space coordinate} \qquad (2.19)$$

$$\theta = \frac{T - T_2}{T_i - T_2} = \text{dimensionless temperature} \qquad (2.20)$$

These dimensionless variables are introduced into Eqs. (2.15)-(2.18).

$$\frac{\partial^2 \theta}{\partial X^2} + \frac{gL^2}{(T_i - T_2)k} = \frac{\partial \theta}{\partial (\alpha t / L^2)} \qquad \text{for t > 0, in region } 0 \le X \le 1 \quad (2.21)$$

$$\theta(X, t = 0) = 1 \text{ in region } 0 \le X \le 1 \tag{2.22}$$

$$\theta(X = 0, t) = \theta_1 \quad \text{for t > 0} \tag{2.23}$$

$$\frac{\partial \theta}{\partial X} + \frac{hL}{k}\theta = 0 \quad \text{at } X = 1, \text{ for t > 0} \tag{2.24}$$

Introducing dimensionless parameters,

$$\text{Bi} \equiv \frac{hL}{k} = \text{Biot number} \tag{2.25}$$

$$\tau \equiv \frac{\alpha t}{L^2} = \text{Fourier number} = \text{Fo} \tag{2.26}$$

$$G \equiv \frac{gL^2}{k(T_i - T_2)} = \text{dimensionless heat generation} \tag{2.27}$$

Eqs. (2.21)-(2.24) become more compact, and are written as

$$\frac{\partial^2 \theta}{\partial X^2} + G = \frac{\partial \theta}{\partial \tau} \quad \text{for } \tau > 0, \text{ in region } 0 \le X \le 1 \tag{2.28}$$

$$\theta(X, \tau = 0) = 1 \text{ in region } 0 \le X \le 1 \tag{2.29}$$

$$\theta(X = 0, \tau) = \theta_1 \quad \text{for } \tau > 0 \tag{2.30}$$

$$\frac{\partial \theta}{\partial X} + Bi\theta = 0 \qquad \qquad \text{at } X = 1, \text{ for } \tau > 0 \tag{2.31}$$

The Fourier number and the Biot number are commonly used heat transfer numbers. The Biot number is the ratio of the heat transfer coefficient to the unit conductance of a solid over the characteristic length.

$$Bi = \frac{hL}{k} = \frac{h}{k/L} = \frac{\text{heat transfer coefficient at the surface of solid}}{\text{internal conductance of solid across length L}}. \quad (2.32)$$

The Fourier number is the ratio of heat conduction across a distance in a given volume to the rate of heat storage in that volume. It can be written as

$$\tau = \frac{\alpha t}{L^2} = \frac{k(1/L)L^2}{\rho C_p L^3 / t} = \frac{\text{rate of heat conduction across L in volume } L^3}{\text{rate of heat storage in volume } L^3}.$$

$$(2.33)$$

PROBLEMS

2.1. Consider the one-dimensional, steady-state heat conduction in a hollow cylinder with constant thermal conductivity in the region $c \leq r \leq d$. Heat generation is a rate of g_r W/m^3. Heat is convected away by fluids flowing on the inside and the outside of the hollow cylinder. Assume that the heat transfer coefficients are h_c and h_d on the inside and outside, and the fluid temperatures on the inside and the outside are T_c and T_d, respectively. Formulate the mathematical expression of this problem.

2.2. Consider the one-dimensional, steady-state heat conduction in a hollow sphere with constant thermal conductivity in the region $c \leq r \leq d$. Heat generation is a rate of g_r W/m^3. Heat is supplied to the inside of the hollow sphere at a rate of q_1 W/m^2. Heat is convected at the surface at $r = d$ into a medium at temperature T_m with a heat transfer coefficient of h_m. Formulate the mathematical expression of this problem.

2.3. In the absence of internal heat sources or sinks, under steady-state conditions, the two surfaces of a slab are kept at constant uniform temperatures T_a and T_b respectively. Show that the rate of heat conduction through the slab is constant.

2.4.   If the thickness of the slab is t and its thermal conductivity is k, derive an expression for the temperature distribution in the slab in Prob. 2.3.

2.5.   In rectangular coordinates, the heat conduction equation with constant thermal conductivity is

$$\frac{\partial^2 T}{\partial x^2} + \frac{\partial^2 T}{\partial y^2} + \frac{\partial^2 T}{\partial z^2} + \frac{g}{k} = \frac{1}{\alpha}\frac{\partial T}{\partial t}$$

Derive the corresponding heat conduction equation in (i) cylindrical coordinates and (ii) spherical coordinates, using coordinate transformations.

2.6.   The steady-state temperature distribution (in °C) in a slab at steady-state is provided by $T = 222 - 250\ x^2$, where x is the distance in meters along the width of the slab and measured from the surface maintained at 222°C. The thermal conductivity of the slab is 35 W/(m.K), and the thickness of the slab is 0.18 m. Calculate the heat fluxes at the two surfaces of the plate.

2.7.   If T_a and T_b are constants, show that the one-dimensional Fourier conduction equation with the following initial and boundary conditions has a unique solution:
$T(x,0) = T_i(x)$ $T(0,t) = T_a$ $T(L,t) = T_b$.

2.8.   Transform the one-dimensional, nonlinear Poisson's equation , with the boundary conditions given, into a linear problem in terms of a new temperature function defined as

$$\theta(x) = \frac{1}{k_r}\int_{T_r}^{T(x)} k(T)dT$$ where $k_r = k(T_r)$.

Given boundary conditions are

$T(0) = T_r$ and $\frac{dT(L)}{dx} = 0$.

2.9.   Consider a slab of thickness L with uniform thermal conductivity and a uniform heat generation of g W/m³. The boundary at x = 0 is kept at a constant temperature T_1 and the boundary at x = L loses heat by convection to a fluid at a constant temperature T_2

with a heat transfer coefficient h. Find the expression for the steady-state temperature distribution in the slab and the heat flux.

Dimensionless Conduction Numbers

Dimensionless numbers help in conduction heat transfer engineering
Used to compare relative values in the practice of engineering
In conduction, there are the Biot number and the Fourier number,
There is also the dimensionless heat generation number.

The Biot number compares the heat transfer coefficient
To unit conductance of a solid with a characteristic length
Fourier compares heat conduction across a distance in given volume
To the rate of heat being stored in that given volume.

K.V. Wong

3

One-Dimensional Steady-State Heat Conduction

This chapter discusses one-dimensional steady-state heat conduction in three different coordinate systems. There is a discussion on temperature-dependent thermal conductivity. Extended surfaces or fins are treated exhaustively.

3.1 The Slab (One-Dimensional Cartesian Coordinates)

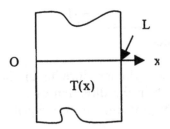

Figure 3.1   The slab.

Consider a slab, infinite in the direction of the y-coordinate, with L the thickness in direction of the x-coordinate. For the steady-state situation, the governing energy equation is

$$\frac{d^2T(x)}{dx^2} + \frac{g(x)}{k} = 0 \quad \text{in } 0 \leq x \leq L \tag{3.1}$$

If $g(x)$ is defined and boundary conditions are defined as first, second or third kind at $x = 0$, $x = L$, the equation can be integrated to solve for $T(x)$. Note that the assumption of steady-state temperatures with Eq.(1), are not consistent for homogenous second kind boundary conditions at both $x = 0$ and $x = L$. Once $T(x)$ is known, then the heat flux can be calculated by

$$q(x) = -k \frac{dT(x)}{dx} \qquad (3.2)$$

3.2 The Cylinder (One-Dimensional Cylindrical Coordinates)

(a) Solid Cylinder

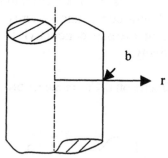

Figure 3.2   Solid cylinder.

Consider a solid cylinder, infinite in the direction of the z-coordinate, with radius b in the direction of the r-coordinate. For the steady-state situation, the governing energy equation is

$$\frac{1}{r} \frac{d}{dr}\left(r \frac{dT}{dr}\right) + \frac{g(r)}{k} = 0 \quad \text{in} \quad 0 \le r \le b. \qquad (3.3)$$

In Figure 3.2, r = 0 is the line of symmetry. It follows that one boundary condition is

$$\left.\frac{dT}{dr}\right|_{r=0} = 0 \quad \text{or} \quad T\big|_{r=0} \text{ is finite.} \qquad (3.4)$$

If g(x) is defined and the boundary condition is defined as first, second or third kind at r = b, the equation can be integrated to solve for T(r). The heat flux can be calculated as $q(r) = -k \dfrac{dT}{dr}$. When q(r)>0, the heat is moving in the positive r-direction. As before, note that the assumption of steady-state temperatures with Eq. (3.3), are not consistent for a homogenous second kind boundary condition at r = b.

(b) Hollow Cylinder

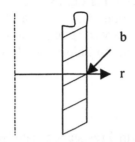

Figure 3.3 Hollow cylinder.

The governing equation is Eq. (3.3), and it is valid for $a \leq r \leq b$. Boundary conditions are required at $r = a$ and $r = b$. As previously noted, the assumption of steady-state temperatures with Eq. (3.3), are not consistent for homogenous second kind boundary conditions at $r = a$ and $r = b$.

3.3 The Sphere (One-Dimensional Spherical Coordinates)

Consider one-dimensional steady-state heat conduction in sphere; that is, there is the temperature has only r dependence. The governing energy equation for a sphere with radius b, is

$$\frac{1}{r^2} \frac{d}{dr}\left[r^2 \frac{dT}{dr} \right] + \frac{g(r)}{k} = 0 \quad \text{in} \quad 0 \leq r \leq b. \qquad (3.5)$$

(a)   Solid Sphere

$$\left.\frac{dT}{dr}\right|_{r=0} = 0 \quad \text{or} \quad T\big|_{r=0} \text{ is finite.} \qquad (3.6)$$

A first, second or third kind boundary condition has to be specified at $r = b$. The assumption of steady state temperatures with Eq. (3.5), are not consistent for a homogenous second kind boundary condition at $r = b$.

(b)   Hollow Sphere

The governing equation is Eq. (3.5), and it is valid for a   r   b. Boundary conditions are required at r = a and r = b. As previously noted, the assumption of steady-state temperatures with Eq. (3.5) is not consistent for homogenous second kind boundary conditions at r = a and r = b.

3.4 Thermal Resistance

For a solid, the thermal resistance is defined as

$$R = \frac{\Delta T}{Q} = \frac{\Delta T}{qA}.$$ (3.7)

(a)   Slab

For one-dimensional steady-state heat conduction with no heat generation, if the first kind boundary conditions are $T|_{x=0} = T_o$ and $T|_{x=L} = T_1$, then $q = \frac{k(T_o - T_1)}{L}$. The thermal resistance R is thus

$$R = \frac{\Delta T}{qA} = \frac{T_o - T_1}{k(T_o - T_1)A/L} = \frac{L}{kA}.$$ (3.8)

(b)   Hollow Cylinder

For one-dimensional steady-state heat conduction with no heat generation, in a cylinder of length H, if the first kind boundary conditions are $T|_{r=a} = T_o$ and $T|_{r=b} = T_1$, then

$$Q = (2\pi H)(k)\frac{(T_o - T_1)}{\ln\left(\frac{b}{a}\right)}$$ (3.9)

The thermal resistance R is thus

$$R = \frac{\ln\left(b/a\right)}{2\pi k H}.$$ (3.10)

The thermal resistance may be arranged to be in a form similar to that for the slab.

$$R_{cyl} = \frac{\ln\left(b/a\right)}{2\pi k H} = \frac{(b-a)\ln\left[\left(\frac{2\pi H}{2\pi H}\right)\left(b/a\right)\right]}{(b-a)2\pi k H}.$$ (3.11)

Note that $L_{cyl} = (b-a) = $ thickness of the cylinder, $A_{cyl}(r) = 2\pi r H$, so

$$R_{cyl} = \frac{L_{cyl} \ln\left(A_1/A_o\right)}{(A_1 - A_o)k}$$

Hence, $R_{cyl} = \dfrac{L_{cyl}}{\vec{A}_{cyl}\,k}$ where $\vec{A}_{cyl} = \dfrac{A_1 - A_o}{\ln\left(A_1/A_o\right)}$ is the log mean area.

(3.12)

(c) Hollow Sphere

For one-dimensional steady-state heat conduction with no heat generation, in a hollow sphere of inner and outer radii a and b respectively, if the first kind boundary conditions are $T|_{r=a} = T_o$ and $T|_{r=b} = T_1$, then

$$Q = 4\pi k\left(\frac{ab}{b-a}\right)(T_o - T_1).$$ (3.13)

The thermal resistance R_{sph} is thus

$$R_{sph} = \frac{b-a}{ab}\frac{1}{4\pi k}.$$ (3.14)

This can be rearranged in a similar form to the resistance of the one-dimensional slab.

$$R_{sph} = \frac{b-a}{k\sqrt{(4\pi a^2)(4\pi b^2)}} = \frac{L_{sph}}{k\sqrt{A_o A_1}}$$

$$R_{sph} = \frac{L_{sph}}{kA_g} \tag{3.15}$$

where A_o, A_1 are the inner and outer sphere areas respectively
   $L_{sph} = (b-a)$
   A_g = geometric mean area.

3.5   Conduction Through a Slab from One Fluid to Another Fluid

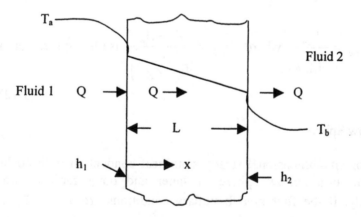

Figure 3.4   Heat transfer through a slab from one fluid to another fluid.

A slab separates two fluids of different temperatures as shown in Fig. 4. There are no heat sources or sinks in the slab. Heat will be transferred from the fluid of higher temperature to the slab, then conducted through the slab, and then transferred from the wall to the fluid of lower temperature. Under steady state conditions, the heat transfer will be the same on both surfaces and through the slab. If the heat transfer coefficients h_1 and h_2 at each side of the slab are constant, the following equations apply:

$$Q = Ah_1(T_a - T_1) \qquad \text{for convection at surface } x = 0 \quad (3.16)$$

$$Q = A\frac{k}{L}(T_1 - T_2) \qquad \text{for conduction through the wall} \quad (3.17)$$

$$Q = Ah_2(T_2 - T_b) \qquad \text{for convection at surface } x = L \quad (3.18)$$

From the above equations, any unknowns may be found by calculation. The heat flux Q through the slab may be written as

$$Q = \frac{(T_a - T_b)}{1/h_1 A + L/kA + 1/h_2 A} \tag{3.19}$$

It may be recognized that the thermal resistances for fluid 1, the slab and fluid 2 are, respectively,

$$R_{f1} = 1/h_1 A, \qquad R_s = L/kA, \qquad \text{and} \qquad R_{f2} = 1/h_2 A \qquad (3.20)$$

3.6 Composite Medium

Consider the one-dimensional, steady-state heat transfer for composite structure consisting of parallel plates, coaxial cylinders, etc., in perfect thermal contact with each other.

(a) Parallel Slabs

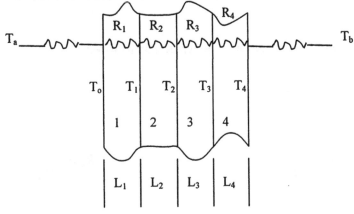

Figure 3.5 Parallel slabs.

Consider a four layer slab. The heat flux through the slab Q may be written as

$$Q = \quad Ah_a\left(T_a - T_o\right) \qquad \text{for convection heat transfer at the leftmost surface}$$

$$= \quad \frac{Ak_1}{L_1}\left(T_o - T_1\right) \qquad \text{for conduction through the first layer}$$

$$= \quad \frac{Ak_2}{L_2}\left(T_1 - T_2\right) \qquad \text{for conduction through the second layer}$$

$$= \quad \frac{Ak_3}{L_3}\left(T_2 - T_3\right) \qquad \text{for conduction through the third layer}$$

$$= \quad \frac{Ak_4}{L_4}\left(T_3 - T_4\right) \qquad \text{for conduction through the fourth layer}$$

$$= \quad Ah_b\left(T_4 - T_b\right) \qquad \text{for convection at the rightmost surface.}$$
(3.21)

The thermal resistances may be written as follows:

$$R_a = \frac{1}{Ah_a} \qquad\qquad R_3 = \frac{L_3}{Ak_3}$$

$$R_1 = \frac{L_1}{Ak_1} \qquad\qquad R_4 = \frac{L_4}{Ak_4}$$

$$R_2 = \frac{L_2}{Ak_2} \qquad\qquad R_b = \frac{1}{Ah_b} \qquad (3.22)$$

The heat flux Q through the slab is then

$$Q = \frac{T_a - T_b}{R_{TOTAL}} \qquad (3.23)$$

where $R_{TOTAL} = R_a + R_1 + R_2 + R_3 + R_4 + R_b$. The units for the thermal resistances are either (hr.°F)/Btu or °C/Watt.

The overall heat transfer coefficient or conductance, U, may be defined as

$$UA = \frac{1}{R} \qquad \text{or} \qquad U = \frac{1}{AR} \qquad (3.24)$$

The total heat transfer rate through an area A of a composite structure, Q, may be defined as

$$Q = UA(T_a - T_b) \qquad (3.25)$$

where U, the conductance, is in Btu/(hr.ft^2.°F) or W/(m^2.°C). The associated area need not be defined.

3.7 Thermal Contact Resistance

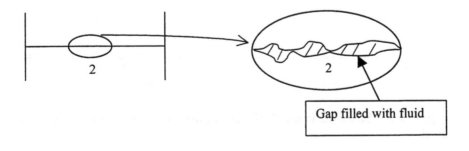

Figure 3.6 Thermal contact resistance between two solid surfaces.

In real contacts between two solid surfaces, direct contact occurs between the two solids at a limited number of spots with voids between these spots filled with some fluid, such as the surrounding medium (air). Heat transfer across the interface occurs by conduction through the solid spots of solid-to-solid direct contact and through the fluid filled gap.

If the thermal conductivity of the fluid is less than that of the two solids, then the interface between the two solids acts as a resistance to heat flow. This resistance is referred to the "thermal contact resistance". It can be rather significant if the contact between the two solid surfaces is poor, and/or the fluid conductivity is much less than the conductivities of the solids.

3.8   Standard Method of Determining the Thermal Conductivity in a Solid

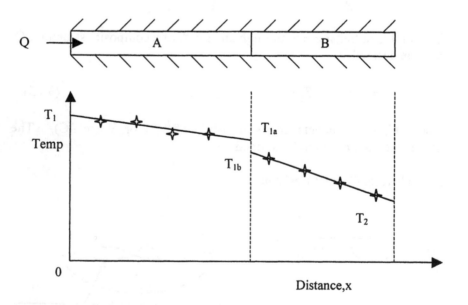

Figure 3.7   Standard method of determining thermal conductivity in a solid.

The standard method of determining the thermal conductivity in a solid is as described in the following. Two bars of the same cross-sectional area and similar lengths are used: one bar made from a metal of known thermal conductivity, and the other made from the material whose thermal conductivity is to be found. In the figure, A is a standard bar of known dimensions and thermal conductivity k_A. B is the bar of known dimensions, the thermal conductivity, k_B, of which is to be measured.

Heat is supplied to one end of A, and the whole system is insulated from heat loss. At steady state, the temperature profiles within A and within B are plotted. The result will be a plot similar to that shown in the figure. Since the heat transfer, Q, in rod A is equal to the heat transfer in rod B, Fourier's law gives

$$Q = \frac{Ak_A}{L_A}(T_o - T_{1a}) = \frac{Ak_B}{L_B}(T_{1b} - T_2) \tag{3.26}$$

Hence, $k_B = k_A \cdot \dfrac{T_o - T_{1a}}{T_{1b} - T_2}$ \hfill (3.27)

Note that the temperature drop ($T_{1a} - T_{1b}$) is caused by the thermal contact resistance between the surfaces of rod A and rod B.

3.9 Temperature-Dependent Thermal Conductivity

(a) One-Dimensional with No Heat Generation

In general, the thermal conductivity of a substance is a function of temperature. For steady-state one-dimensional heat conduction in a solid with variable conductivity, eg., in a slab,

$$\frac{d}{dx}\left(k(T)\frac{dT}{dx}\right) = 0 \quad \text{in} \quad 0 \le x \le L \tag{3.28}$$

where $T\big|_{x=0} = T_o$ and $T\big|_{x=L} = T_1$.

Integrating yields

$$k(T)\frac{dT}{dx} = c, \text{ a constant, or } k(T)dT = cdx.$$ (3.29)

The heat-flow rate Q through an area A is given by

$$Q = qA = -A.k(T)\frac{dT}{dx} = -A.c$$ (3.30)

Integrating to find c,

$$\int_{T_o}^{T_1} k(T)dT = c.L \text{ or } c = \frac{1}{L}\int_{T_o}^{T_1} k(T)dT$$ (3.31)

Hence,

$$Q = \frac{A}{L}\int_{T_o}^{T_1} k(T)dT.$$ (3.32)

Example 3.1

Problem: The thermal conductivity of a plane wall varies with temperature according to the relation

$$k(T) = k_o(1 + \beta T^2)$$

The surfaces at x = 0 and x = L are maintained at uniform temperatures T_1 and T_2, respectively. Find the relation for the heat flow through the slab per unit area.

Solution

From Eq. (3.28),

$$\frac{d}{dx}\left(k(T)\frac{dT}{dx}\right) = 0 \text{ in } 0 \le x \le L$$

$$T\big|_{x=0} = T_1 \text{ and } T\big|_{x=L} = T_2$$

$$k(T) = k_o(1 + \beta T^2)$$

$$k(T)dT = c_1 dx$$

$$k_o(1 + \beta T^2)dT = c_1 dx$$

$$c_1 = \frac{k_o}{L} \int_{T_1}^{T_2} (1 + \beta T^2)dT$$

$$= \frac{k_o}{L}\left[(T_2 - T_1) + \frac{1}{3}\beta(T_2^3 - T_1^3)\right]$$

$$= \frac{k_o}{L}(T_2 - T_1)\left[1 + \frac{1}{3}\beta(T_2^2 + T_1 T_2 + T_1^2)\right]$$

$$q = -k(T)\frac{dT}{dx} = -c_1.$$

Hence, $q = \dfrac{T_1 - T_2}{L}.k_o\left[1 + \dfrac{1}{3}\beta(T_2^2 + T_1 T_2 + T_1^2)\right].$

(b) Poisson's Equation, with Heat Generation

When the thermal conductivity is dependent on temperature, the Poisson equation is

$$\nabla.(k\nabla T) + g = 0. \tag{3.33}$$

Equation (3.33) is a nonlinear equation and difficult to solve. Equation (3.33) may be reduced to a linear differential equation by introducing a new temperature function θ by means of the Kirchhoff transformation as

$$\theta = \frac{1}{k_r} \int_{T_r}^{T} k(\hat{T})d\hat{T} \tag{3.34}$$

where T_r is a convenient reference temperature and $k_r = k(T_r)$. It follows from Eq. (3.34) that

$$\nabla\theta = \frac{k}{k_r}\nabla T. \tag{3.35}$$

Thus, Eq. (3.33) can be written as

$$\nabla^2 \theta + \frac{g}{k_r} = 0 \qquad\qquad (3.36)$$

which is not dissimilar to the heat conduction equation with constant k. The transformed equation may be solved with the usual techniques, as long as the boundary conditions can also be transformed. Boundary conditions of the first and second kind can be transformed; boundary conditions of the third kind usually cannot be transformed. Boundary conditions of the third kind are generally solved using numerical methods.

Consider a long rod of radius r_1, and the surface is kept at a uniform temperature T_w. The internal heat generation is at a uniform rate of g per unit volume. The governing equation is

$$\frac{1}{r}\frac{d}{dr}\left[rk(T)\frac{dT}{dr}\right] + g = 0 \qquad\qquad (3.37)$$

with $\quad \left(\dfrac{dT}{dr}\right)_{r=0} = 0 \quad$ and $\quad T(r_1) = T_w.$ $\qquad\qquad (3.38)$

Employing the Kirchhoff transformation as

$$\theta = \frac{1}{k_w}\int_{T_w}^{T} k(\hat{T})d\hat{T} \qquad\qquad (3.34)$$

where $k_w = k(T_w)$, Eq. (3.37) and the boundary conditions, Eq. (3.38) are transformed to

$$\frac{1}{r}\frac{d}{dr}\left(r\frac{d\theta}{dr}\right) + \frac{g}{k_w} = 0. \qquad\qquad (3.39)$$

with $\left(\dfrac{d\theta}{dr}\right)_{r=0} = 0$ and $\theta(r_1) = 0.$ (3.40)

The solution of Eq. (3.39) with boundary conditions Eq. (3.40) is

$$\theta(r) = \frac{gr_1^2}{4k_w}\left[1-\left(\frac{r}{r_1}\right)^2\right].$$ (3.41)

From Eq. (3.40) and Eq. (3.34), we get

$$\int_{T_w}^{T(r)} k(T)dT = \frac{gr_1^2}{4}\left[1-\left(\frac{r}{r_1}\right)^2\right]$$ (3.42)

If the relation $k = k(T)$ is known, then the relation can be written entirely in terms of T. At $r = 0$, the equation gives

$$\int_{T_w}^{T_o} k(T)dT = \frac{gr_1^2}{4}$$ (3.43)

where T_o is the centerline temperature. For the situation where k is a constant, Eq. (3.42) reduces to

$$T(r) - Tw = \frac{gr_1^2}{4k}\left[1-\left(\frac{r}{r_1}\right)^2\right]$$ (3.44)

It should be observed that the following relation exists between the heat generation rate and the surface heat flux:

$$q_sA=gV$$ (3.45)

where q_s = surface heat flux
    A = total surface area
    V = volume

Thus,

$$q_s = \frac{gV}{A} \tag{3.46}$$

It can be seen that the surface heat flux is proportional to the ratio of volume to surface area, and to the strength of the internal heat generation.

Example 3.2

Problem: Calculate the rate of heat generation per unit volume in a rod that will produce a centerline temperature of 1000°C for the following values of the parameters:

$$T_w = 300°C, \qquad r_1 = 2 \text{ cm}, \qquad k = \frac{1000}{273 + T}$$

where k is in W/(m.K) and T is in °C. In addition, find the surface heat flux.

Solution

From Eq. (3.43),

$$g = \frac{4}{r_1^2} \int_{300}^{1000} \frac{1000}{273 + T} dT$$

$$= \frac{4 \times 1000}{(0.02)^2} \ln \frac{1273}{573} = 7.982 \times 10^6 \, W/m^3$$

From Eq. (3.46), the surface heat flux is given by

$$q_s = \frac{g \pi r_1^2 L}{2 \pi r_1 L} = \frac{g r_1}{2} = \frac{7.982 \times 10^6 \times 0.02}{2} = 7.982 \times 10^4 \text{ W/m}^2.$$

3.10 Critical Radius of Insulation

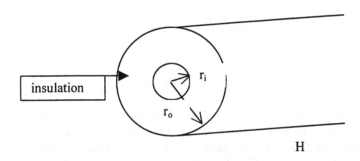

Figure 3.8   Insulation around a cylindrical system.

Consider the insulation around a cylindrical system, as shown in Fig. 3.8. The boundary conditions are the first kind at $r = r_i$ and the third kind at $r = r_o$.

At $r = r_i$, $T = T_i$

At $r = r_o$, $k\dfrac{\partial T}{\partial r} + h_o T = h_o T_\infty$

The heat flux, Q, is given by

$$Q = \frac{T_i - T_\infty}{R_{in} + R_o} \tag{3.47}$$

where R_{in} = insulation resistance = $\dfrac{1}{2\pi H k_{in}} \ln \dfrac{r_o}{r_i}$

and R_o = outside surface resistance = $\dfrac{1}{2\pi H h_o} \dfrac{1}{r_o}$.

Note that Q_{max} is at $r_o = r_{ocritical}$, which is determined by putting $\dfrac{d\theta}{dr_o} = 0$. It can be shown that

$$\frac{dQ}{dr_o} = \frac{2\pi H(T_i - T_\infty)}{\left[\ln\dfrac{r_o}{r_i} + \dfrac{k}{h_o r_o}\right]^2}\left(\frac{1}{r_o} - \frac{k}{h_o r_o^2}\right) \tag{3.48}$$

Hence, the critical radius of insulation is $r_{ocritical} = \dfrac{k}{h_o}$. Any insulation which results in a radius less than or greater than this value, will cause the heat flux to be less than this maximum. For a practical example, consider insulation being added to a wire whose surface is kept at a uniform temperature. The heat loss from the wire will increase as the insulation is added until the outside radius of the insulation equals to the critical radius $r_{ocritical}$. As the insulation thickness is increased past this value, the heat loss from the wire will begin to decrease.

Example 3.3

Problem: Insulating material with k = 0.2 W/(m².K), is added to a 0.02 m outer diameter pipe. The heat transfer coefficient on the outer surface is h = 6.6 W/(m².K). Will the heat loss increase or decrease?

Solution
The critical radius is $r_{ocritical}$ = k/h = 0.030 m.
The outer radius of the pipe is 0.02 m < 0.03 m; hence, heat loss from the pipe will decrease until the outer radius is 0.03 m, after which it will increase.

3.11   Effects of Radiation

The preceding discussion in this subsection does not include the effects of thermal radiation. If the radiation effects can be linearized (that is, assuming small temperature differences), the heat transfer coefficient at the outer surface, h_o, takes the form

$h_o = h_c + h_r$ (3.49)

where h_c is the convective component and h_r is the radiative component and approximately equal to $4\sigma F T^3$. The expression for the critical radius then becomes

$$r_{ocritical} = \frac{k}{h} + \frac{k}{h_c + h_r} \qquad (3.50)$$

The effect of including the thermal radiation effects is to reduce the value of the critical radius.

3.12   One-Dimensional Extended Surfaces

3.12.1   Introduction

Heat transfer in regular heat exchangers from one fluid to another through a conducting wall takes place at a rate that is directly proportional to the surface area of the wall and the temperature difference between the fluids. One way to increase the rate of heat transfer is to increase the effective area for heat transfer. This may be done by adding fins or spines to the surface of the conducting wall. These are thin conducting strips that can be a variety of shapes and sizes. The average surface temperature of the fins will not be the same as the original surface temperature, but will be nearer to that of the surrounding fluid. Because of this fact, the rate of heat transfer will not be proportionately increased even though the surface area has been increased by the fins.

In the discussions in this section, the following assumptions are used:
(i)   Heat flow in the extended surface is steady.
(ii)   There are no heat sources or sinks within the extended surface.
(iii)   The thermal conductivity of the solid is constant.
(iv)   The fluid is at a uniform and constant temperature.
(v)   The heat transfer coefficient between the extended surface and the fluid is constant.
(vi)   Temperature in the extended surface is one-dimensional. This can be achieved if the cross-section of the fin is small compared to its length.
(vii)   The temperature of the base of the fins is constant.

Asssumption (v) may be challenged, but an average value in analytical studies gives heat transfer results with reasonable agreement to experimental measurements. For most engineering applications, assumption (vii) provides a true picture.

3.12.2 One-Dimensional Fin Equations

To augment the effective heat transfer, fins (extended surfaces) are often used in practical applications. To understand the heat flow through a fin requires a knowledge of the temperature distribution in the fin. Consider the variable cross section fin shown in Fig. 3.9.

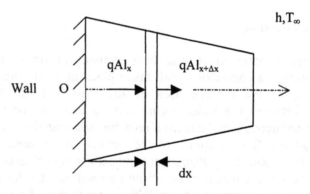

Figure 3.9   Variable cross-section fin.

We assume that the temperature of the fin, T, is only a function of the coordinate x. In other words, the temperature is uniform at any cross section. The one-dimensional steady-state fin energy conservation equation gives

[Net rate of heat gain   +   [Net rate of heat gain
  by conduction in x   by convection through
  direction into volume]   lateral surfaces into volume] = 0 (3.51)

        Conduction   +   Convection   = 0

Here,   Conduction = $-\dfrac{d}{dx}(qA)\Delta x$

and   Convection= $h[T_\infty - T(x)]\Delta a$

where h and T_∞ are assumed to be constant. Letting $\Delta x \to dx$,

$$-\frac{d}{dx}(qA) + h(T_\infty - T)\frac{da}{dx} = 0 \quad \text{where} \quad q = -k\frac{dT}{dx} \tag{3.52}$$

$$\frac{d}{dx}\left(A\frac{dT}{dx}\right) + \frac{h}{k}(T_\infty - T)\frac{da}{dx} = 0$$

Let $\quad \theta(x) = T - T_\infty$

Hence, $\quad \dfrac{d}{dx}\left(A\dfrac{d\theta}{dx}\right) - \dfrac{h}{k}\dfrac{da}{dx}\theta = 0$ $\qquad\qquad$ (3.53)

Equation (3.53) is the one-dimensional fin equation for fins with variable cross section. This special case occurs when A is constant. Let this constant be equal to a = Px, where P is the perimeter. In this case, da/dx = P. The one-dimensional fin equation then becomes

$$\frac{d^2\theta}{dx^2} - \frac{Ph}{Ak}\theta = 0 \quad \text{or} \quad \frac{d^2\theta}{dx^2} - m^2\theta = 0 \quad \text{where} \quad m^2 = \frac{Ph}{Ak} \tag{3.54}$$

Boundary conditions are needed to solve the equation.

3.12.3 Temperature Distribution and Heat Flow in Fins of Uniform Cross Section

The governing equation for heat flow in fins of uniform cross section is Eq. (4.4). Different boundary conditions will result in different temperature distributions in the fin. The temperature distributions are thus classified under the different boundary conditions.

(a) Long Fin

For the long fin, the boundary conditions are

$$\theta(x) = T_0 - T_\infty = \theta_0 \qquad \text{at} \qquad x = 0 \qquad\qquad (3.55a)$$

$$\theta(x) \to 0 \ (\text{ i.e. } T \to T_\infty) \text{ as } x \to \infty. \qquad\qquad (3.55b)$$

The general solution for this case is $\theta(x) = c_1 e^{-mx} + c_2 e^{mx}$ (3.56) where c_1, c_2 are constants of integration to be determined by the boundary conditions.

As $x \to \infty$, $\qquad \theta \to 0 = \lim_{x \to \infty}(c_1 e^{-mx} + c_2 e^{mx})$

Hence, $c_2 = 0$ to satisfy the boundary condition. At x = 0,

$$\theta(0) = \theta_o = T_o - T_\infty = c_1 e^{-m.0}$$

Thus, $c_1 = T_0 - T_\infty$. So,

$$\frac{\theta}{\theta_o} = \frac{T - T_\infty}{T_o - T_\infty} = e^{-mx} \quad \text{or} \quad \theta = \theta_o e^{-mx}. \qquad\qquad (3.57)$$

The heat flux, Q, can be found from

$$Q = \int_{x=0}^{\infty} h\theta(x)P\,dx \qquad \text{(Btu/hr or W/hr).} \qquad\qquad (3.58)$$

Since $\quad \theta = \frac{1}{m^2}\frac{d^2\theta}{dx^2}$ from the governing equation,

$$Q = \frac{hP}{m^2}\int_o^\infty \frac{d^2\theta}{dx^2}\,dx$$

$$= \frac{hP}{Ph\!\!\Big/\!Ak}\int_o^\infty \frac{d^2\theta}{dx^2}\,dx = Ak\left[\frac{d\theta}{dx}\bigg|_\infty - \frac{d\theta}{dx}\bigg|_o\right].$$

As $x \to \infty$, $\theta \to 0$ and $\dfrac{d\theta}{dx}\bigg|_\infty \to 0$.

Recall that since $\theta = \theta_o e^{-mx}$,

$$\frac{d\theta}{dx} = -m\theta_o e^{-mx} \quad \text{and} \quad \frac{d\theta}{dx}\bigg|_{x=0} = -m\theta_o e^{-m0} = -m\theta_o .$$

So, $\quad Q = -Ak(-m\theta_o) = \theta_o \sqrt{PhAk} \quad$ since $\quad m^2 = \dfrac{Ph}{Ak} \quad$ (3.59)

Example 3.4

Problem: A long fin of 0.02 m diameter is one of the fins conducting heat away from a heat exchanger. The steady-state temperature at two different locations along the fin 0.09 m apart are 130°C and 100°C, respectively. The environment is at 20°C. If the thermal conductivity of the fin is 100 W/(m.K), calculate the heat transfer coefficient between the fin and the environmental fluid.

Solution
    Since the fin is long, the temperature distribution is given by Eq. (3.57),

$$\frac{\theta}{\theta_o} = \frac{T - T_\infty}{T_o - T_\infty} = e^{-mx}$$

Hence, $\quad \dfrac{T(x_1) - T_\infty}{T(x_2) - T_\infty} = e^{m(x_2 - x_1)}$

Substituting values, $\quad \dfrac{130 - 20}{100 - 20} = e^{0.09m}$

which gives $\quad\quad\quad m = 3.5384 \ \text{m}^{-1}$

Since $\quad m = \sqrt{\dfrac{hP}{kA}} = \sqrt{\dfrac{4h\pi D}{k\pi D^2}} = \sqrt{\dfrac{4h}{kD}}$

$h = km^2 D/4$

$$h = 100 \frac{W}{m.K}(3.5384)^2 \frac{1}{m^2}(0.02)m\frac{1}{4}$$

$h = 6.26$ W/(m^2.K).

(b) Fin with Negligible Heat Flow at the Tip

      For a fin with negligible heat flow at the tip, the boundary conditions are

$$\theta(x) = T_o - T_\infty = \theta_o \qquad \text{at} \qquad x = 0 \qquad\qquad (3.60a)$$

$$\frac{d\theta}{dx} = 0 \text{ at } x = L. \qquad\qquad (3.60b)$$

The general solution for this case is
$$\theta(x) = c_1 \cosh m(L-x) + c_2 \sinh m(L-x) \qquad\qquad (3.61)$$

Since $\sinh u = 0.5(e^u - e^{-u})$, $\cosh u = 0.5(e^u + e^{-u})$,

      $\sinh 0 = 0$, $\cosh 0 = 1$, and $c_2 = 0$.

From Eq. (3.60a), $\theta = \theta_o = c_1 \cosh mL$ and $c_1 = \dfrac{\theta_o}{\cosh mL}$.

Thus, $\dfrac{\theta(x)}{\theta_o} = \dfrac{T(x) - T_\infty}{T_o - T_\infty} = \dfrac{\cosh m(L-x)}{\cosh mL}$ $\qquad\qquad$ (3.62)

Now, $Q = \displaystyle\int_0^L h\theta P dx$

$$= \int_0^L \frac{hP}{m^2}\frac{d^2\theta}{dx^2}dx$$

$$0$$

$$= Ak\left(\frac{d\theta}{dx}\bigg|_L - \frac{d\theta}{dx}\bigg|_o\right)$$

$$= -Ak\left(\frac{d\theta}{dx}\Big|_{o}\right)$$

$$= -Ak\left(\frac{-m\theta_{o}\sinh mL}{\cosh mL}\right) = \theta_{o}Akm\tanh mL$$

$$Q = \theta_{0}\sqrt{PhAk}\tanh mL \quad \text{where} \quad m = \sqrt{\frac{Ph}{Ak}}. \qquad (3.63)$$

To check the correctness of Eq. (3.63), we can take the limit of $mL \rightarrow \infty$. When we take the limit, $\tanh mL \rightarrow 1$, and $Q \rightarrow \theta_{o}\sqrt{PhAk}$, which is the same result as that for the long fin. This gives us confidence in Eq.(3.63) since the answer in the limit is an answer we expect. In research, this particular feature is important. Since the scholar is typically in an unknown realm in research, the scholar needs to know that the results obtained make sense in the limit where the answer corresponds to known solutions. This test is often used for the validity of new relations and equations.

Equation (3.62) may be used for a thermocouple in a fluid stream that is at a different temperature from that of the plate supporting the thermocouple. This is a useful correction for temperature measurement devices employed by experimental engineers.

Example 3.5

Problem: A thermocouple well mounted through the wall of a gas pipe may be considered as a metal rod of 0.01 m outer diameter and 0.06 m length, with a thermal conductivity of 22 W/(m.K). The thermocouple reads 160°C and the temperature of the pipe is 70°C. If the gas heat transfer coefficient to the well is 115 W/(m².K), find the average gas temperature.

Solution
The thermocouple well may be modeled as a long fin with negligible heat flow at the tip. Hence, Eq.(3.62) applies to the temperature distribution.

$$m = \sqrt{\frac{hP}{kA}} = \sqrt{\frac{4h\pi D}{k\pi D^2}} = \sqrt{\frac{4h}{kD}}$$

Hence, $m = \sqrt{\dfrac{4 \times 115}{22 \times 0.01}\left(\dfrac{W}{m^2 .k} \cdot \dfrac{m.K}{W} \cdot \dfrac{1}{m}\right)} = 45.73 m^{-1}$

From Eq. (3.62), $\dfrac{\theta(x)}{\theta_o} = \dfrac{T(x) - T_\infty}{T_o - T_\infty} = \dfrac{\cosh m(L-x)}{\cosh mL}$

Substituting 70°C for the temperature at the base of the fin and writing the expression for the temperature at $x = L$,

$$\frac{160 - T_\infty}{70 - T_\infty} = \frac{\cosh m(L - L)}{\cosh(45.73 \times 0.06)}$$

$T_\infty = 173.2°C$

Here, it has been assumed that the thermocouple reading is the same as the temperature at the tip of the well. The average temperature of the gas is 173.2°C.

(c)   Fin with Convection at the Tip

For a fin with convection at the tip, the boundary conditions are

$$\theta(x) = T_o - T_\infty = \theta_o \qquad \text{at} \qquad x = 0 \qquad\qquad (3.64a)$$

$$k\frac{d\theta}{dx} + h_e \theta(x) = 0 \text{ at } x = L. \qquad\qquad (3.64b)$$

The general solution for this case is

$$\theta(x) = c_1 \cosh m(L-x) + c_2 \sinh m(L-x) \qquad\qquad (3.65)$$

At $x = L$,

$$k\left[-c_1 m \sinh m.0 - c_2 m \cosh m.0\right] + h_e\left[c_1 \cosh o + c_2 \sinh 0\right] = 0$$

(with annotations: 0, $=1$, $=1$, 0 above the respective terms)

$$0 = -k.c_2 m + h_e c_1 \quad \text{or} \quad c_1 = c_2 \frac{km}{h_e}$$

At $x = 0$,

$$\theta_o = \frac{c_2 km}{h_e}\cosh(mL) + c_2 \sinh mL, \quad \text{so} \quad c_2 = \frac{\theta_o}{\sinh mL + \frac{km}{h_e}\cosh(ml)}.$$

Therefore, $\quad c_1 = \frac{km}{h_e}\left(\dfrac{\theta_o}{\sinh mL + \frac{km}{h_e}\cosh mL}\right).$

Hence, $\quad \dfrac{\theta(x)}{\theta_o} = \dfrac{T(x) - T_\infty}{T_o - T_\infty} = \dfrac{\cosh m(L - x) + \frac{c_2}{c_1}\sinh m(L - x)}{\frac{\theta_o}{c_1}}.$

Thus, $\quad \dfrac{\theta(x)}{\theta_o} = \dfrac{\cosh m(L - x) + \frac{h_e}{km}\sinh m(L - x)}{\cosh mL + \frac{h_e}{km}\sinh mL}.$ \hfill (3.66)

With negligible heat flow at the tip, this would be the same as if $h_e = 0$. If $h_e = 0$, Eq. (3.66) gives $\dfrac{\theta(x)}{\theta_o} = \dfrac{\cosh m(L - x)}{\cosh mL}.$ This temperature distribution is the same as that given by Eq. (3.62). Hence, the solution checks out at this limit. This particular case embodies the two previous cases. The solution for this case gives the other two previous cases when the appropriate limits are taken. In research, the aspiration is to obtain a general relation or solution, which encompasses many cases. This situation clearly illustrates the accomplishment of this feature.

The rate of heat transfer from the extended surface to the surrounding fluid is

$$Q = \sqrt{PhAk}\,\theta_o \, \frac{\sinh mL + \dfrac{h_e}{km}\sinh mL}{\cosh ml + \dfrac{h_e}{km}\sinh mL}. \tag{3.67}$$

In addition, Eq. (3.67) reduces to Eq. (3.63) when $h_e = 0$.

Example 3.6

Problem: A nickel fin, k = 20 W/(m.K), 0.04 m in diameter and 0.2 m in length, juts out from a plane wall which is at 300°C. The rod is cooled by a fluid at 10°C with an average heat transfer coefficient of 12 W/(m².K). Find the rate of heat loss from the fin.

Solution

$$\text{Evaluating the parameter, } m = \sqrt{\frac{hP}{kA}} = \sqrt{\frac{4h}{kD}} = 7.746m^{-1}$$

Hence, mL = 7.746 x 0.2 = 1.55
$\qquad\qquad h_e/(km) = 0.07746$
Perimeter $P = \pi D = 0.04\pi$ m = 0.1257 m
Cross-area $A = \pi D^2/4 = 0.001257$ m²

$$\sqrt{PhAk} = (0.1257 \times 12 \times 0.001257 \times 20)^{0.5} = 0.1947 \text{ W/K}$$

Rate of heat loss from the fin, Q = 0.1947(300 – 10) x
$$\frac{2.2476 + 0.07746 \times 2.46}{2.46 + 0.07746 \times 2.2476} = 52.3W$$

3.12.4 Fin Efficiency

Extended surfaces are used to augment heat transfer from the base area. To compare and evaluate these extended surfaces, two performance factors are used: fin efficiency and fin effectiveness. In the practice of engineering, the fin efficiency is more widely used. The fin effectiveness is defined as the ratio of the rate of heat transfer from an extended surface to the rate of heat transfer that would take place from the same base area without the extended surface.

The fin efficiency, η, is defined as

$$\eta = \frac{\text{actual heat transfer in fin}}{\text{ideal heat transfer in fin if entire fin were at } T_o}. \quad (3.67)$$

In other words, $\eta = \dfrac{Q_{fin}}{Q_{ideal}}$

$$Q_{ideal} = a_f h \theta_o = a_f h (T_o - T_\infty) \quad (3.68)$$

$$Q_{fin} = \eta Q_{ideal} = \eta a_f h \theta_o \quad (3.69)$$

Example 3.7

Problem: For a fin with a uniform cross section with x = L, find the fin efficiency.

Solution

$$a_f = PL$$
$$Q_{ideal} = PLh\theta_o$$
$$Q_{fin} = \theta_o \sqrt{PhkA} \tanh mL$$

Hence,

$$\eta = \frac{\theta_o \sqrt{PhkA} \tanh mL}{PLh\theta_o} = \frac{\sqrt{PhkA}}{PhL} \tanh mL = \frac{1}{mL} \tanh mL.$$

In practice,
$$Q_{total} = Q_{fin} + Q_{unfinned} \quad (3.70)$$

$$Q_{total} = \eta a_f h \theta_o + (a - a_f) h \theta_o \quad \text{where} \quad a = \text{total area} = a_f + a_{uf}$$

$$Q_{total} = (\eta\beta + 1 - \beta)ha\theta_o = \eta'ah\theta_o$$

$$\text{where} \quad \eta' = \eta\beta + 1 - \beta \text{ and } \beta = \frac{a_f}{a}.$$

In practice, $\dfrac{Pk/A}{h}$ should be > 1 to justify the use of fins. In other words, $Pk/A > h$ or the internal conductance should be larger than the convective film coefficient to justify the use of fins.

Consider a fin of given shape and material. Its efficiency decreases as h decreases. In other words, a fin that is very efficient when used with a gas like air is not as efficient when used in a liquid like alcohol because the h in alcohol generally has a much higher value.

3.12.5 Heat Transfer from a Finned Surface

The rate of heat transfer from the fins on an extended surface would be

$$Q_f = \eta h a_f (T_b - T_\infty) \tag{3.71}$$

where a_f is the total heat transfer surface area of the fins. The rate of heat removed from the surface between the fins is given by

$$Q_s = h a_s (T_b - T_\infty) \tag{3.72}$$

where a_s is the total surface area between the fins. Hence, the total rate of heat transfer is

$$Q_t = Q_f + Q_s = h(a_s + \eta a_f)(T_b - T_\infty) = h a_{eff}(T_b - T_\infty) \tag{3.73}$$

The effective heat transfer area of the surface, a_{eff} is equal to $(a_s + \eta a_f)$. If h is constant, the rate of heat transfer is increased by a factor of $(a_s + \eta a_f)/(a_s + a_{ba})$, where a_{ba} is the base area of the fins.

PROBLEMS

3.1.   A furnace is made of walls comprising 0.1 m thick fire brick on the inside and 0.3 m thick regular brick on the outside. Under steady-state conditions, the high surface temperature of the wall is 720°C and the low surface temperature is 100°C. A 0.05 m layer of insulation, k = 0.1 W/(m.K), is added onto the outside of the regular brick to reduce the heat loss. With the added layer of insulation, the steady-state temperatures are as follows: 740°C at the flame side of the fire brick, 680°C at the junction between the fire brick and the regular brick, 520°C at the junction between the regular brick and the insulation, and 80°C on the outer surface of the insulation. Find the rate of heat loss from the furnace, expressed as a fraction of the original rate of heat loss.

3.2.   Determine an expression for the steady-state temperature distribution T(x) in a plane wall, $0 \leq x \leq L$, having uniform heat generation of strength g W/m^3. The thermal conductivity of the wall, k, is a constant. At x = 0, the wall surface is at a constant temperature of T_1 while at x = L, it is at T_2.

3.3.   A plane wall, 0.15 m thick, internally generates heat at a rate of 6 x 10^4 W/m^3. One side of the wall is insulated and the other side is exposed to an environment at 25°C. The heat transfer coefficient between the wall and the environment is 750 W/(m^2.K). The thermal conductivity of the wall is 20 W/(m.K). Calculate the maximum temperature in the wall.

3.4.   Insulating material with k = 0.1 W/(m^2.K), is added to a 0.02 m outer radius pipe. The heat transfer coefficient on the outer surface is h = 6.6 W/(m^2.K). Will the heat loss increase or decrease?

3.5.   Insulating material with k = 0.15 W/(m^2.K), is added to a 0.03 m outer radius pipe. The heat transfer coefficient on the outer surface is h = 4.6 W/(m^2.K). Will the heat loss increase or decrease?

3.6.   Derive the expression for the critical radius of insulation for a sphere.

3.7.   Insulation with k = 0.1 W/(m.K) is added to a steam pipe of outer radius 0.01 m. The heat transfer coefficient with and without insulation may be assumed to be the same at h = 7 W/(m²K). What is the thickness of the insulation when the heat loss is the same as that without insulation?

3.8.   A long fin 0.03 m in diameter is one of the fins conducting heat away from a heat exchanger. The steady-state temperatures at two different locations along the fin 0.1 m apart, are 110°C and 80°C, respectively. The environment is at 25°C. If the thermal conductivity of the fin is 80 W/(m.K), calculate the heat transfer coefficient between the fin and the environmental fluid.

3.9.   A long fin 0.01 m in diameter is part of an array of fins for a radiator. The steady-state temperatures at two different positions along the fin 0.15 m apart are 150°C and 75°C, respectively. The environment is at 30°C. If the heat transfer coefficient between the fin and the environmental air is 9 W/(m².K), find the thermal conductivity of the fin.

3.10. A thermocouple well mounted through the wall of a boiler may be considered as a metal rod of 0.02 m outer diameter and 4 m length, with a thermal conductivity of 30 W/(m.K). The thermocouple reads 600°C, and the temperature of the boiler wall where the well is located is 200°C. If the gas heat transfer coefficient to the well is 200 W/(m².K), find the average gas temperature.

3.11. A thermocouple well mounted through the wall of a vapor pipe may be considered as a metal rod of 0.015 m outer diameter and 0.05 m length, with a thermal conductivity of 18 W/(m.K). The thermocouple reads 225°C, and the temperature of the pipe is 100°C. If the vapor heat transfer coefficient to the well is 90 W/(m².K), find the average vapor temperature.

3.12. A fin, k = 30 W/(m.K), 0.03 m in diameter and 0.18 m in length, protrudes from a plane wall which is at 450°C. The rod is cooled by a fluid at 100°C with an average heat transfer coefficient of 20 W/(m².K). Find the rate of heat loss from the fin.

3.13. A fin, k = 100 W/(m.K), 0.05 m in diameter and 0.28 m in length, protrudes from a plane wall which is at 150°C. The rod is cooled by a fluid at T_f with an average heat transfer coefficient of 10 W/(m^2.K). The rate of heat loss from the fin is 50 W. Find T_f.

Fins

Thermal conductivity of the solid is constant
Fluid is at temperature that is uniform and constant
Heat transfer coefficient between fin and fluid is constant
The temperature of the base of the fin is constant.

There are not any heat sources or sinks within the fin
The heat energy flow is steady throughout the fin
Temperature distribution is only in one dimension
Cross-section of fin is small compared to length dimension.

K.V. Wong

4

Two-Dimensional Steady and One-Dimensional Unsteady Heat Conduction

In this chapter, two-dimensional steady and one-dimensional unsteady heat conduction is discussed. The main method introduced to solve this group of problems is the method of separation of variables.

4.1 Method of Separation of Variables

The basis behind separation of variables is the orthogonal expansion technique. The method of separation of variables produces a set of auxiliary differential equations. One of these auxiliary problems is called the eigenvalue problem with its eigenfunction solutions.

Consider the second-order ordinary differential equation in the domain $0 \leq x \leq L$:

$$\frac{d^2 \psi(x)}{dx^2} + \lambda^2 \psi(x) = 0 . \tag{4.1}$$

The boundary conditions are:

$$-k\frac{d\psi}{dx} + h\psi = 0 \qquad \text{at } x = 0 \tag{4.2a}$$

$$-k\frac{d\psi}{dx} + h\psi = 0 \qquad \text{at } x = L \tag{4.2b}$$

where λ, h and k are constants. This is called an *eigenvalue* problem. This problem has solutions for certain values of the parameter $\lambda = \lambda_n$ where n = 1,2,3,...where λ_n's are called the eigenvalues. The nontrivial solutions $\psi(\lambda_n, x)$ are called the *eigenfunctions*.

Let $\psi(\lambda_m, x)$ and $\psi(\lambda_n, x)$ denote two different eigenfunctions corresponding to eigenvalues λ_m and λ_n. The orthogonality principal means that

61

Table 4.1   Eigenfunctions and Eigenvalues.

Boundary Condition at		Eigen-functions	Eigenvalues	
$x = 0$	$x = L$	$\psi(\lambda, x)$	λ 's are roots	$\dfrac{1}{N} = \dfrac{1}{\displaystyle\int_0^L \psi^2(\lambda, x)dx}$
1. $\psi = 0$	$\psi = 0$	$\sin \lambda x$	$\sin \lambda L = 0$	$\dfrac{2}{L}$
2. $\psi = 0$	$\dfrac{d\psi}{dx} = 0$	$\sin \lambda x$	$\cos \lambda L = 0$	$\dfrac{2}{L}$
3. $\psi = 0$	$k\dfrac{d\psi}{dx} + h\psi = 0$	$\sin \lambda x$	$\lambda \cot \lambda L = -H$	$\dfrac{2(\lambda^2 + H^2)}{L(\lambda^2 + H^2) + H}$
4. $\dfrac{d\psi}{dx} = 0$	$\psi = 0$	$\cos \lambda x$	$\cos \lambda L = 0$	$\dfrac{2}{L}$
5. $\dfrac{d\psi}{dx} = 0$	$\dfrac{d\psi}{dx} = 0$	$\cos \lambda x$	$\sin \lambda L = 0$	$\dfrac{2}{L} \; for \lambda \neq 0$ $\dfrac{1}{L} \; for \lambda = 0$
6. $\dfrac{d\psi}{dx} = 0$	$k\dfrac{d\psi}{dx} + h\psi = 0$	$\cos \lambda x$	$\lambda \tan \lambda L = H$	$\dfrac{2(\lambda^2 + H^2)}{L(\lambda^2 + H^2) + H}$
7. $-k\dfrac{d\psi}{dx} + h\psi = 0$	$\psi = 0$	$\sin \lambda(L - x)$	$\lambda \cot \lambda L = -H$	$\dfrac{2(\lambda^2 + H^2)}{L(\lambda^2 + H^2) + H}$
8. $-k\dfrac{d\psi}{dx} + h\psi = 0$	$\dfrac{d\psi}{dx} = 0$	$\cos \lambda(L - x)$	$\lambda \tan \lambda L = H$	$\dfrac{2(\lambda^2 + H^2)}{L(\lambda^2 + H^2) + H}$
9. $-k\dfrac{d\psi}{dx} + h\psi = 0$	$k\dfrac{d\psi}{dx} + h\psi = 0$	$\lambda \cos \lambda x + H \sin \lambda x$	$\tan \lambda L = \dfrac{2\lambda H}{\lambda^2 - H^2}$	$\dfrac{2}{L(\lambda^2 + H^2) + 2H}$

$$\int_0^L \psi(\lambda_m, x)\psi(\lambda_n, x)dx = 0 \quad \text{for} \quad \lambda_m \neq \lambda_n$$

$$= N \quad \text{for} \quad \lambda_m = \lambda_n \tag{4.3}$$

where N, the normalization integral, is

$$N = \int_0^L \psi^2(\lambda_m, x)dx . \tag{4.4}$$

The reader can consult any standard text on mathematics for the proof of the above orthogonality principal. The general case shown above is the eigenvalue problem with homogeneous boundary conditions of the third kind. There are nine different combinations of such boundary conditions (first, second and third kinds) for a finite region $0 \leq x \leq L$, and any of these nine combinations may be derived from the boundary conditions Eqs. (4.2a,b). The eigenfunctions, eigenvalues and the normalization integrals for all nine cases are listed in Table 4.1.

Example 4.1

Problem: Solve $\dfrac{d^2\psi}{dx^2} + \lambda^2\psi = 0$ in $0 \leq x \leq L$ subject to the boundary conditions $\psi(0) = 0$ and $\psi(L) = 0$.

Solution
The solution is $\psi(x) = c_1 \sin \lambda x + c_2 \cos \lambda x$.
At $x = 0$, $c_2 = 0 \Rightarrow \psi(x) = c_1 \sin \lambda x$.
At $x = L$, $\sin \lambda x = 0$ since $c_1 \neq 0$. Hence, $L\lambda_n = n\pi$ where $n = 1,2,3....$

Therefore, $\psi(\lambda_n, x) = \sin \lambda_n x$ where $\lambda_n = \dfrac{n\pi}{L}$.

Thus, $N = \int_0^L \sin^2 \lambda_n x \, dx = \int_0^L \left(\sin \dfrac{n\pi}{L}x\right)^2 dx = \dfrac{L}{2}$.

4.2   Steady-State Heat Conduction in a Rectangular Region

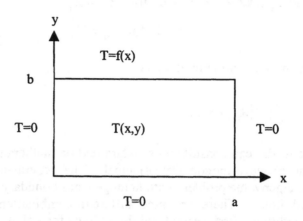

Figure 4.1   A rectangular region for heat conduction.

Consider a rectangular region for heat conduction as shown in Fig. 4.1. The assumptions about the problem are usually as follows:

• Steady state heat conduction takes place.
• Thermal conductivity k is constant.
• There is no heat generation.
• There is only one nonhomogeneous boundary condition.

With the assumptions above, the governing heat conduction equation is the Laplace equation.

$$\frac{\partial^2 T}{\partial x^2} + \frac{\partial^2 T}{\partial y^2} = 0 \ \text{ in } 0 \le x \le a, 0 \le y \le b \tag{4.5}$$

The boundary conditions shown in the figure are expressed mathematically as follows:

At $x = 0$, $T(x,y) = 0$.
At $x = a$, $T(x,y) = 0$.
At $y = 0$, $T(x,y) = 0$.
At $y = b$, $T(x,y) = f(x)$.

This problem may be solved by the method called separation of variables. The first step in this method is to assume that the temperature distribution function T(x,y) is a product of X(x) and Y(y) where X(x) is only a function of x, and Y(y) is only a function of y.

$$T(x,y)=X(x)Y(y) \tag{4.6}$$

Thus, $\dfrac{1}{X}\dfrac{d^2X}{dx^2} = -\dfrac{1}{Y}\dfrac{d^2Y}{dy^2}.$ (4.7)

Since the left-hand side of Eq. (4.7) is a function of x only, and the right-hand side of Eq. (4.7) is a function of y only, then both sides should equal a constant. A constant, $-\lambda^2$, is selected as in Eq.(4.8) such that the equation in x (with both boundary conditions homogeneous) will give an eigenvalue equation.

$$\frac{1}{X}\frac{d^2X}{dx^2} = -\lambda^2 = -\frac{1}{Y}\frac{d^2Y}{dy^2} \tag{4.8}$$

The eigenvalue equation in x is

$$\frac{d^2X}{dx^2} + \lambda^2 X = 0 \quad \text{in} \quad 0 \le x \le a \tag{4.9}$$

At x = 0, X = 0. At x = a, X = 0.

The differential equation for Y separation is

$$\frac{d^2Y}{dx^2} - \lambda^2 Y = 0 \quad \text{in} \quad 0 \le y \le b \tag{4.10}$$

At y = 0, T(x,0) = 0 and X(0) = 0. Thus Y(0) = 0.
At y = b, T(x,b) = f(x) and X(0) = X(a) = 0.

The solution to Eq.(4.9) is the eigenfunction X(x) = sin λ_n x, where $\lambda_n = \dfrac{n\pi}{a}$, n = 1,2,3.... The normalized integral N = a/2.

The solution for the Y-separation equation may be assumed to take the form

$$Y(y) = c_1 \sinh \lambda_n y + c_2 \cosh \lambda_n y.$$ (4.11)

The boundary condition at y = 0 makes $c_2 = 0$. So,

$$Y(y) = c_1 \sinh \lambda_n y.$$ (4.12)

Since T(x,y) = X(x)Y(y),

$$T(x,y) = c_1 \sin(\lambda_n x) \sinh(\lambda_n y).$$ (4.13)

Since there are multiple values of λ_n, the complete solution for the temperature should be taken as a linear combination of all these possible solutions. Therefore,

$$T(x,y) = \sum_{n=1}^{\infty} c_n \sinh \lambda_n y \sin \lambda_n x$$ (4.14)

where c_n's are unknown expansion coefficients. They will be determined by the constraint at y = b, T(x,b) = f(x).

So, $$f(x) = \sum_{n=1}^{\infty} c_n \sinh(\lambda_n b) \sin \lambda_n x$$ (4.15)

Since $\sin \lambda_n x$ has an orthogonality property, we multiply Eq.(4.15) by $\sin \lambda_m x$ and integrate from x = 0 to x = a.

$$\int_0^a f(x') \sin \lambda_m x' \, dx' = \sum_{n=1}^{\infty} c_n \sinh \lambda_n b \int_0^a \sin \lambda_m x' \sin \lambda_n x' dx'$$ (4.16)

where x' is the dummy variable of integration.

$$\int_0^a f(x')\sin\lambda_m x'\,dx' = c_m \sinh(\lambda_m b).N \quad \text{where} \quad N = \frac{a}{2} \qquad (4.17)$$

Hence, $c_n = \dfrac{2}{a\sinh(\lambda_n b)} \int_0^a f(x')\sin(\lambda_n x')\,dx'$ (4.18)

Therefore,

$$T(x,y) = \frac{2}{a}\sum_{n=1}^{\infty}\{\ \frac{\sinh(\lambda_n y)}{\sinh(\lambda_n b)}\sin(\lambda_n x)\ \}.\int_0^a f(x')\sin(\lambda_n x')\,dx'$$

(4.19)

where $\lambda_n = \dfrac{n\pi}{a}$.

4.3 Heat Flow

Since the heat flow or heat flux at the boundaries is often of interest, we can calculate it as shown below.

$$q(x,y)\Big|_{x=0} = -k\frac{\partial T}{\partial x}(x,y)\Big|_{x=0} \quad \frac{Btu}{hr.ft^2}\left(\frac{W}{m^2}\right). \qquad (4.20)$$

Using the temperature distribution, Eq. (4.19), as an example, at $x = 0$ the heat flux

$$q(x,y) = -\frac{2k}{a}\sum_{n=1}^{\infty}\frac{\sinh(\lambda_n y)}{\sinh(\lambda_n b)}\lambda_n \overset{=1}{\cos(\lambda_n \cancel{0})}\int_0^a f(x')\sin(\lambda_n x')\,dx'.$$

(4.21)

At $y = b$, $q(x)\Big|_{y=b}$ = function of x. Thus Q is obtained by integrating over the length for a unit depth.

Example 4.2

Problem: From Eq. (4.18), obtain the steady-state temperature distribution T(x,y) in a rectangular cross section with the following boundary conditions:

   At x = 0, T(x,y) = 0.
   At x = a, T(x,y) = 0.
   At y = 0, T(x,y) = f(x).
   At y = b, T(x,y) = 0.

Solution

  Use the transformation $\gamma = b - y$. Hence the boundary conditions in y become

   At $\gamma = b$, T(x,γ) = f(x).
   At $\gamma = 0$,   T(x,γ) = 0.

From Eq. (4.18), the temperature distribution is thus

$$T(x,\gamma) = \frac{2}{a}\sum_{n=1}^{\infty}\left\{\frac{\sinh(\lambda_n\gamma)}{\sinh(\lambda_n b)}\sin(\lambda_n x).\int_0^a f(x')\sin(\lambda_n x')dx'\right.$$

which can be written as

$$T(x,y) = \frac{2}{a}\sum_{n=1}^{\infty}\left\{\frac{\sinh(\lambda_n\{b-y\})}{\sinh(\lambda_n b)}\sin(\lambda_n x).\int_0^a f(x')\sin(\lambda_n x')dx'\right..$$

Example 4.3

Problem: Obtain the steady-state temperature distribution T(x,y) in a rectangular cross-section with k constant and the following boundary conditions:

   At x = 0, T(x,y) = 0.
   At x = a, T(x,y) = $T_1 \sin(\pi y/b)$.
   At y = 0, T(x,y) = 0.
   At y = b, T(x,y) = 0.

Solution

Assuming that $T(x,y) = X(x)Y(y)$, the governing equation is separated as

$$\frac{1}{X}\frac{d^2 X}{dx^2} = -\frac{1}{Y}\frac{d^2 Y}{dy^2} = +\lambda^2,$$ where $+\lambda^2$ has been chosen so that the eigenfunction equation is in y. The solution for $Y(y)$ is obtained from Table 4.1, case 1, as $Y(\lambda,y) = \sin \lambda y$ where $\lambda_n = \frac{n\pi}{b}, n = 1,2,3,...$ and $1/N = 2/b$.

The solution for $X(\lambda,x)$ function satisfying the boundary condition at $x = 0$ is $X(\lambda,x) = \sinh \lambda x$.

The formal solution for $T(x,y)$ is expressed as

$$T(x,y) = \sum_{n=1}^{\infty} c_n \sinh \lambda_n x \sin \lambda_n y.$$

Using the boundary condition at $x = a$,

$$T_1 \sin \lambda_1 y = \sum_{n=1}^{\infty} c_n \sinh \lambda_n a \sin\lambda_n y, \text{ since } \lambda_1 = \frac{\pi}{b}.$$

Operating both sides by $\int_0^b \sin \lambda_m y \, dy$,

$$T_1 \int_0^b \sin \lambda_m y . \sin \lambda_1 y \, dy = \sum_{n=1}^{\infty} c_n \sinh \lambda_n a . \delta_{mn} . N$$

where $\delta_{mn} = \begin{cases} 1 & m = n \\ 0 & m \neq n \end{cases}$.

Hence, $c_n = \dfrac{T_1}{N \sinh \lambda_n a} \cdot \delta_{n1}$

where $\delta_{n1} = \begin{cases} 1 & n = 1 \\ 0 & n \neq 1 \end{cases}$.

The complete solution for the temperature becomes

$$T(x,y) = \sum_{n=1}^{\infty} \delta_{n1} \cdot T_1 \cdot \left(\frac{\sinh \lambda_n x}{\sinh \lambda_n a} \right) \sin \lambda_n y$$

$$T(x,y) = T_1 \cdot \left(\frac{\sinh \lambda_1 x}{\sinh \lambda_1 a} \right) \sin \lambda_1 y \qquad \text{where } \lambda_1 = \frac{\pi}{b}.$$

It is noted that the expression of the temperature distribution contains one term, rather than a summation of an infinite number of terms. This may be expected from the following practical consideration. The medium is passive, without heat generation; it is at steady state. Only one boundary has a sinusoidal temperature distribution imposed on it, the other three being homogeneous. A modified sinusoidal temperature distribution within the medium is not surprising. Hence, the single term in the expression is adequate to describe this modified sinusoidal temperature distribution.

Contrast this case with one where the nonhomogeneous boundary condition is a constant temperature. One should expect a summation of an infinite number of terms in the final expression for temperature because it takes a large number of sinusoidal terms to describe the approximately constant temperature along lines parallel to this boundary.

4.4 Separation into Simpler Problems

When the problem has more than one nonhomogeneous boundary condition, the principle of superposition can be used. For instance, the Laplacian equation in the region $0 \leq x \leq a$ and $0 \leq y \leq b$ is subjected to the following boundary conditions:

$$T\big|_{x=0} = 0 \qquad\qquad T\big|_{y=0} = 0$$

$$T\big|_{x=a} = f_1(y) \qquad\qquad T\big|_{y=b} = f_2(x)$$

There are two nonhomogeneous boundary conditions. We can use the superposition theorem, and write the solution of T(x,y) in terms of $T_1(x,y)$ and $T_2(x,y)$. Hence,

$$T(x,y) = T_1(x,y) + T_2(x,y) \tag{4.22}$$

where $T_1(x,y)$ is the solution to the Laplacian equation subjected to the boundary conditions

$$T\big|_{x=0} = 0 \qquad\qquad T\big|_{y=0} = 0$$

$$T\big|_{x=a} = f_1(y) \qquad\qquad T\big|_{y=b} = 0 \tag{4.23}$$

and $T_2(x,y)$ is the solution to the Laplacian equation subjected to the boundary conditions

$$T\big|_{x=0} = 0 \qquad\qquad T\big|_{y=0} = 0$$

$$T\big|_{x=a} = 0 \qquad\qquad T\big|_{y=b} = f_2(x). \tag{4.24}$$

4.5 Summary of Steps Used in the Method of Separation of Variables

The following are the steps used in the method of separation of variables in solving the Laplacian equation.

(1)   Assume the temperature function, T, is a product of X and Y, where X is a function of x only, and Y is a function of y only.

(2)   Substitute XY into the Laplacian equation and select a constant such that the equation with the coordinate that has

homogeneous boundary conditions is the eigenfunction
equation.

(3)   Say Y is the non-eigenfunction equation, use the
homogeneous boundary condition to obtain the constant of
integration.

(4)   The eigenfunction X and Y gives the formal solution for the
temperature T.

(5)   Use the nonhomogeneous boundary condition and the
orthogonality principle to obtain the expansion coefficient.

4.6 Steady-State Heat Conduction in a Two-Dimensional Fin

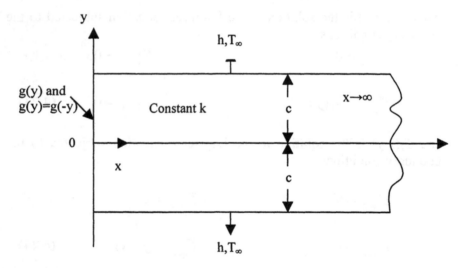

Figure 4.2   Infinitely long two-dimensional fin with convective
boundary conditions.

Consider the infinitely long two-dimensional fin with convective
boundary conditions, as shown in Fig. 4.2. The temperature of the
surrounding fluid is T_∞, and the heat transfer coefficient is h. The x-axis
is taken to be the axis of symmetry. The governing equations of the
problem may be stated as below.

$$\frac{\partial^2 T}{\partial x^2} + \frac{\partial^2 T}{\partial y^2} = 0 \qquad\qquad (4.25)$$

$$T(0,y) = g(y) \quad \text{and} \quad T(x \to \infty, y) = T_\infty \qquad (4.26a,b)$$

$$\frac{\partial T(x,0)}{\partial y} = 0 \quad \text{and} \quad -k\frac{\partial T(x,c)}{\partial y} = h\{T(x,c) - T_\infty\}. \qquad (4.26c,d)$$

The problem as mathematically expressed above does not satisfy the condition that three of the four boundary conditions be homogeneous. Thus, the method of separation of variables is not applicable with this expression of the problem. However, with the introduction of

$$\psi(x,y) = T(x,y) - T_\infty \qquad (4.27)$$

the expression of the problem becomes

$$\frac{\partial^2 \Psi}{\partial x^2} + \frac{\partial^2 \Psi}{\partial y^2} = 0 \qquad (4.28)$$

$$\psi(0,y) = g(y) - T_\infty = G(y) \quad \text{and} \quad \psi(x \to \infty, y) = 0 \qquad (4.29a,b)$$

$$\frac{\partial \psi(x,0)}{\partial y} = 0 \quad \text{and} \quad -k\frac{\partial \psi(x,c)}{\partial y} = h\psi(x,c) \qquad (4.29c,d)$$

which satisfies all the requirements of the method of separation of variables. First, assume a product solution of the form

$$\psi(x, y) = X(x)Y(y). \qquad (4.30)$$

Introduce this expression into Eq. (4.28) and divide each term of the resulting expression by XY,

$$\frac{1}{X}\frac{d^2 X}{dx^2} = -\frac{1}{Y}\frac{d^2 Y}{dy^2} = \lambda^2 \qquad (4.31)$$

or $\quad \dfrac{d^2 Y}{dy^2} + \lambda^2 Y = 0 \qquad (4.32)$

and $\qquad \dfrac{d^2X}{dx^2} - \lambda^2 X = 0$ (4.33)

The solution of Eq. (4.27) is given by

$$\psi(x,y) = (C_1 e^{-\lambda x} + C_2 e^{\lambda x})(D_1 \cos \lambda y + D_2 \sin \lambda y)$$ (4.34)

The sign of the separation constant λ^2 is selected such that the homogeneous y direction results in an eigenvalue problem. In other words,

$$\dfrac{d^2Y}{dy^2} + \lambda^2 Y = 0$$ (4.35)

$$\dfrac{dY(0)}{dy} = 0$$ (4.36a)

$$k\dfrac{dY(c)}{dy} + hY(c) = 0$$ (4.36b)

This is a Sturm-Liouville (eigenfunction) problem with the following characteristic eigenfunctions:

$$\varphi_n(x) = \cos \lambda_n y.$$ (4.37)

The eigenvalues are the positive roots of the transcendental equation:

$$\lambda_n \tan(\lambda_n c) = \dfrac{h}{k} \qquad \text{where } n = 1,2,3,\dots$$ (4.38)

that is obtained from the use of the boundary condition (4.36b).

The use of the boundary conditions (4.29b-d) gives the following result:

$$\psi(x, y) = \sum_{n=1}^{\infty} c_n e^{-\lambda_n x} \cos \lambda_n y. \qquad (4.39)$$

The nonhomogeneous boundary condition (4.29a) provides that

$$G(y) = g(y) - T_\infty = \sum_{n=1}^{\infty} c_n \cos \lambda_n y \qquad (4.40)$$

where the expansion coefficients c_n are given by

$$c_n = \frac{2\lambda_n}{\lambda_n c + \sin(\lambda_n c)\cos(\lambda_n c)} \int_0^c \{g(y) - T_\infty\}\cos(\lambda_n y)dy. \qquad (4.41)$$

Therefore, the temperature distribution is

$$\psi(x, y) = T(x, y) - T_\infty$$
$$= 2\sum_{n=1}^{\infty} \frac{\lambda_n e^{-\lambda_n x} \cos \lambda_n y}{\lambda_n c + \sin(\lambda_n c)\cos(\lambda_n c)} \int_0^c \{g(y') - T_\infty\}\cos(\lambda_n y')dy'. $$
$$(4.42)$$

For the special case of $g(y) = T_0 = $ constant, the temperature distribution may be expressed as follows:

$$\frac{T(x, y) - T_\infty}{T_0 - T_\infty} = 2\sum_{n=1}^{\infty} \frac{\sin(\lambda_n c)}{\lambda_n c + \sin(\lambda_n c)\cos(\lambda_n c)} e^{-\lambda_n x} \cos(\lambda_n y).$$
$$(4.43)$$

The characteristic values λ_n in Eqs. (4.42) and (4.43) can be obtained from the transcendental Eq. (4.38), which may be written as

$$\tan(\lambda_n c) = \frac{Bi}{\lambda_n c} \qquad \text{or} \qquad \cot(\lambda_n c) = \frac{\lambda_n c}{Bi} \qquad (4.44)$$

where $Bi = hb/k$. The roots of either form of Eq. (4.44) are infinite in number. They should be determined numerically or graphically.

4.7 Transient Heat Conduction in a Slab

For one-dimensional, time dependent heat conduction with constant properties, we can select a slab for convenience of illustration. The typical problem is governed by the energy equation:

$$\frac{\partial^2 T(x,t)}{\partial x^2} = \frac{1}{\alpha}\frac{\partial T(x,t)}{\partial t} \quad \text{in } 0 \leq x \leq t \text{ for } t > 0. \tag{4.45}$$

Boundary conditions, for example, can be

$$\text{At } x = 0, t > 0, \qquad \frac{\partial T(x,t)}{\partial x} = 0 \tag{4.46}$$

$$\text{At } x = L, t > 0, \quad k\frac{\partial T(x,t)}{\partial x} + hT(x,t) = 0. \tag{4.47}$$

The initial condition is

$$\text{At } t = 0, \qquad T(x,t) = F(x) \text{ in } 0 \leq x \leq t. \tag{4.48}$$

It is assumed that $T(x,t) = X(x)\Gamma(t)$.

Hence,
$$\frac{1}{X}\frac{d^2 X}{dx^2} = \frac{1}{\alpha\Gamma}\frac{d\Gamma}{dt} \equiv -\lambda^2 \tag{4.49}$$

Here, "$-\lambda^2$" is selected to allow for the temperature decay with time. This later behavior is expected from physics.

The eigenvalue problem becomes

$$\frac{d^2 X}{dx^2} + \lambda^2 X = 0 \text{ in } 0 \leq x \leq L. \tag{4.50}$$

The boundary conditions are as follows:

At x = 0, $\dfrac{dX}{dx} = 0$ (4.51)

At x = L, $k\dfrac{dX}{dx} + hX = 0$.   (4.52)

The ordinary differential equation in time t is

$$\dfrac{d\Gamma}{dt} + \lambda^2 \alpha \Gamma = 0 \quad \text{for } t > 0 \tag{4.53}$$

The solution for Eq.(4.53) is $\Gamma(t) = \exp(-\alpha\lambda_n^2 t)$ and so on to complete the solution for the temperature.

Example 4.4

Problem: Solve the one-dimensional, transient heat conduction problem with the following boundary conditions:

At x = 0, t > 0, $\dfrac{\partial T(x,t)}{\partial x} = 0$ (i)

At x = L, t > 0, $k\dfrac{\partial T(x,t)}{\partial x} + hT(x,t) = 0$.   (ii)

The initial condition is

At t = 0,   T(x,t) = T_o in $0 \le x \le t$ (iii)

Solution

    The solution for $\Gamma(t)$ separation is $\Gamma(t) = \exp(-\alpha\lambda_n^2 t)$. The solution for X(x) separation is $X(\lambda_n, x) = \cos \lambda_n x$ where λ_n are the roots of $\lambda\tan(\lambda L) = h/k = H$ and the normalization integral N is given by:

$$\dfrac{1}{N} = \dfrac{2(\lambda^2 + H^2)}{L(\lambda^2 + H^2) + H}. \tag{iv}$$

The solution for T(x,t) is written in the form

$$T(x,t) = \sum_{n=1}^{\infty} c_n e^{-\alpha \lambda_n^2 t} \cos \lambda_n x. \qquad \text{(v)}$$

The initial conditions give $\qquad T_o = \sum_{n=1}^{\infty} c_n \cos \lambda_n x \qquad \text{(vi)}$

where c_n is determined as $c_n = \dfrac{T_o}{N} \int_0^L \cos \lambda_n x \, dx = \dfrac{T_o}{N \lambda_n} \sin \lambda_n L$.(vii)

Then, the complete solution becomes

$$T(x,t) = T_o \sum_{n=1}^{\infty} \frac{\sin \lambda_n L}{N \lambda_n} e^{-\alpha \lambda_n^2 t} \cos \lambda_n x \qquad \text{(viii)}$$

where

$$N = \frac{L(\lambda_n^2 + H^2) + H}{2(\lambda_n^2 + H^2)} \quad \text{and} \quad \lambda_n \text{ are the roots of } \lambda \tan \lambda L = \frac{h}{k} = H.$$

PROBLEMS

4.1.   Solve for the eigenfunctions and eigenvalues in Table 4.1.

4.2.   Derive an expression for the steady-state temperature distribution $T(x,y)$ by solving the differential equation

$$\frac{\partial^2 T}{\partial x^2} + \frac{\partial^2 T}{\partial y^2} = 0$$

in a rectangular region $0 \le x \le a$, $0 \le y \le b$, subject to the boundary conditions $T(0,y) = 0$, $T(a,y) = 0$, $T(x,0) = 0$, $T(x,b) = Ax + f(x)$.

4.3.   Consider two-dimensional, steady-state heat transfer without heat generation in a rectangular region. Solve using an exact analytical method for the problem where the boundaries are of the first-kind homogeneous everywhere except at $x = 0$, where $T = T_1 \cos(\pi y / b)$ and at $y=0$, $T = T_2 \cos(\pi x / a)$. The region has dimensions a x b.

4.4.   A long iron bar has a rectangular cross-section with the following temperature boundary conditions: $T(0,y) = T_1e(y)$, $T(a,y) = T_2f(y)$, $T(x,0) = 0$, $T(x,b) = 0$ where T_1 and T_2 are constants and $e(y)$ and $f(y)$ are functions of y. Find the steady-state temperature distribution.

4.5.   Solve the problem of the Laplacian equation of two-dimensional, steady-state heat conduction, with the following boundary conditions: $T(0,y) = T_1$, $T(a,y) = T_2$, $T(x,0) = T_3$, $T(x,b) = T_4$.

4.6.   Derive an expression for the steady-state temperature distribution $T(x,y)$ by solving the differential equation

$$\frac{\partial^2 T}{\partial x^2} + \frac{\partial^2 T}{\partial y^2} = 0$$

in a rectangular region $0 \le x \le a$, $0 \le y \le b$, subject to the boundary conditions $T(0,y) = 0$, $T(a,y) = By^2+g(y)$, $T(x,0) = 0$, $T(x,b) = 0$.

4.7.   A straight rectangular fin has a thickness c in the direction and is extremely long in the y direction. If the thermal conductivity is constant, determine the steady-state temperature distribution in this fin for the following boundary conditions:

$$T(0,y) = 0, \qquad \frac{\partial T(c,y)}{\partial x} = 0 \quad \text{and} \quad T(x,0) = e(x).$$

4.8.   Obtain the steady-state temperature distribution $T(x,y)$ in a rectangular cross section with constant k and the following boundary conditions:
At $x = 0$, $T(x,y) = T_o\cos(\pi y/b)$.
At $x = a$, $T(x,y) = 0$.
At $y = 0$, $T(x,y) = 0$.
At $y = b$, $T(x,y) = 0$.

4.9.   Obtain the steady-state temperature distribution $T(x,y)$ in a rectangular cross section with constant k and the following boundary conditions:
At $x = 0$, $T(x,y) = 0$.
At $x = a$, $T(x,y) = T_1 \cos(\pi y/2b)$.
At $y = 0$, $dT/dy = 0$.

At $y = b$, $T(x,y) = 0$.

4.10. Consider steady-state heat transfer in a rectangular cross section, with constant thermal conductivity. Design the four nonhomogeneous boundary conditions such that the final expression of the temperature distribution consists only of four terms and not an infinite series of terms. Write down this expression. (Hint: Look at the solutions of Ex. 4.3 and Prob. 4.9.)

4.11. The initial temperature of a wall is T_1, and it extends from $x = 0$ to $x = L$. For times $t \geq 0$, the surface at $x = 0$ is kept insulated and at $x = L$ is kept at a constant T_2. Derive
  (i) the unsteady temperature distribution in the wall for $t > 0$,
  (ii) the mean temperature across the wall as a function of time,
  (iii) the instantaneous rate of heat transfer from the slab.

4.12. A wall, $0 \leq x \leq L$, is initially at a temperature $K(x)$, for times $t > 0$ the boundaries at $x = 0$ and $x = L$ are kept insulated. In other words, $\partial T/\partial x = 0$ at $x = 0$ and $x = L$. Find the temperature distribution $T(x,t)$ in the wall.

4.13. A piece of beef steak is cooked either in a microwave oven or a radiant heating oven. Sketch temperature distributions at specific times during the heating and the cooling processes in each oven.

Separation of Variables

To solve Laplacian by separation of variables
Assume that x and y are the two variables
X is a function of x variable only,
Y is a function of y variable only.

Put XY into the Laplacian equation
So that the following is the resulting condition
The variable with homogeneous boundary condition
Produces the equation with the eigenfunctions.

Say capital Y is the non-eigenfunction equation
Use associated homogeneous boundary condition
To find expression for the constant of integration
XY is for temperature T, formal solution.

Use the nonhomogeneous boundary condition
And orthogonality principle of eigenfunctions
To obtain coefficient accompanying expansion
Hence temperature T has a final expression.

K.V. Wong

5

Numerical Analysis in Conduction

5.1   Introduction

With the development of high-speed personal computers, it is very convenient to use numerical techniques to solve heat transfer problems. The finite-difference method and the finite-element method are two popular and useful methods. The finite-element method is not as direct, conceptually, as the finite-difference method. It has some advantages over the finite-difference method in solving heat transfer problems, especially for problems with complex geometries.

We will discuss the solution of steady-state and unsteady-state heat conduction problems in this chapter, using the finite-difference method. The finite-difference method comprises the replacement of the governing equations and corresponding boundary conditions by a set of algebraic equations. The discussion here is not meant to be exhaustive in its mathematical rigor. The basics are presented, and the solution of the finite-difference equations by numerical methods are discussed. The solution of convection problems using the finite-difference method is discussed in a later chapter.

5.2   Finite Difference of Derivatives

The finite difference of derivatives involves the approximation of a differential equation or a boundary condition by algebraic equations. Consider the function T(x) shown in Fig. 5.1. The definition of the derivative of T(x) at x_i is given by

$$\left.\frac{dT}{dx}\right|_{x_i} = \lim_{\Delta x \to 0} \frac{T(x_i + \Delta x) - T(x_i)}{\Delta x}. \tag{5.1}$$

In taking the limit, we obtain

$$\left.\frac{dT}{dx}\right|_{x_i} \approx \frac{T(x_i + \Delta x) - T(x_i)}{\Delta x} = \frac{T_{i+1} - T_i}{\Delta x}. \qquad (5.2)$$

This approximate relation is an algebraic expression for the derivative at x_i. It is called the forward difference form of the first derivative since it involves the value of x at the point I and the point forward of i, that is, at i+1.

Another approximate relation for the gradient of T at x_i may be written as

$$\left.\frac{dT}{dx}\right|_{x_i} \approx \frac{T_i - T_{i-1}}{\Delta x} \qquad (5.3)$$

This approximate relation is called the backward-difference form of the first derivative at x_i. It involves the value of x at the point i and the point backward of i, that is, at i-1.

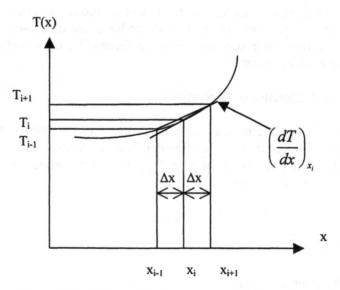

Figure 5.1   Finite-difference approximation of a derivative.

As illustrated in Fig. 5.1, an approximation that is more accurate than either the forward-difference form or the backward-difference form may be written as

$$\left.\frac{dT}{dx}\right|_{x_i} \approx \frac{T_{i+1} - T_{i-1}}{2\Delta x}. \tag{5.4}$$

This approximate relation is called the central-difference form of the first derivative.

The central-difference form has a truncation error (or discretization error) of the order of the magnitude of $(\Delta x)^2$, whereas the truncation error of both the forward-difference and backward-difference forms have a truncation error of the order of the magnitude of Δx.

The second derivative of $T(x)$ can be written in central-difference approximation as

$$\left.\frac{d^2T}{dx^2}\right|_{x_i} \approx \frac{\left(\frac{dT}{dx}\right)\Big|_{x_i+\Delta x/2} - \left(\frac{dT}{dx}\right)\Big|_{x_i-\Delta x/2}}{\Delta x}. \tag{5.5}$$

Substitute the central-difference forms of $(dT/dx)|_{x_i+\Delta x/2}$ and $(dT/dx)|_{x_i-\Delta x/2}$ into Eq. (5.5) and get

$$\left.\frac{d^2T}{dx^2}\right|_{x_i} \approx \frac{T_{i+1} + T_{i-1} - 2T_i}{(\Delta x)^2}. \tag{5.6}$$

The relation Eq. (5.6) has a truncation error of order $(\Delta x)^2$.

5.3   Finite Difference Equations for 2-D Rectangular Steady-State Conduction

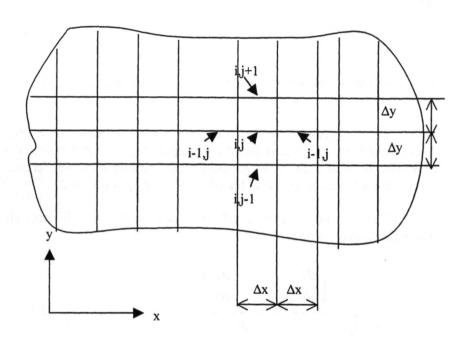

Figure 5.2   Grid points for a rectangular grid system.

In a 2-D rectangular region undergoing heat transfer, the first step is to divide the region into a rectangular grid system. For a solid undergoing steady-state heat conduction, with constant thermal conductivity, Laplace's equation applies.

$$\frac{\partial^2 T}{\partial x^2} + \frac{\partial^2 T}{\partial y^2} = 0 \qquad\qquad (5.7)$$

This equation is pointwise continuous; that is, it is applicable throughout the region. In setting up the rectangular grid system, we are deriving finite difference equations that are only valid at the grid points. The resulting solutions obtained for the finite difference equations are therefore valid at these points only.

The grid points are identified by two subscripts, say i and j; i is the number of Δx increments and j the number of Δy increments. At grid point (i,j), apply Eq. (5.6) and the corresponding equation in y to each second-order derivative, giving the expression

$$\frac{T_{i+1,j} + T_{i-1,j} - 2T_{i,j}}{(\Delta x)^2} + \frac{T_{i,j+1} + T_{i,j-1} - 2T_{i,j}}{(\Delta y)^2} = 0. \qquad (5.8)$$

If the grid is square rather than rectangular, $\Delta x = \Delta y$, this expression reduces to

$$T_{i+1,j} + T_{i-1,j} + T_{i,j+1} + T_{i,j-1} - 4T_{i,j} = 0. \qquad (5.9)$$

In other words,

$$T_{i,j} = 0.25(\, T_{i+1,j} + T_{i-1,j} + T_{i,j+1} + T_{i,j-1}). \qquad (5.10)$$

Equation (5.10) expresses the temperature of grid point (i,j) in terms of the temperatures of four neighbouring grid points (i+1,j), (i-1,j), (i,j+1) and (i,j-1).

The finite difference equation (5.9) is valid for all interior points. If the temperature is known throughout the boundaries (i.e., first kind boundary conditions), then the application of Eq. (5.9) to all the interior points is adequate to allow the temperatures at these points to be solved. If there are n interior points, the procedure will give n simultaneous algebraic equations, which can then be solved.

The number of grid points determine the detail to which the temperature distribution is calculated. A smaller grid gives a more detailed solution, but results in more algebraic equations to solve. If done manually, it involves more time. But with fast personal computers nowadays, a 2-D problem seldom presents a problem. A coarser grid will provide a less detailed solution of the temperature distribution. However, sometimes that is all that is necessary to get an idea of the temperatures in the solid.

5.4 Finite Difference Representation of Boundary Conditions

The expression of the heat conduction problem in finite difference form is completed by also expressing the boundary conditions in finite differences. If the first kind of boundary condition exists in the whole boundary, then the known boundary temperatures enter into the finite difference equations. Each equation for the grid points next to the boundary will have a prescribed term in it. If any part or the whole boundary is not at the first kind of boundary condition, then the boundary conditions have to be finite differenced. The following subsections consider the finite differencing of these different boundary conditions.

Boundary with Convective Heat Transfer

Consider a grid point (i,j) on a boundary subjected to convective heat exchange with an environment at temperature T_∞, and with a heat transfer coefficient h. The finite difference formulation of the boundary condition can either be obtained by converting the corresponding differential boundary condition, or by writing an energy balance on the shaded volume element shown in the figure. The conservation of energy for the element gives

Rate of heat entering volume
element through boundaries $= 0$ (5.11)

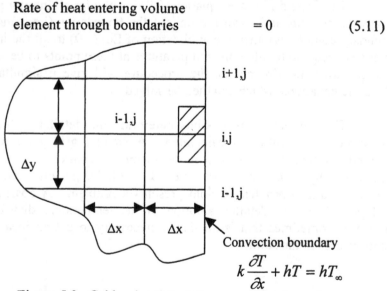

$$k \frac{\partial T}{\partial x} + hT = hT_\infty$$

Figure 5.3   Grid point (i,j) at the convection boundary.

Consider a volume with unit depth into the plane of Fig 5.3. The energy balance gives

$$k\frac{T_{i-1,j}-T_{i,j}}{\Delta x}\Delta y + k\frac{T_{i,j-1}-T_{i,j}}{\Delta y}\frac{\Delta x}{2}+k\frac{T_{i,j+1}-T_{i,j}}{\Delta y}\frac{\Delta x}{2}+h(T_\infty-T_{i,j})\Delta y=0$$

$$(5.12)$$

If the grid is square, that is, $\Delta x = \Delta y$, then Eq. (5.12) reduces to

$$0.5\left(2T_{i-1,j}+T_{i,j+1}+T_{i,j-1}\right)+\frac{h\Delta x}{k}T_\infty-\left(2+\frac{h\Delta x}{k}\right)T_{i,j}=0 \qquad (5.13)$$

Hence, Eq. (5.13) is used for the grid points on the boundary, and Eq. (5.9) is used for the interior points. Equation (5.13) is, however, not applicable for a grid point in the corner, Fig. 5.4. Consider the corner section in the figure. The conservation of energy gives

$$0.5\left(T_{i-1,j}+T_{i,j-1}\right)+\frac{h\Delta x}{k}T_\infty-\left(1+\frac{h\Delta x}{k}\right)T_{i,j}=0.$$

$$(5.14)$$

Equation (5.14) should be used for corner grid points undergoing convective heat transfer, when the convective heat transfer coefficient is prescribed.

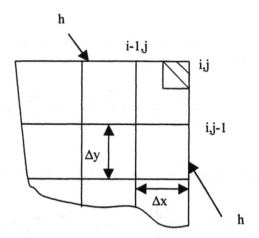

Figure 5.4   Grid point (i,j) on a corner with convective boundary
conditions.

Insulated Boundary

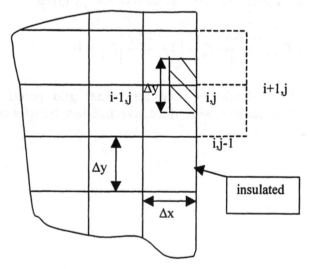

Figure 5.5   Grid point on an insulated boundary.

When the boundary is insulated, as in Fig. 5.5, the finite difference equations for the grid points on the insulated boundary may be obtained as previously. The conservation of energy gives

$$T_{i,j+1} + T_{i,j-1} + 2T_{i-1,j} - 4T_{i,j} = 0. \qquad (5.15)$$

Since $\dfrac{\partial T}{\partial x} = 0$ on the boundary, the temperatures at the boundary may be set equal to the temperatures of the grid points on the penultimate column. In other words, $T_{k-1,j} = T_{k,j}$, where k designates the x-value of the column of grid points on the boundary. A similar action may be taken for the insulated boundary which is parallel to the x-axis. In that case, $\dfrac{\partial T}{\partial y} = 0$ on the boundary, and the temperatures at the boundary may be set equal to the temperatures of the grid points on the penultimate row. In other words, $T_{i-1,m} = T_{i,m}$, where m designates the y-value of the row of grid points on the boundary.

Irregular Boundaries

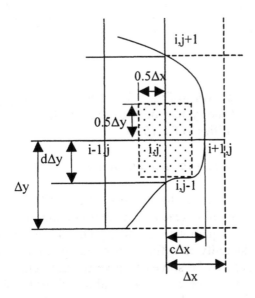

Figure 5.6   Grid point next to an irregular boundary.

For simple geometries, grid points may be made to lie on the boundaries exactly. For irregular boundaries, the boundary does not fall on regular grid points. Numerical methods are an excellent way to treat such physical boundaries. Consider the irregular boundary shown in Fig. 5.6, and assume that the temperatures are known at the boundaries. Equation (5.9) cannot be used for the grid point (i,j) next to the boundary. Using the conservation of energy to the system (shaded rectangle) shown in the figure,

$$k\frac{(1+d)\Delta y}{2}\frac{T_{i-1,j}-T_{i,j}}{\Delta x}+k\frac{(1+d)\Delta y}{2}\frac{T_{i+1,j}-T_{i,j}}{c\Delta x}$$

$$+k\frac{(1+c)\Delta x}{2}\frac{T_{i,j-1}-T_{i,j}}{d\Delta y}+k\frac{(1+c)\Delta x}{2}\frac{T_{i,j+1}-T_{i,j}}{\Delta y}=0. \qquad (5.16)$$

If there is an additional condition that $\Delta x = \Delta y$, then Eq. (5.16) simplifies to

$$\frac{1}{c(1+c)}T_{i+1,j}+\frac{1}{1+c}T_{i-1,j}+\frac{1}{1+d}T_{i,j+1}+\frac{1}{d(1+d)}T_{i,j-1}-\left(\frac{1}{c}+\frac{1}{d}\right)T_{i,j}=0$$

$$(5.17)$$

where c and d are shown in Fig. 5.6. The grid point (i,j) is not at the geometric center of the system defined around (I,j). If $c = d = \Delta x = \Delta y$, then Eq. (5.17) simplifies to Eq. (5.9).

5.5 Solution of Finite Difference Equations

The finite difference formulation of the differential equation and boundary conditions of a steady-state heat conduction problem gives a system of algebraic equations. These equations are linear except when the thermal conductivity is a function of temperature. The system of algebraic equations can be solved to give the temperature values at the various grid points. For a small number of grid points, the finite difference equations may be solved using a calculator, but a larger number require the use of a computer.

Gaussian Elimination Method

This method can be used to solve a coupled system of linear algebraic equations. The finite differencing of a two-dimensional, steady-state heat conduction equation gives a set of algebraic equations which form a banded matrix, as shown in Eq. (5.18). The nonzero elements of the matrix are in a band on either side of the diagonal. In general, a banded matrix is solved efficiently on a computer using the gaussian elimination method. As an illustration, consider the banded matrix shown in Eq. (5.18). The matrix is transformed into an upper diagonal form in the following manner. The first equation in the system of Eq. (5.18) is used to eliminate the nonzero elements a_{21} and a_{31} in the first column. In other words, the first equation is multiplied by a_{21}/a_{11}, and the resulting equation is subtracted from the second equation in order to eliminate a_{21}. Similarly, a_{31} is eliminated from the third equation. The second equation is then used to eliminate a_{32} and a_{42}. The third equation is used to eliminate a_{43}, and so on. When this procedure is carried out to the last equation, the result is an upper diagonal matrix as shown in Eq. (5.19). The last equation in Eq. (5.19) directly gives T_n. With T_n known, the temperature T_{n-1} is determined from the (n-1)th equation, and the computations are carried out until T_1 is found from the first equation. Computer programs are easily available to solve a system of simultaneous algebraic equations using the Gaussian elimination method.

$$
\begin{bmatrix}
a_{11} & a_{12} & a_{13} & 0 & .. & .. & 0 & 0 \\
a_{21} & a_{22} & a_{23} & a_{24} & .. & .. & .. & 0 \\
a_{31} & a_{32} & a_{33} & a_{34} & .. & .. & .. & 0 \\
 & & & a_{44} & .. & .. & .. & 0 \\
\multicolumn{8}{c}{\dotfill} \\
\multicolumn{8}{c}{\dotfill} \\
\multicolumn{8}{c}{\dotfill} \\
\multicolumn{8}{c}{\dotfill} \\
0 & 0 & 0 & .. & .. & a_{n,n-2} & a_{n,n-1} & a_{m,n}
\end{bmatrix}
\begin{bmatrix}
T_1 \\ T_2 \\ T_3 \\ T_4 \\ . \\ . \\ . \\ . \\ T_n
\end{bmatrix}
=
\begin{bmatrix}
C_1 \\ C_2 \\ C_3 \\ C_4 \\ . \\ . \\ . \\ . \\ C_n
\end{bmatrix}
\qquad (5.18)
$$

$$
\begin{bmatrix}
a_{11}^* & a_{12}^* & a_{13}^* & 0 & .. & .. & .. & 0 \\
0 & a_{22}^* & a_{23}^* & a_{24}^* & .. & .. & .. & 0 \\
0 & 0 & a_{33}^* & a_{34}^* & .. & .. & .. & 0 \\
0 & 0 & 0 & a_{44}^* & .. & .. & .. & 0 \\
\multicolumn{8}{c}{\dotfill} \\
\multicolumn{8}{c}{\dotfill} \\
\multicolumn{8}{c}{\dotfill} \\
.. & .. & .. & .. & a_{n-1,n-1}^* & a_{n-1,n}^* & & \\
0 & 0 & .. & .. & 0 & 0 & 0 & a_{n,n}^*
\end{bmatrix}
\begin{bmatrix}
T_1 \\ T_2 \\ T_3 \\ T_4 \\ . \\ . \\ . \\ T_{n-1} \\ T_n
\end{bmatrix}
=
\begin{bmatrix}
C_1^* \\ C_2^* \\ C_3^* \\ C_4^* \\ . \\ . \\ . \\ C_{n-1}^* \\ C_n^*
\end{bmatrix}
\qquad (5.19)
$$

Example 5.1

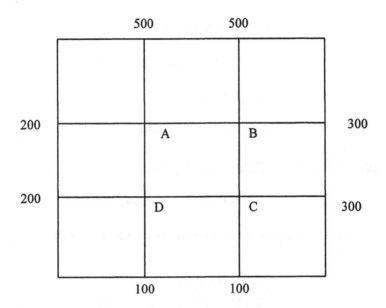

Figure 5.7   Figure for Example 5.1.

Problem: Find the steady-state temperatures at the grid points A, B, C and D of the two-dimensional solid with the boundary conditions shown in degrees centigrade.

Solution
         Using Eq. (5.9), the banded matrix equations we obtain are given by Eqs. (i-iv).

$$-4T_A + T_B \qquad +T_D = -700 \qquad\qquad (i)$$
$$+T_A - 4T_B + T_C \qquad = -800 \qquad\qquad (ii)$$
$$+T_B - 4T_C + T_D = -400 \qquad\qquad (iii)$$
$$+T_A \qquad + T_C - 4T_D = -300 \qquad\qquad (iv)$$

Divide Eq. (i) by 4 and add that to Eq. (ii):
$$-3.75T_B + T_C + 0.25T_D = -975. \qquad\qquad (v)$$

Divide Eq. (i) by 4 and add that to Eq. (iv):

$$0.25T_B + T_C - 3.75T_D = -475. \qquad\qquad \text{(vi)}$$

Multiply Eq. (v) by 0.2667 and add Eq. (iii):

$$-3.7333T_C + 1.067T_D = -660. \qquad\qquad \text{(vii)}$$

Multiply Eq. (vi) by 15 and add Eq.(v):

$$16T_C - 56\,T_D = -8100. \qquad\qquad \text{(viii)}$$

Multiply Eq. (viii) by 0.233 and add Eq. (vii):

$$-11.999T_D = -2547.3. \qquad\qquad \text{(ix)}$$

The upper tridiagonal matrix equations we obtain are given by Eqs. (v), (vii) and (ix),

$$
\begin{aligned}
-3.75T_B + T_C + 0.25T_D &= -975 &\qquad \text{(v)}\\
-3.7333T_C + 1.067T_D &= -660 &\qquad \text{(vii)}\\
-11.999T_D &= -2547.3 &\qquad \text{(ix)}
\end{aligned}
$$

Hence, from Eq. (ix), $T_D = 212.3$.
Substituting this value of T_D in Eq.(vii), $T_C = 237.5$.
Substituting these values of T_D and T_C in Eq. (v), $T_B = 337.5$.
Lastly, by substituting these calculated values of T_B and T_D in Eq. (i),
$\qquad T_A = 312.5$.

Matrix Inversion

As we discussed in the previous section, the two-dimensional, steady-state heat conduction problem gives a set of algebraic equations implicitly involving the unknown temperatures, which form a banded matrix, as shown in Eqs. (5.18). The matrix equations may be written as

$$\mathbf{A} \bullet \mathbf{T} = \mathbf{C} \qquad\qquad (5.20)$$

where **A** is the coefficient matrix, **C** is the column vector of known constants, and **T** is the vector of unknown temperatures. The temperature vector is thus given by

$$\mathbf{T} = A^{-1} \bullet \mathbf{C}. \tag{5.21}$$

The inversion of the coefficient matrix is not complicated because it is a banded matrix; furthermore, computer programs for carrying out this procedure are commonplace.

Example 5.2

Problem: Re-solve the problem in Example 5.1 using the matrix inversion method.

Solution
        The finite difference equations, shown in Eqs.(i-iv) in Example 5.1, can be written as

$$\mathbf{A} \bullet \mathbf{T} = \mathbf{C}. \tag{i}$$

where

$$A = \begin{bmatrix} -4 & 1 & 0 & 1 \\ 1 & -4 & 1 & 0 \\ 0 & 1 & -4 & 1 \\ 1 & 0 & 0 & -4 \end{bmatrix} \tag{ii}$$

$$T = \begin{bmatrix} T_A \\ T_B \\ T_C \\ T_D \end{bmatrix} \quad \text{and} \quad C = \begin{bmatrix} -700 \\ -800 \\ -400 \\ -300 \end{bmatrix}. \tag{iii}$$

Inversion of matrix A gives

$$
A^{-1} = \begin{bmatrix}
-\dfrac{7}{24} & -\dfrac{1}{12} & -\dfrac{1}{24} & -\dfrac{1}{12} \\
-\dfrac{1}{12} & -\dfrac{7}{24} & -\dfrac{1}{12} & -\dfrac{1}{24} \\
-\dfrac{1}{24} & -\dfrac{1}{12} & -\dfrac{7}{24} & -\dfrac{1}{12} \\
-\dfrac{1}{12} & -\dfrac{1}{24} & -\dfrac{1}{12} & -\dfrac{7}{24}
\end{bmatrix}. \tag{iv}
$$

Thus,

$$
\begin{bmatrix} T_A \\ T_B \\ T_C \\ T_D \end{bmatrix} =
\begin{bmatrix}
-0.29167 & -0.08333 & -0.04167 & -0.08333 \\
-0.08333 & -0.29167 & -0.08333 & -0.04167 \\
-0.04167 & -0.08333 & -0.29167 & -0.08333 \\
-0.08333 & -0.04167 & -0.08333 & -0.29167
\end{bmatrix}
$$

which gives

$T_A = 0.29167(700) + 0.08333(800) + 0.04167(400) + 0.08333(300) = 312.5$

$T_B = 0.08333(700) + 0.29167(800) + 0.08333(400) + 0.04167(300) = 337.5$

$T_C = 0.04167(700) + 0.08333(800) + 0.29167(400) + 0.08333(300) = 237.5$

$T_D = 0.08333(700) + 0.04167(800) + 0.08333(400) + 0.29167(300) = 212.5$.

These are essentially the same answers as obtained by the Gaussian elimination method in Example 5.1.

Relaxation Method

If there are only several grid points and thus several equations, the system of finite-difference equations can be solved by hand using a simple calculator by the relaxation method. When there is a large number of grid points, then a computer program may be written to solve the equations using a personal computer.

The following steps outline the relaxation method:
1.   Set the right-hand side of Eq. (5.9) to a residual $R_{i,j}$ as
$$T_{i+1,j} + T_{i-1,j} + T_{i,j+1} + T_{i,j-1} - 4T_{i,j} = R_{i,j}. \qquad (5.22)$$

2.   In the heat conduction problem with only first kind boundary conditions, only the temperatures at the interior grid points are unknown. Guess temperatures for these unknowns.
3.   Compute the residual $R_{i,j}$ at each unknown grid point using the guessed values. When the guessed values of the temperatures are the true values at each grid point, the residuals will be zero. You would not expect the first calculation of the residuals to be all zero.
4.   Select the largest residual, and try to reduce it to zero by changing the guessed temperature of the corresponding grid point while keeping the temperatures of the other grid points constant.
5.   Calculate the new residuals. This calculation may be done by using the computational module shown in Fig. 5.8. Repeat Step 4 above.
6.   Continue the relaxation procedure until all the residuals are as close to zero as required.

This relaxation method is demonstrated by Example 5.3, where the problem in Examples 5.1 and 5.2 is solved using the relaxation method.

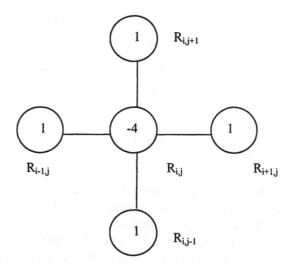

Figure 5.8   Two-dimensional computational
module for relaxation method.

Example 5.3

Problem: Re-solve the problem in Example 5.1 using the relaxation
method.   In addition, calculate the rate of heat loss from the 100°C
surface.

Solution
            The relaxation calculations for this problem are shown in Table
5.1. The computation is stopped while some of the residuals still have
nonzero values. The precision is acceptable because all the temperatures
are within 1°C of their true values.

Table 5.1   Relaxation Table for Example 5.3

Point A		Point B		Point C		Point D	
R	T_A	R	T_B	R	T_C	R	T_D
-200	400	-100	400	-100	300	-200	300
0	350	-150	400	-100	300	-250	300
-62.5	350	-150	400	-162.5	300	0	237.5
-62.5	350	-190.6	400	0	259.4	-40.6	237.5
-110.5	350	0	352	-48.1	259.4	52	237.5
1.5	322	-26.6	352	-48.1	259.4	24	237.5
1.5	322	-39	352	1.5	247	-81	237.5
-19	322	-39	352	-19	247	1	217
-29	322	1	342	-29	247	1	217
-1	315	-14	342	3	239	-14	217
-9	315	2	338	-5	239	2	213
-1	313	-1.5	338	1	237.5	-1.5	213

Note that we have overrelaxed the residuals to speed the calculations.

The heat loss rate from the 100°C surface is calculated as

q = kd{ 0.5(200 −100) + (213 − 100) + (237.5 − 100) + 0.5(300 − 100) }
  = 400.5 kd

where k is the thermal conductivity and d is the thickness of the solid. The rate of heat transfer at the other surfaces may be calculated in the same way.

5.6   Finite Difference Equations for 1-D Unsteady-State Conduction in Rectangular Coordinates

For a system with constant thermal properties, the one-dimensional, unsteady-state conduction problem is governed by the equation

$$\frac{\partial^2 T}{\partial x^2} = \frac{1}{\alpha}\frac{\partial T}{\partial t}.$$ (5.23)

This equation is pointwise continuous, that is, it is applicable throughout the region and space considered. Let us develop a network of grid points by dividing x and t domains into small intervals of Δx and Δt, such that T_i^n represents the temperature at location $x = i\Delta x$ at $t = n\Delta t$. At time t (= $n\Delta t$), the left-hand side of Eq. (5.23) can be written in finite-difference form as

$$\left.\frac{\partial^2 T}{\partial x^2}\right|_i \cong \frac{T_{i+1}^n + T_{i-1}^n - 2T_i^n}{(\Delta x)^2}. \tag{5.24}$$

The time derivative in Eq. (5.23) may be approximated in terms of forward, backward, or central differences as

$$\left.\frac{\partial T}{\partial t}\right|_x \cong \frac{T_i^{n+1} - T_i^n}{\Delta t}, \tag{5.25}$$

$$\left.\frac{\partial T}{\partial t}\right|_i \cong \frac{T_i^n - T_i^{n-1}}{\Delta t}, \tag{5.26}$$

$$\left.\frac{\partial T}{\partial t}\right|_i \cong \frac{T_i^{n+1} - T_i^{n-1}}{2\Delta t}. \tag{5.27}$$

These three finite difference approximations have different error levels and different stability properties. From the forward-difference approximation of time derivative, we obtain finite-difference equations that are uncoupled and thus easy to solve, but their solutions are not stable for all situations. The backward-difference approximation gives equations that are coupled and thus difficult to solve, but their solutions are stable for all situations. We now discuss the solution of the finite difference equations by different methods.

Explicit Method

If the forward-difference approximation is used in Eq. (5.23), the finite-difference equation is represented by

$$\frac{T_{i+1}^n + T_{i-1}^n - 2T_i^n}{(\Delta x)^2} = \frac{1}{\alpha} \frac{T_i^{n+1} - T_i^n}{\Delta t} \qquad (5.28)$$

which may be arranged as

$$T_i^{n+1} = \left[1 - \frac{2\alpha\Delta t}{(\Delta x)^2}\right]T_i^n + \frac{\alpha\Delta t}{(\Delta x)^2}\left(T_{i+1}^n + T_{i-1}^n\right) \qquad (5.29)$$

This is an explicit development since the temperature T_i^{n+1} (at location i, at time t +Δt) is expressed explicitly in terms of the temperatures at time t and at locations i-1, i and i+1. When the temperatures of the grid points are known at any particular time t, the temperatures after a time increment Δt can be computed by writing an equation similar to Eq. (5.28) for each grid point and calculating the values of T_i^{n+1}. The computation continues from one time increment to the next until the temperature distribution is found at the required time.

For discussion and better understanding of the method, let us consider the space increment Δx and time increment Δt are chosen such that $\alpha\Delta t/(\Delta x)^2 = 0.5$. Equation (5.29) then becomes

$$T_i^{n+1} = 0.5\left(T_i^{n+1} + T_{i-1}^n\right) \qquad (5.30)$$

In this simple case, the temperature at grid point i after one time increment is given by the arithmetic average of the temperatures of the adjacent points at the start of the time increment.

The coefficient of T_i^n may be positive, zero, or negative depending on the values of Δx and Δt. For a better physical feel for stability, let the temperatures T_{i+1}^n and T_{i-1}^n be zero, and the temperature T_i^n be positive. In addition, if the coefficient of T_i^n is negative, the temperature T_i^{n} at location i becomes negative. This is not allowed since it would violate the second law of thermodynamics, as heat will only flow in the direction of a negative temperature gradient. Hence, for stable solutions the following condition has to be satisfied:

$$\frac{\alpha \Delta t}{(\Delta x)^2} \le \frac{1}{2}.$$ (5.31)

Once Δx has been selected, this restriction limits the choice of Δt.

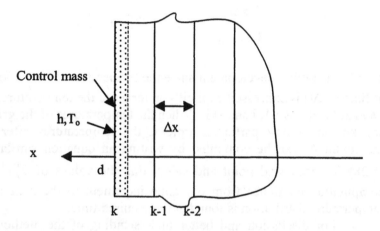

Figure 5.9   Control mass system defined next to the boundary at $x = d$.

When the boundary temperatures are known, the finite difference equation (5.29) is employed to calculate the temperatures of the internal grid points as a function of time. If a convection boundary condition exists, then the boundary has to be considered separately. For the one-dimensional system shown in Fig. 5.9, the boundary condition at $x = d$ is

$$-k\frac{\partial T}{\partial x}\bigg|_{x=d} = h\{T(d) - T_o\}.$$ (5.32)

The finite difference form of the boundary condition (5.32) may be expressed as

$$k\frac{T_{k-1}^{n+1} - T_k^{n+1}}{\Delta x} = h\{T_k^{n+1} - T_o\}$$ (5.33)

or $\qquad T_k^{n+1} = \dfrac{1}{1+\dfrac{h\Delta x}{k}}\left\{T_{k-1}^{n+1} + \dfrac{h\Delta x}{k}T_o\right\}.$ $\qquad\qquad$ (5.34)

Substituting T_{k-1}^{n+1} from Eq. (5.29) into Eq. (5.34),

$$T_k^{n+1} = \frac{1}{1+\dfrac{h\Delta x}{k}}\left\{\left[1-\frac{2\alpha\Delta t}{(\Delta x)^2}\right]T_{ki-1}^n + \frac{\alpha\Delta t}{(\Delta x)^2}\left(T_k^n + T_{k-2}^n\right)+\frac{h\Delta x}{k}T_o\right\}.$$

$$(5.35)$$

$\qquad$ The effect of the heat capacity of the system next to the boundary is not included in Eq. (5.35). If the Δx used is small, then this approximation is pretty good since the heat capacity of the system becomes insignificant. An improvement is to take into consideration the heat capacity of the system shown. The conservation of energy principle gives

$$k\frac{T_{k-1}^n - T_k^n}{\Delta x} + h\left(T_o - T_k^n\right) = \rho c\,\frac{\Delta x}{2}\frac{T_k^{n+1} - T_k^n}{\Delta t}.$$ $\qquad$ (5.36)

This equation may be written as

$$T_k^{n+1} = \frac{\alpha\Delta t}{(\Delta x)^2}\left\{\left[\frac{(\Delta x)^2}{\alpha\Delta t}-2\frac{h\Delta x}{k}-2\right]T_k^n + 2T_{k-1}^n + 2\frac{h\Delta x}{k}T_o\right\}.$$ $\quad$ (5.37)

$\qquad$ The stability of the temperatures of the grid points on the surface must also be guaranteed by the proper selection of the parameter $(\Delta x)^2/\alpha\Delta t$. This parameter can be selected so that the coefficients of T_i^n for both interior and boundary grid points are either positive or zero:

$$\frac{(\Delta x)^2}{\alpha\Delta t} \geq 2\left(\frac{h\Delta x}{k}+1\right)$$ $\qquad\qquad$ (5.38)

Example 5.4

Problem: In Figure 5.10 is shown the initial temperature distribution
inside a uniform flat plate of thickness 40 cm. The plate experiences
convection of both sides from a coolant at 200°C, with a heat transfer
coefficient h of 18.89 W/(m².K). The properties of the solid medium are
$\rho = 1700$ kg/m³, c = 0.8 kJ/(kg.K) and k = 3.778 W/(m.K). Determine
the temperatures at the grid points as a function of time.

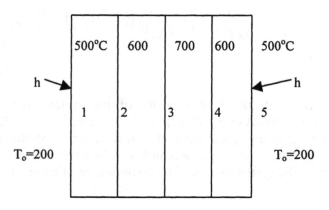

Figure 5.10 Figure for Example 5.4.

Solution
          The temperatures of the interior grid points may be calculated
using Eq. (5.29),

$$T_i^{n+1} = \left[1 - \frac{2\alpha\Delta t}{(\Delta x)^2}\right]T_i^n + \frac{\alpha\Delta t}{(\Delta x)^2}\left(T_{i+1}^n + T_{i-1}^n\right) \qquad (5.29)$$

The boundary temperatures may be found with Eq. (5.37).

          Since a convection boundary condition exists on both
boundaries, for a stable solution

$$\frac{(\Delta x)^2}{\alpha\Delta t} \ge 2\left(\frac{h\Delta x}{k} + 1\right).$$

Also, $h\Delta x/k = 0.5$, so

$$\frac{(\Delta x)^2}{\alpha \Delta t} \geq 3.$$

If we choose $(\Delta x)^2 \alpha/\Delta t = 3$, the computational equations become

$$T_i^{n+1} = \frac{1}{3}\left(T_i^n + T_{i+1}^n + T_{i-1}^n\right) \qquad\qquad i = 2,3,4,$$

$$T_1^{n+1} = \frac{1}{3}\left(2T_2^n + T_o\right),$$

and $\qquad T_5^{n+1} = \frac{1}{3}\left(2T_4^n + T_o\right).$

In addition,

$$\Delta t = \frac{(\Delta x)^2}{3\alpha} = \frac{(0.1)^2 \times 1700 x 0.8}{3 \times 3.778 \times 10^{-3}} = 1200 \sec s = 20 \min.$$

The temperatures at the grid points can be computed using the above relationships as a function of time with 20-min time intervals, as shown in Table 5.2.

Table 5.2   Results for Example 5.4

n	t(min)	T_1	T_2	T_3	T_4	T_5
0	0	500	600	700	600	500
1	20	466.7	600	633.3	600	466.7
2	40	466.7	566.7	611.1	566.7	466.7
3	60	444.5	548.2	581.5	548.2	444.5
4	80	432.1	524.7	559.3	524.7	432.1
5	100	416.5	505.4	536.2	505.4	416.5
6	120	403.6	486.0	515.7	486.0	403.6

Implicit Method

     In the explicit method, the requirement that $\alpha \Delta t / (\Delta x)^2 \leq 0.5$ for interior grid points place a severe restriction on the time increment Δt. For problems involving large values of time, this may result in excessive computation. The implicit method overcomes this shortcoming by using a slightly more complex calculation. In this implicit method, $\partial^2 T / \partial x^2$ is converted to the finite difference form evaluated at $t + \Delta t$, and $\partial T / \partial t$ is replaced by the backward finite difference form. The governing finite difference equation is thus

$$\frac{T_{i+1}^{n+1} + T_{i-1}^{n+1} - 2T_i^{n+1}}{(\Delta x)^2} = \frac{1}{\alpha} \frac{T_i^{n+1} - T_i^n}{\Delta t}. \tag{5.39}$$

This equation may be rewritten as

$$T_i^n = \left\{ 1 + \frac{2\alpha \Delta t}{(\Delta x)^2} \right\} T_i^{n+1} - \frac{\alpha \Delta t}{(\Delta x)^2} \left(T_{i+1}^{n+1} + T_{i-1}^{n+1} \right) \tag{5.40}$$

Notice that the above formulation does not allow the explicit evaluation of T_i^{n+1} in terms of T_i^n. At any time level, writing Eq. (5.40) for all the grid points produces a family of algebraic equations that must be solved simultaneously to find the temperatures T_i^{n+1}. These finite difference equations can be solved by the methods discussed previously, that is, by using Gaussian elimination, matrix inversion or relaxation methods. The advantage of this method is that there is no restriction on the step size Δx or the time increment Δt for stable solutions. It is obviously more complex than the explicit method of solution.

The Crank-Nicolson Method

     In this method, the arithmetic average of Eqs. (5.28) and (5.39) is taken, giving

$$\frac{1}{2} \left[\frac{T_{i+1}^n + T_{i-1}^n - 2T_i^n}{(\Delta x)^2} + \frac{T_{i+1}^{n+1} + T_{i-1}^{n+1} - 2T_i^{n+1}}{(\Delta x)^2} \right] = \frac{1}{\alpha} \frac{T_i^{n+1} - T_i^n}{\Delta t}. \tag{5.41}$$

This equation may be re-written as

$$2\left[1+\frac{\alpha\Delta t}{(\Delta x)^2}\right]T_i^{n+1}-\frac{\alpha\Delta t}{(\Delta x)^2}\left[T_{i+1}^{n+1}+T_{i-1}^{n+1}\right]=2\left[1-\frac{\alpha\Delta t}{(\Delta x)^2}\right]T_i^n+\frac{\alpha\Delta t}{(\Delta x)^2}\left[T_{i+1}^n+T_{i-1}^n\right]$$

(5.42)

Equation (5.42) is similar to the implicit method Eq. (5.40) since the Crank-Nicolson method is a variation of the implicit method. It can be shown that for all values of $\alpha\Delta t/(\Delta x)^2$, the Crank-Nicolson method is stable and converges. In addition, its truncation error is of the order of $\{(\Delta x)^2 + (\Delta t)^2 \}$. Both the explicit and implicit methods discussed previously have truncation errors of the order of magnitude of $\{\Delta t + (\Delta x)^2\}$. So, the Crank-Nicolson method shows significant improvement over both those two other methods. However, the finite difference equations obtained by applying Eq. (5.42) are a little more complex than the equations of even the fully implicit method.

Another method can be designed by using a weighted average of Eq. (5.28) and Eq. (5.39). In the literature, there are many other methods of finite differencing derived in a similar manner.

5.7 Finite Difference of 2-D Unsteady-State Problems in Rectangular Coordinates

With constant thermal properties, the governing equation for a 2-D solid under heat conduction is

$$\frac{\partial^2 T}{\partial x^2}+\frac{\partial^2 T}{\partial y^2}=\frac{1}{\alpha}\frac{\partial T}{\partial t}.$$

(5.43)

At time $t = 0$, the initial temperatures at all interior grid points and points on the boundaries are known. After time $t = 0$, the boundaries are at the first, second or third kind of boundary conditions. The solid domain is divided into a rectangular grid, as was done before, in Fig. 5.11.

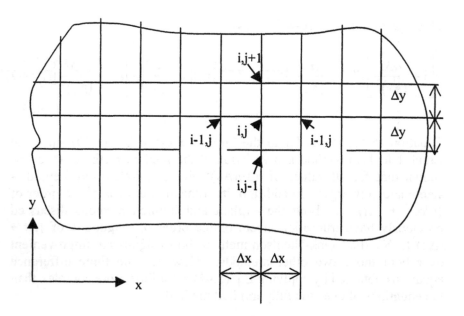

Figure 5.11   Grid points for a rectangular grid system.

As shown in the one-dimensional, unsteady-state case, the finite difference form of Eq. (5.43) can be written in a number of different ways. The explicit finite difference of Eq. (5.43) is

$$\frac{T_{i+1,j}^{n} + T_{i-1,j}^{n} - 2T_{i,j}^{n}}{(\Delta x)^2} + \frac{T_{i,j+1}^{n} + T_{i,j-1}^{n} - 2T_{i,j}^{n}}{(\Delta y)^2} = \frac{1}{\alpha} \frac{T_{i,j}^{n+1} - T_{i,j}^{n}}{\Delta t}. \quad (5.44)$$

The temperature at any specific time is represented in terms of known values of the previous time step. If the increments in x and y are selected to be the same, then Eq. (5.44) can be written as

$$T_{i,j}^{n+1} = \frac{\alpha \Delta t}{(\Delta x)^2} \left\{ T_{i+1,j}^{n} + T_{i-1,j}^{n} + T_{i,j+1}^{n} + T_{i,j-1}^{n} \right\} + \left[1 - 4 \frac{\alpha \Delta t}{(\Delta x)^2} \right] T_{i,j}^{n}.$$

$$(5.45)$$

As in the case of a one-dimensional, unsteady-state problem, the explicit method is only stable for limited values of Δx and Δt. In the case of the first kind boundary condition, for t 0, the limit of stability is given by

$$\frac{\alpha \Delta t}{(\Delta x)^2} \le \frac{1}{4} \qquad (5.46)$$

Equation (5.45) can be used to find the temperatures of the interior grid points. If the boundary conditions are other than the first kind, then finite difference relations have to be developed for the points on the boundary. This procedure has been illustrated for the 1-D case. For example, if there is a convection boundary condition, the finite difference equation for a point on the boundary is

$$T_{i,j}^{n+1} = \frac{\alpha \Delta t}{(\Delta x)^2} \left[2\frac{h\Delta x}{k}T_o + 2T_{i-1,j}^n + T_{i,j+1}^n + T_{i,j-1}^n + \left\{ \frac{(\Delta x)^2}{\alpha \Delta t} - 2\frac{h\Delta x}{k} - 4 \right\} T_{i,j}^n \right] \qquad (5.47)$$

where T_o is the temperature of the surrounding fluid. For the current problem, the condition for stability is

$$\frac{(\Delta t)^2}{\alpha \Delta t} \ge 2\left(\frac{h\Delta x}{k} + 2 \right). \qquad (5.48)$$

An implicit formulation of Eq. (5.43) is given by

$$\frac{T_{i+1,j}^{n+1} + T_{i-1,j}^{n+1} - 2T_{i,j}^{n+1}}{(\Delta x)^2} + \frac{T_{i,j+1}^{n+1} + T_{i,j-1}^{n+1} - 2T_{i,j}^{n+1}}{(\Delta y)^2} = \frac{1}{\alpha}\frac{T_{i,j}^{n+1} - T_{i,j}^n}{\Delta t}. \qquad (5.49)$$

If we put $\Delta x = \Delta y$, then Eq. (5.49) gives

$$T_{i,j}^n = \left\{ 1 + \frac{4\alpha \Delta t}{(\Delta x)^2} \right\} T_{i,j}^{n+1} - \frac{\alpha \Delta t}{(\Delta x)^2} \left\{ T_{i+1,j}^{n+1} + T_{i-1,j}^{n+1} + T_{i,j+1}^{n+1} + T_{i,j-1}^{n+1} \right\} \qquad (5.50)$$

Although this implicit formulation is stable for all values of Δx and Δt, it is a little more difficult to solve than the explicit formulation. At each time step, a number of simultaneous algebraic equations have to be solved, depending on the number of grid points.

Example 5.5

Problem: The 2-D body shown in Fig.5.12 is initially at a uniform temperature of 100°C. For times   0, the boundary temperatures are kept at the levels shown in Fig. 5.12.   Compute the temperatures at points A, B, C and D as a function of time.

Solution

Put $\dfrac{\alpha \Delta t}{(\Delta x)^2} = \dfrac{1}{4}$.   Using the explicit formulation, the temperatures T_A, T_B, T_C and T_D are given by the equations written below.

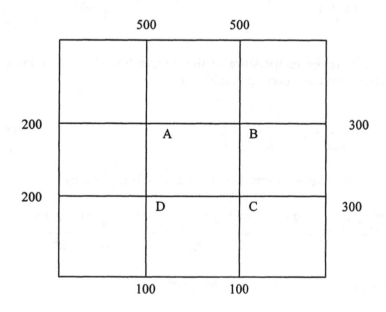

Figure 5.12 Figure for Example 5.5.

$$T_A^{n+1} = \frac{1}{4}\left(700 + T_B^n + T_D^n\right) \qquad\qquad T_B^{n+1} = \frac{1}{4}\left(800 + T_A^n + T_C^n\right)$$

$$T_C^{n+1} = \frac{1}{4}\left(400 + T_B^n + T_D^n\right) \qquad\qquad T_D^{n+1} = \frac{1}{4}\left(300 + T_A^n + T_C^n\right)$$

Table 5.3 Results for Example 5.5

n	t	T_A	T_B	T_C	T_D
0	0	100	100	100	100
1	$\Delta t = (\Delta x)^2/4\alpha$	225	250	150	125
2	$2\Delta t$	268.8	293.8	193.8	168.8
3	$3\Delta t$	290.7	315.7	215.7	190.7
4	$4\Delta t$	301.6	326.6	226.6	201.6
5	$5\Delta t$	307	332	232	207
6	$6\Delta t$	310	335	235	210
7	$7\Delta t$	311	336	236	211
∞	∞	312	337	237	212

5.8 Finite Difference Method Applied in Cylindrical Coordinates

The finite difference methods that have been applied to rectangular coordinates may be similarly applied to cylindrical coordinates and spherical coordinates. The discussion for the problem in cylindrical coordinates is done in this section. Without energy sources or sinks, the governing steady-state equation for a medium with constant thermal conductivity in cylindrical coordinates is

$$\frac{\partial^2 T}{\partial r^2} + \frac{1}{r}\frac{\partial T}{\partial r} + \frac{1}{r^2}\frac{\partial^2 T}{\partial \theta^2} + \frac{\partial^2 T}{\partial z^2} = 0. \tag{5.51}$$

When the temperature has no dependence on the z-coordinate, the equation becomes

$$\frac{\partial^2 T}{\partial r^2} + \frac{1}{r}\frac{\partial T}{\partial r} + \frac{1}{r^2}\frac{\partial^2 T}{\partial \theta^2} = 0. \tag{5.52}$$

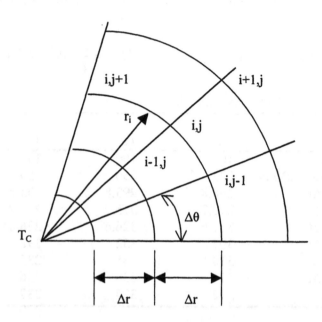

Figure 5.13   Finite difference grid for cylindrical coordinates.

Equation (5.52) may be represented in finite difference form at the point (i,j) shown in Fig. 5.13 by

$$\frac{T_{i+1,j}+T_{i-1,j}-2T_{i,j}}{(\Delta r)^2} + \frac{1}{r_i}\frac{T_{i+1,j}-T_{i-1,j}}{2\Delta r} + \frac{1}{r_i^2}\frac{T_{i,j+1}+T_{i,j-1}-2T_{i,j}}{(\Delta \theta)^2}$$

(5.53)

where $r_i = i\Delta r$. At $r = 0$, Eq. (5.53) becomes $T_C = T_m$ where $T_C =$ temperature at $r = 0$, and T_m = mean temperature of the grid points that surround $r = 0$.

When Eq. (5.53) is applied to all the grid points, with the appropriate boundary conditions, the result is a system of simultaneous linear algebraic equations not dissimilar to those obtained in rectangular coordinates.

For problems with no θ dependence, that is, T is only a function of r and z, then the governing equation is

$$\frac{\partial^2 T}{\partial r^2} + \frac{1}{r}\frac{\partial T}{\partial r} + \frac{\partial^2 T}{\partial z^2} = 0. \qquad (5.54)$$

The finite difference form of Eq.(5.54) may be written as

$$\frac{T_{i+1,j} + T_{i-1,j} - 2T_{i,j}}{(\Delta r)^2} + \frac{1}{r_i}\frac{T_{i+1,j} - T_{i-1,j}}{2\Delta r} + \frac{T_{i,j+1} + T_{i,j-1} - 2T_{i,j}}{(\Delta z)^2} = 0$$

$$(5.55)$$

where j is the number of increments in the z direction, and i the number of increments in the r direction. At r = 0, the Laplacian in Eq. (5.54) reduces to

$$2\frac{\partial^2 T}{\partial r^2} + \frac{\partial^2 T}{\partial z^2} = 0 \qquad (5.56)$$

which be represented in finite difference form quite readily.

The unsteady-state problems in cylindrical coordinates can be handled in ways that are similar to the ones discussed here.

5.9 Truncation Errors and Round-Off Errors in the Finite Difference Method

Consider a function T(x) and its derivatives to be single-valued, finite, and continuous with respect to x. The Taylor series expansion of $T(x + \Delta x)$ about $T(x)$ may be written as

$$T(x + \Delta x) = T(x) + \frac{dT}{dx}\bigg|_x \Delta x + \frac{1}{2!}\frac{d^2 T}{dx^2}\bigg|_x (\Delta x)^2 + \frac{1}{3!}\frac{d^3 T}{dx^3}\bigg|_x (\Delta x)^3 +$$

... Higher order terms (H.O.T.)

$$(5.57)$$

$$T(x - \Delta x) = T(x) - \frac{dT}{dx}\bigg|_x \Delta x + \frac{1}{2!}\frac{d^2T}{dx^2}\bigg|_x (\Delta x)^2 - \frac{1}{3!}\frac{d^3T}{dx^3}\bigg|_x (\Delta x)^3$$

$$+ ... \text{H.O.T.}$$

(5.58)

Adding Eqs. (5.57) and (5.58),

$$T(x + \Delta x) + T(x - \Delta x) = 2T(x) + \frac{d^2T}{dx^2}\bigg|_x (\Delta x)^2 + O\{(\Delta x)^4\}$$ (5.59)

where the last term in Eq. (5.59) represents terms in the fourth and higher powers of Δx. Equation (5.59) may be written as

$$\frac{d^2T}{dx^2}\bigg|_x = \frac{T(x + \Delta x) + T(x - \Delta x) - 2T(x)}{(\Delta x)^2} + O\big[(\Delta)^2\big]$$ (5.60)

Comparing Eq. (5.60) to Eq. (5.6), the finite difference approximation of $\left(\dfrac{d^2T}{dx^2}\right)$ has a truncation error of the order of magnitude of $(\Delta x)^2$.

If we subtract Eq. (5.58) from Eq. (5.57), and rearrange, we obtain

$$\frac{dT}{dx}\bigg|_x = \frac{T(x + \Delta x) - T(x - \Delta x)}{2\Delta x} + O\{(\Delta x)^2\}$$ (5.61)

Comparing Eq. (5.61) to Eq. (5.4), we can see that the finite difference approximation of dT/dx has a truncation error of the order of magnitude of $(\Delta x)^2$.

From Eq.(5.57), the first derivative of T with respect to x may be written as

$$\frac{dT}{dx}\bigg|_x = \frac{T(x + \Delta x) - T(x)}{\Delta x} + O\{\Delta x\}. \tag{5.62}$$

Similarly, from Eq. (5.58),

$$\frac{dT}{dx}\bigg|_x = \frac{T(x) - T(x - \Delta x)}{\Delta x} + O\{\Delta x\}. \tag{5.63}$$

The other two approximations for the first derivative of T with respect to x have truncation errors of the order of magnitude of Δx. Thus, the central difference formulation is more precise than the other two.

The truncation errors are inherent in the finite difference method and cannot be eliminated. The errors may be reduced by selecting a finer grid. In other words, smaller increments for space and time will reduce the truncation errors.

Numerical solutions are carried out to a finite number of significant figures; the numbers are rounded-off and thus, round-off errors are introduced. Round-off errors compound, and this may result in a large cumulative error. It is difficult to estimate the order of magnitude of the cumulative round-off errors. The use of smaller increments in space and time increases the accumulation of round-off errors, even though they lead to less truncation errors.

5.10   Stability and Convergence

The precision of a finite difference numerical method depends on its "stability" and "convergence". The precision is dependent on the step sizes employed, and an increase in precision is attained with increased labor.

A numerical method is convergent if the solution obtained approaches the exact solution as the increments in time and space approach zero. The numerical solution has to converge to the exact solution in the limit; otherwise, it is not convergent.

As discussed in the previous section, there are truncation errors and round-off errors associated with finite difference methods. If these errors increase as the solution method proceeds and this increase is unbounded, the solution is said to be unstable. A numerical method that does not allow the increase in errors is said to be stable. Stability is thus necessary for convergence. It is easy to see that if the errors increase at a faster rate than that at which convergence is approached, than instability also exists. The next few paragraphs outline the method to determine the stability of finite difference methods, first presented by O'Brien, Hyman and Kaplan in the Journal of Mathematical Physics, 1951.

Consider the explicit finite difference form of the one-dimensional, unsteady-state heat conduction equation, Eq. (5.29),

$$T_i^{n+1} = \left[1 - \frac{2\alpha\Delta t}{(\Delta x)^2}\right]T_i^n + \frac{\alpha\Delta t}{(\Delta x)^2}\left(T_{i+1}^n + T_{i-1}^n\right) \tag{5.29}$$

The solution of the governing equation, Eq. (5.23), can be expanded at any specified time t into a Fourier series in x. Neglecting the coefficient, a typical term in this expansion will be of the form $\psi(t)e^{i\gamma x}$. Substituting this term into Eq. (5.29), the form of $\Psi(t)$ may be found, and a criterion determined as to whether it remains bounded in the limit as t becomes very large. Carrying out the substitution,

$$\psi(t+\Delta t)e^{i\gamma x} = \left[1 - \frac{2\alpha\Delta t}{(\Delta x)^2}\right]\psi(t)e^{i\gamma x} + \frac{\alpha\Delta t}{(\Delta x)^2}\left[e^{i\gamma(x+\Delta x)} + e^{i\gamma(x-\Delta x)}\right]\psi(t). \tag{5.64}$$

This may be written as

$$\frac{\psi(t+\Delta t)}{\Delta t} = 1 - \frac{4\alpha\Delta t}{(\Delta x)^2}\sin^2\frac{\gamma\Delta x}{2}. \tag{5.65}$$

For stability, $\Psi(t)$ must be bounded as Δt and Δx approach zero. This requirement becomes

$$\max\left|1 - \frac{4\alpha\Delta t}{(\Delta x)^2}\sin^2\frac{\gamma\Delta x}{2}\right| \le 1 \quad \text{for all values of } \gamma.. \tag{5.66}$$

In the actual case, components of all frequencies of γ can be present. Even if they are not present in the initial conditions and are not introduced by the boundary conditions, they may be introduced by the round-off errors. For condition Eq. (5.66) to be satisfied,

$$\frac{\alpha\Delta t}{(\Delta x)^2} \le \frac{1}{2}. \tag{5.67}$$

Hence, Eq. (5.67) is the necessary (and sufficient) stability criterion for the explicit finite difference equation, Eq. (5.29).

PROBLEMS

5.1. Find the steady-state temperatures at the grid points A, B, C and D of the two-dimensional solid with boundary conditions shown in degrees centigrade. The grid shown is square. Use the Gaussian elimination method.

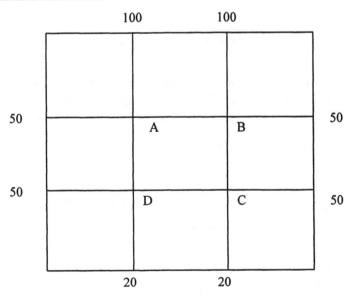

Figure for Problem 5.1.

5.2. The diagram below represents a 2-D conduction system with steady state boundary conditions in °C as shown. Calculate the temperatures at the internal node points, A, B, C and D. The grid shown is a square grid. State assumptions.

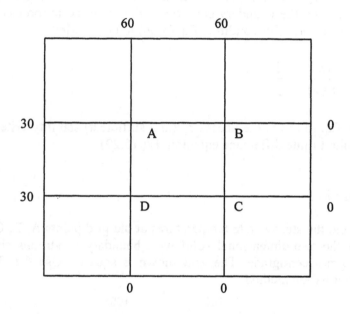

Figure for Problem 5.2.

5.3. For some situations in heat conduction it is more convenient to use the grid shown rather than the conventional square or rectangular grid. All the triangles are equilateral with sides of length l.

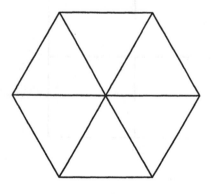

Figure for Problem 5.3.

Establish a generalized notation for the nodes of the grid, and write the Laplace equation in finite difference form using the notation you have established.

5.4. A turbine blade profile (in dashed lines) may be approximated by the square grid mesh as shown in the figure. With the boundary conditions as shown in °F, find by the relaxation method the internal temperatures at points 1, 2, 3, 4, 5, 6, 7 and 8.

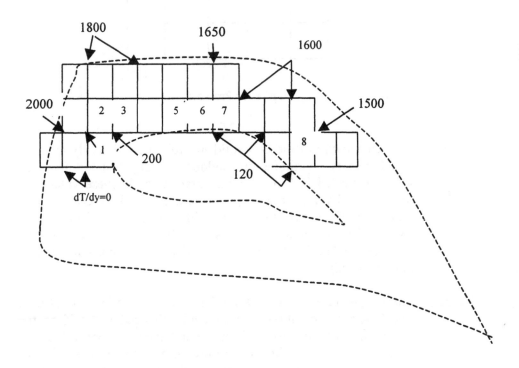

Figure for Problem 5.4.

5.5. The cross-area of a metallic bar is shown in the figure. The boundaries are insulated except for the two faces shown, which are maintained at 150°C and 25°C respectively. The grid shown is square, i.e., $\Delta x = \Delta y$. Estimate the steady-state nodal temperatures of the metal.

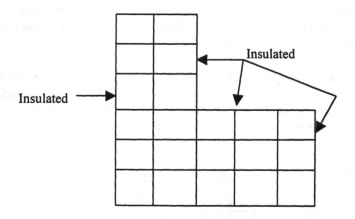

Figure for Prob.5.5.

5.6.   The temperature of the 2-D solid in Prob. 5.1 is initially at 20°C.
        The boundary conditions are suddenly changed at t = 0 to the
        values indicated in Prob. 5.1 and maintained at these values for
        times t > 0. Estimate the temperatures at points A, B, C and D as a
        function of time. Take the thermal diffusivity α = 0.01 m²/s, Δx =
        Δy = 0.316 m, and Δt = 1 sec.   By using the steady-state
        temperature distribution of Prob. 5.1, deduce the approximate time
        to reach steady state.

5.7.   The temperature of the 2-D solid in Prob. 5.2 is initially at 0°C.
        The boundary conditions are suddenly changed at t = 0 to the
        values indicated in Prob. 5.2 and maintained at these values for
        times t > 0. Estimate the temperatures at points A, B, C and D as a
        function of time. Take the thermal diffusivity α = 0.2 m²/s, Δx =
        Δy = 1 m, and Δt = 1 sec.   By using the steady-state temperature
        distribution of Prob. 5.2, deduce the approximate time to reach
        steady-state.

5.8.   Solve Prob. 5.6 using an implicit method.

5.9.   Solve Prob. 5.7 using an implicit method.

5.10. Solve numerically the two-dimensional transient heat conduction problem of a square plate. First write down the Fourier equation in finite difference form with second-order accuracy in space and first-order accuracy in time. Divide the square plate into 3 x 3 squares, i.e., with 4 internal points. Consider the plate to be at 20°C initially. At time t = 0 sec, the top boundary is kept at 100°C, and the left boundary is kept at 100°C. The other two boundaries are maintained at 20°C. Take the thermal diffusivity α = 0.1 m²/s, Δx = Δy = 1 m, and Δt = 1 sec. Solve for the first five temperature profiles within the square plate,i.e., from zero up to and including five seconds.

5.11. Find the steady-state temperatures at the grid points of the two-dimensional solid with boundary conditions shown in degrees centigrade. The long sides of the solid are insulated. The grid shown is square. The solid is a long rod made up of two materials, A and B, such that k_A is 0.01 the value of k_B. The contact interface is 45° to the horizontal.

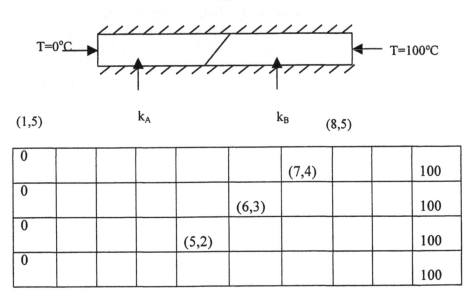

Figure for Problem 5.11.

The contact interface is the straight line going through the points (4,1), (5,2), (6,3), (7,4) and (8,5).

Gaussian Elimination Method

Finite difference of the two-dimensional, Laplacian equation
Produces a banded coefficient matrix for the conduction equation
The first step is to transform the matrix into an upper diagonal form
The first operation is to multiply the first equation by a21/a11.

The resulting equation subtracted from the second to eliminate a21.
Similarly, the third equation is rid of the term in a31
The second equation is then used to eliminate a32 and a42
The third equation is used to eliminate a43, and so on.

The upper diagonal form of the coefficient matrix is thus obtained
The last equation directly gives the temperature Tn
With Tn known, Tn-1 is found from the (n-1)th equation
Computations are carried out until T1 is found from first equation.

K.V. Wong

6

Equations for Convection

The equations for convection are the continuity or conservation of mass equation, the momentum equations and the energy equation. From the dimensionless equation of energy, useful dimensionless numbers are obtained.

6.1   Continuity

The continuity equation is the conservation of mass equation. It is derived by a mass balance of the fluid entering and exiting a volume element taken in the flow field. In Fig. 6.1, consider a differential volume element $\Delta x \Delta y \Delta z$. For ease of understanding, we shall consider steady, two-dimensional flow with velocity components $u(x,y)$ and $v(x,y)$ in the x and y directions, respectively.

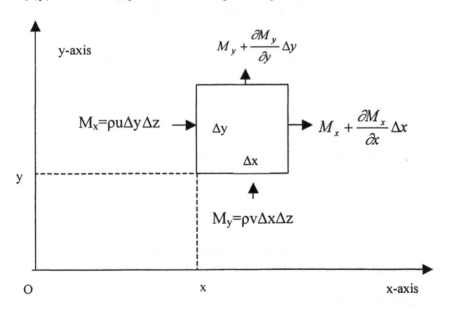

Figure 6.1   Control volume for continuity equation.

The conservation of mass may be stated as

$$\begin{pmatrix} \text{Net rate of mass flow entering} \\ \text{volume element in x direction} \end{pmatrix} + \begin{pmatrix} \text{Net rate of mass flow entering} \\ \text{volume element in y direction} \end{pmatrix} = 0$$

$$(6.1)$$

If the mass flow rate into the volume element in the x direction through the surface x is $M_x = \rho u \Delta y \Delta z$, then

$$\begin{pmatrix} \text{Net rate of mass flow entering} \\ \text{volume element in x direction} \end{pmatrix} = -\frac{\partial M_x}{\partial x} \Delta x = -\frac{\partial(\rho u)}{\partial x} \Delta x \Delta y \Delta z$$

$$(6.2)$$

If the mass flow rate into the volume element in the y direction through the surface y is $M_y = \rho v \Delta x \Delta z$, then

$$\begin{pmatrix} \text{Net rate of mass flow entering} \\ \text{volume element in y direction} \end{pmatrix} = -\frac{\partial M_y}{\partial y} \Delta y = -\frac{\partial(\rho u)}{\partial y} \Delta x \Delta y \Delta z$$

$$(6.3)$$

Substituting Eqs. (6.2) and (6.3) in Eq. (6.1) and simplifying,

$$\frac{\partial(\rho u)}{\partial x} + \frac{\partial(\rho v)}{\partial y} = 0.$$

$$(6.4)$$

If the density is constant, Eq. (6.4) simplifies to

$$\frac{\partial u}{\partial x} + \frac{\partial v}{\partial y} = 0.$$

$$(6.5)$$

Equation (7.5) is the continuity equation for a two-dimensional, steady, incompressible flow in rectangular coordinates.

6.2 Momentum Equations

The momentum equations are derived using Newton's second law of motion. This law states that the external forces acting on a body in a given direction is equal to the mass times the acceleration in the same direction. The external forces may be classified as body forces and surface forces. The surface forces are from the stresses acting on the surface of the volume element. The body forces include gravitational, magnetic and electric fields acting on the body of fluid. Newton's second law may be stated as

$$
(\text{Mass}) \begin{pmatrix} \text{acceleration in} \\ \text{direction j} \end{pmatrix} = \begin{pmatrix} \text{Body forces acting} \\ \text{in direction j} \end{pmatrix} + \begin{pmatrix} \text{Surface forces acting} \\ \text{in direction j} \end{pmatrix}.
$$

(6.6)

For three-dimensional flow, for instance, in rectangular coordinates, Eq.(6.6) gives three independent momentum equations. For ease of understanding, we consider steady, two-dimensional, incompressible flow with constant properties having velocity components u(x,y) and v(x,y) in the x and y directions, respectively.

The mass of a differential volume element is given by

$$
M = \rho \Delta x \Delta y \Delta z
$$

(6.7)

For a three-dimensional, unsteady flow field, with velocity components u,v and w in x, y and z directions, respectively, the rate of change of a property θ in the flow field is provided by the substantial or total derivative $D\theta/Dt$ defined as

$$
\frac{D\theta}{Dt} = \frac{\partial \theta}{\partial t} + u\frac{\partial \theta}{\partial x} + v\frac{\partial \theta}{\partial y} + w\frac{\partial \theta}{\partial z}.
$$

(6.8)

We consider steady, two-dimensional flow with velocity components u and v. The corresponding acceleration in the x direction is

$$
\frac{Du}{Dt} = u\frac{\partial u}{\partial x} + v\frac{\partial u}{\partial y}.
$$

(steady, two-dimensional flow) (6.9)

and the corresponding acceleration in the y direction is

$$\frac{Dv}{Dt} = u\frac{\partial v}{\partial x} + v\frac{\partial v}{\partial y} \qquad \text{(steady, two-dimensional flow). (6.10)}$$

Without specifying the nature of the body forces, if B_x and B_y are the components of the body forces acting per unit volume of the fluid in the x and y directions, respectively, then

$$\begin{pmatrix} \text{Body forces acting on} \\ \Delta x \Delta y \Delta z \text{ in x direction} \end{pmatrix} = B_x \Delta x \Delta y \Delta z \qquad (6.11)$$

and

$$\begin{pmatrix} \text{Body forces acting on} \\ \Delta x \Delta y \Delta z \text{ in y direction} \end{pmatrix} = B_y \Delta x \Delta y \Delta z \qquad (6.12)$$

In Fig. 6.2 are shown the surface stresses on a differential volume element. The normal stresses in the x and y directions are shown by σ_x and σ_y, respectively. The shear stresses are shown by τ_{xy} and τ_{yx}, where the first subscript denotes the axis to which the surface is perpendicular and the second subscript denotes the direction of the shear stress. Hence, τ_{xy} is the shear stress acting on the surface $\Delta y \Delta z$ (the surface at right angles to the x axis) at x in the direction y. The net normal surface force acting on the element in the positive x direction is $(\partial/\partial y)(\sigma_x \Delta y \Delta z)\Delta x$, and the net shear force acting on the element in the positive x direction is $(\partial/\partial y)(\tau_{yx}\Delta x \Delta z)\Delta y$. Thus, the net surface forces acting on the element in the positive x direction is

$$\begin{pmatrix} \text{Net surface forces} \\ \text{acting in x direction} \end{pmatrix} = \left(\frac{\partial \sigma_x}{\partial x} + \frac{\partial \tau_{yx}}{\partial y}\right)\Delta x \Delta y \Delta z. \qquad (6.13)$$

The net surface force acting in the y direction can be similarly found to be

$$\begin{pmatrix} \text{Net surface forces} \\ \text{acting in y direction} \end{pmatrix} = \left(\frac{\partial \sigma_y}{\partial y} + \frac{\partial \tau_{xy}}{\partial x}\right)\Delta x \Delta y \Delta z \qquad (6.14)$$

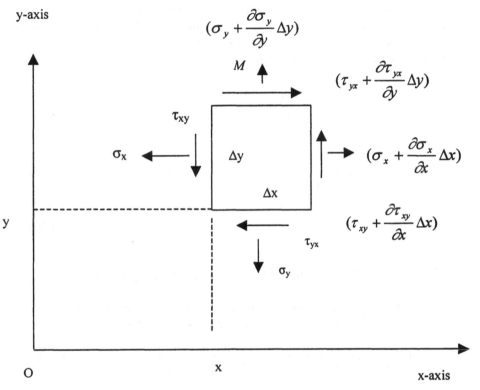

Figure 6.2   Surface stresses on the volume element.

Equations (6.7), (6.9), (6.11) and (6.13) are substituted into Eq. (6.6), and the x-momentum equation is

$$\rho(u\frac{\partial u}{\partial x} + v\frac{\partial u}{\partial y}) = B_x + \frac{\partial \sigma_x}{\partial x} + \frac{\partial \tau_{yx}}{\partial y}.$$
(6.15)

Equations (6.7), (6.10), (6.12) and (6.14) are substituted into Eq. (6.6), and the y-momentum equation is

$$\rho(u\frac{\partial v}{\partial x} + v\frac{\partial v}{\partial y}) = B_y + \frac{\partial \sigma_y}{\partial y} + \frac{\partial \tau_{xy}}{\partial x}.$$
(6.16)

The various stresses have to be related to the velocity components; a discussion of this matter is provided by Schlichting, 1979 [1]. For the two-dimensional, incompressible, constant property Newtonian fluid flow under consideration, the stresses in Eqs. (6.15) and (6.16) are related to the velocity components by

$$\tau_{xy} = \tau_{yx} = \mu\left(\frac{\partial u}{\partial y} + \frac{\partial v}{\partial x}\right) \tag{6.17}$$

$$\sigma_x = -p + 2\mu\frac{\partial u}{\partial x} \tag{6.18}$$

$$\sigma_y = -p + 2\mu\frac{\partial v}{\partial y} \tag{6.19}$$

where p is the pressure and μ is the dynamic viscosity of the fluid.

When the stresses in Eqs. (6.17) to (6.19) are substituted into Eqs. (6.15) and (6.16), the x- and y-momentum equations are obtained as

$$\rho(u\frac{\partial u}{\partial x} + v\frac{\partial u}{\partial y}) = B_x - \frac{\partial P}{\partial x} + \mu(\frac{\partial^2 u}{\partial x^2} + \frac{\partial^2 u}{\partial y^2}) \tag{6.20}$$

$$\rho(u\frac{\partial v}{\partial x} + v\frac{\partial v}{\partial y}) = B_y - \frac{\partial P}{\partial y} + \mu(\frac{\partial^2 v}{\partial x^2} + \frac{\partial^2 v}{\partial y^2}). \tag{6.21}$$

Equations (6.20) and (6.21) are the x- and y-momentum equations for the steady, two-dimensional flow of an incompressible fluid with constant properties.

In Eqs. (6.20) and (6.21), the terms on the left-hand side are the inertia forces, the first term on the right-hand side is the body force, the second term is the pressure force, and the last term within brackets is the viscous forces acting on the fluid element. With known body forces B_x and B_y, the continuity equation (6.5) and the two momentum equations (6.20) and (6.21) are three independent equations for the solution of the three unknown quantities u, v and p for the steady, two-dimensional flow of an incompressible fluid.   The solution of these equations are not simple except for a few special cases. The various terms are relevant in convective heat transfer. In the following chapters on convective heat transfer, the governing equations for velocity distribution will be obtained from these equations with the appropriate simplification in each situation.

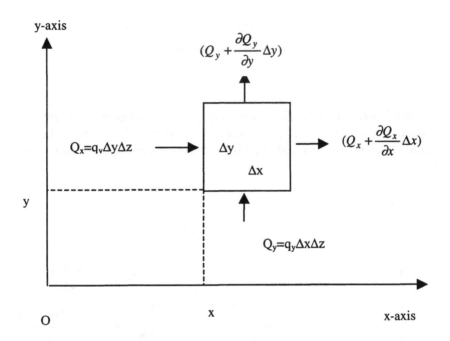

Figure 6.3   Heat addition by conduction.

6.3   Energy Equation

The energy equation may be derived using the first law of thermodynamics for a differential volume element in a flow field. In the absence of radiation and heat sources or sinks in the fluid, the energy balance on a differential volume element $\Delta x \Delta y \Delta z$ about a point (x,y,z) may be expressed as

$$\begin{pmatrix} \text{Rate of heat addition} \\ \text{into element by conduction} \end{pmatrix} + \begin{pmatrix} \text{Rate of energy input into element} \\ \text{due to work done by surfaces stresses} \\ \text{and body forces} \end{pmatrix}$$

$$\qquad\qquad\qquad\quad \text{H} \qquad\qquad\qquad\qquad\qquad\qquad\qquad \text{I}$$

$$= \begin{pmatrix} \text{rate of increase of energy} \\ \text{stored in element} \end{pmatrix}. \qquad\qquad (6.22)$$

$$\qquad\qquad\qquad\qquad\qquad \text{J}$$

The following is the derivation of Eq. (6.22) in mathematical terms, for a steady, two-dimensional, constant property flow in which the temperature variation and velocity components are in the x and y

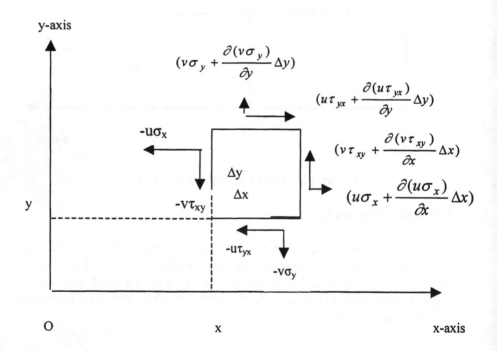

Figure 6.4 Frictional work done by the surface forces.

directions only. In other words, it is assumed that there is no flow or temperature variation in the z direction.

Referring to Fig. 6.3, if q_x and q_y are the heat fluxes in the x and y directions, the net rate of heat addition into the volume element is

$$H = -\left(\frac{\partial Q_x}{\partial x}\Delta x + \frac{\partial Q_y}{\partial y}\Delta y\right) = -\left(\frac{\partial q_x}{\partial x} + \frac{\partial q_y}{\partial y}\right)\Delta x \Delta y \Delta z \qquad (6.23)$$

where the heat fluxes are provided by the Fourier law. Assuming constant thermal conductivity,

$$H = k\left(\frac{\partial^2 T}{\partial x^2} + \frac{\partial^2 T}{\partial y^2}\right)\Delta x \Delta y \Delta z. \qquad (6.24)$$

If B_x and B_y are the body forces acting per unit volume of the fluid in the x and y directions, respectively, while u and v are the corresponding velocity components, respectively, the increase of the potential energy is

$$(uB_x + vB_y)\Delta x \Delta y \Delta z. \qquad (6.25)$$

The rate of energy input into the volume element $\Delta x \Delta y \Delta z$ due to work done by the normal stress σ_x is given by

$$\left[-u\sigma_x + \{u\sigma_x + \frac{\partial}{\partial x}(u\sigma_x)\Delta x\}\right]\Delta y \Delta z = \Delta x \Delta y \Delta z \frac{\partial}{\partial x}(u\sigma_x) \qquad (6.26)$$

and that done by the normal stress σ_y is

$$\left[-v\sigma_y + \{v\sigma_y + \frac{\partial}{\partial y}(u\sigma_y)\Delta y\}\right]\Delta x \Delta z = \Delta x \Delta y \Delta z \frac{\partial}{\partial y}(v\sigma_y). \qquad (6.27)$$

In addition, the rate of work done by the shear stress τ_{yx} and τ_{xy} are respectively given by

$$-u\tau_{yx} + \{u\tau_{yx} + \frac{\partial}{\partial y}(u\tau_{yx})\Delta y\}\Delta x\Delta z = \Delta x\Delta y\Delta z\frac{\partial}{\partial y}(u\tau_{yx}) \qquad (6.28)$$

$$-v\tau_{xy} + \{v\tau_{xy} + \frac{\partial}{\partial x}(v\tau_{xy})\Delta x\}\Delta y\Delta z = \Delta x\Delta y\Delta z\frac{\partial}{\partial x}(v\tau_{xy}). \qquad (6.29)$$

Hence, the rate of energy input owing to the frictional work done by the stresses on the volume element (sum Eqs. (6.26) to (6.29)), is given by

$$\{\frac{\partial}{\partial x}(u\sigma_x) + \frac{\partial}{\partial y}(v\sigma_y) + \frac{\partial}{\partial x}(u\tau_{xy}) + \frac{\partial}{\partial y}(v\tau_{yx})\}\Delta x\Delta y\Delta z. \qquad (6.30)$$

The total rate of energy input into the volume element owirg to the work done by the body forces and the surface stresses is

$$I =$$
$$\left\{uB_x + vB_y + \frac{\partial}{\partial x}(u\sigma_x) + \frac{\partial}{\partial y}(v\sigma_y) + \frac{\partial}{\partial x}(v\tau_{xy}) + \frac{\partial}{\partial y}(u\tau_{yx})\right\}\Delta x\Delta y\Delta z.$$
$$(6.31)$$

The energy of the fluid volume element comprises the specific internal energy e per unit mass and the kinetic energy per unit mass which is $0.5(u^2+v^2)$. The internal energy of the volume element $\Delta x\Delta y\Delta z$ is

$$\rho\{e + 0.5(u^2 + v^2)\}\Delta x\Delta y\Delta z. \qquad (6.32)$$

The rate of increase of the energy contained in the volume element is given by the total derivative of the quantity in Eq. (6.32),

$$J = \rho\{\frac{De}{Dt} + \frac{1}{2}\frac{D}{Dt}(u^2 + v^2)\}\Delta x\Delta y\Delta z \qquad (6.33)$$

where the total derivative D/Dt for two-dimensional, steady flow is defined as

$$\frac{D}{Dt} \equiv u\frac{\partial}{\partial x} + v\frac{\partial}{\partial y}.$$

Equations (6.24), (6.31) and (6.33) are substituted into Eq. (6.22), and the resulting expression is simplified.

$$\rho\frac{De}{Dt} + \frac{\rho}{2}\frac{D}{Dt}\left(u^2 + v^2\right) =$$

$$k\left(\frac{\partial^2 T}{\partial x^2} + \frac{\partial^2 T}{\partial y^2}\right) + \left[uB_x + vB_y + \frac{\partial}{\partial x}(u\sigma_x) + \frac{\partial}{\partial y}(v\sigma_y) + \frac{\partial}{\partial x}(v\tau_{xy}) + \frac{\partial}{\partial y}(u\tau_{yx})\right]$$

(6.34)

Add Eq. (6.15) multiplied by u to Eq. (6.16) multiplied by v:

$$\frac{\rho}{2}\frac{D}{Dt}\left(u^2 + v^2\right) = uB_x + vB_y + u\frac{\partial\sigma_x}{\partial x} + v\frac{\partial\sigma_y}{\partial y} + v\frac{\partial\tau_{xy}}{\partial x} + u\frac{\partial\tau_{yx}}{\partial y}$$

(6.35)

Subtract Eq. (6.35) from Eq. (6.34),

$$\rho\frac{De}{Dt} = k\left(\frac{\partial^2 T}{\partial x^2} + \frac{\partial^2 T}{\partial y^2}\right) + \left[\sigma_x\frac{\partial u}{\partial x} + \sigma_y\frac{\partial v}{\partial y} + \tau_{xy}\frac{\partial v}{\partial x} + \tau_{yx}\frac{\partial u}{\partial y}\right] \qquad (6.36)$$

since $\quad \dfrac{\partial}{\partial x}(u\sigma_x) - u\dfrac{\partial\sigma_x}{\partial x} = \sigma_x\dfrac{\partial u}{\partial x}$

$$\frac{\partial}{\partial x}(v\sigma_y) - v\frac{\partial\sigma_y}{\partial y} = \sigma_y\frac{\partial v}{\partial y}$$

$$\frac{\partial}{\partial y}(u\tau_{yx}) - u\frac{\partial\tau_{yx}}{\partial y} = \tau_{yx}\frac{\partial u}{\partial y}.$$

$$\frac{\partial}{\partial y}\left(v\tau_{xy}\right) - v\frac{\partial \tau_{xy}}{\partial x} = \tau_{xy}\frac{\partial v}{\partial x}$$

When the stresses in Eqs. (6.17) to (6.19) are substituted in Eq. (6.36),

$$\rho\frac{De}{Dt} = k\left(\frac{\partial^2 T}{\partial x^2} + \frac{\partial^2 T}{\partial y^2}\right) + \left[-p\frac{\partial u}{\partial x} + 2\mu\left(\frac{\partial u}{\partial x}\right)^2 - p\frac{\partial v}{\partial y} + 2\mu\left(\frac{\partial v}{\partial y}\right)^2 + \mu\left(\frac{\partial u}{\partial y} + \frac{\partial v}{\partial x}\right)^2\right]$$

$$= k\left(\frac{\partial^2 T}{\partial x^2} + \frac{\partial^2 T}{\partial y^2}\right) + \mu\left[2(\frac{\partial u}{\partial x})^2 + 2(\frac{\partial v}{\partial y})^2 + \left(\frac{\partial u}{\partial y} + \frac{\partial v}{\partial x}\right)^2\right]. \qquad (6.37)$$

Thus, the energy equation for the two-dimensional, steady, incompressible, constant property flow is

$$\rho\frac{De}{Dt} = k\left(\frac{\partial^2 T}{\partial x^2} + \frac{\partial^2 T}{\partial y^2}\right) + \mu\phi \qquad (6.38)$$

where the viscous-energy dissipation term   is defined as

$$\phi \equiv 2\{\left(\frac{\partial u}{\partial x}\right)^2 + \left(\frac{\partial v}{\partial y}\right)^2\} + \left(\frac{\partial v}{\partial x} + \frac{\partial u}{\partial y}\right)^2. \qquad (6.39)$$

If the density is constant, the term De/Dt may be approximated as

$$\frac{De}{Dt} \approx C_p\frac{DT}{Dt} \qquad (6.40)$$

The energy equation for constant density flow becomes

$$\rho C_p\left(u\frac{\partial T}{\partial x} + v\frac{\partial T}{\partial y}\right) = k\left(\frac{\partial^2 T}{\partial x^2} + \frac{\partial^2 T}{\partial y^2}\right) + \mu\phi \qquad (6.41)$$

where ϕ is given by Eq. (6.39).

The left-hand side of the equation represents the convective heat transfer, the first bracketed term on the right-hand side represents the conductive heat transfer, and the second term represents the viscous energy dissipation owing to friction in the fluid.

For many practical engineering cases, the flow velocities are moderate and the viscous energy dissipation term may be neglected. The Eq. (6.41) simplifies to

$$\rho C_P\left(u\frac{\partial T}{\partial x}+v\frac{\partial T}{\partial y}\right)=k\left(\frac{\partial^2 T}{\partial x^2}+\frac{\partial^2 T}{\partial y^2}\right) \tag{6.42}$$

When there is no flow, Eq. (6.42) reduces to the conduction equation with no heat generation, which is expected.

6.4   Summary of Governing Equations

Table 6.1   The continuity equation in different coordinate systems.

Vectorial	Compressible	$\frac{\partial \rho}{\partial t}+\nabla.(\rho\vec{V})=0$
	Incompressible	$\nabla.\vec{V}=0$
Rectangular (x,y)	Compressible	$\frac{\partial \rho}{\partial t}+\frac{\partial}{\partial x}(\rho u)+\frac{\partial}{\partial y}(\rho v)=0$
	Incompressible	$\frac{\partial u}{\partial x}+\frac{\partial v}{\partial y}=0$
Cylindrical (r,z)	Compressible	$\frac{\partial \rho}{\partial t}+\frac{1}{r}\frac{\partial}{\partial r}(\rho r v_r)+\frac{\partial}{\partial z}(\rho w)=0$
	Incompressible	$\frac{\partial v_r}{\partial r}+\frac{v_r}{r}+\frac{\partial v_z}{\partial z}=0$

Table 6.2 The momentum equations for a steady-flow, two-dimensional, incompressible, Newtonian fluid with constant properties in different coordinate systems.

Rectangular	
x-momentum	$$\rho\left(u\frac{\partial u}{\partial x}+v\frac{\partial u}{\partial y}\right)=B_x-\frac{\partial p}{\partial x}+\mu\left(\frac{\partial^2 u}{\partial x^2}+\frac{\partial^2 u}{\partial y^2}\right)$$
y-momentum	$$\rho\left(u\frac{\partial v}{\partial x}+v\frac{\partial v}{\partial y}\right)=B_y-\frac{\partial p}{\partial y}+\mu\left(\frac{\partial^2 v}{\partial x^2}+\frac{\partial^2 v}{\partial y^2}\right)$$
Cylindrical	
r-momentum	$$\rho\left(v_r\frac{\partial v_r}{\partial r}+v_z\frac{\partial v_r}{\partial z}\right)=B_r-\frac{\partial p}{\partial r}+\mu\left(\frac{\partial^2 v_r}{\partial r^2}+\frac{1}{r}\frac{\partial v_r}{\partial r}-\frac{v_r}{r^2}+\frac{\partial^2 v_r}{\partial z^2}\right)$$
z-momentum	$$\rho\left(v_r\frac{\partial v_z}{\partial r}+v_z\frac{\partial v_z}{\partial z}\right)=B_z-\frac{\partial p}{\partial z}+\mu\left(\frac{\partial^2 v_z}{\partial r^2}+\frac{1}{r}\frac{\partial v_z}{\partial r}+\frac{\partial^2 v_z}{\partial z^2}\right)$$

Table 6.3 The energy equations for a two-dimensional, incompressible, Newtonian fluid with constant properties in different coordinate systems.

Rectangular	
Vectorial	$$\rho C_P\frac{DT}{Dt}=k\nabla^2 T+\mu\phi$$
2-dimensional	$$\rho C_P\left(u\frac{\partial T}{\partial x}+v\frac{\partial T}{\partial y}\right)=k\left(\frac{\partial^2 T}{\partial x^2}+\frac{\partial^2 T}{\partial y^2}\right)+\mu\phi \ \text{where}$$ $$\phi\equiv 2\{\left(\frac{\partial u}{\partial x}\right)^2+\left(\frac{\partial v}{\partial y}\right)^2\}+\left(\frac{\partial v}{\partial x}+\frac{\partial u}{\partial y}\right)^2$$

Cylindrical	
Vectorial	$\rho C_P \dfrac{DT}{Dt} = k\nabla^2 T + \mu\phi$
2-dimensional	$\rho C_P \left(v_r \dfrac{\partial T}{\partial r} + v_z \dfrac{\partial T}{\partial z} \right) = k \left(\dfrac{\partial^2 T}{\partial r^2} + \dfrac{1}{r}\dfrac{\partial T}{\partial r} + \dfrac{\partial^2 T}{\partial z^2} \right) + \mu\phi$ where $$\phi \equiv 2\left\{ \left(\frac{\partial v_r}{\partial r} \right)^2 + \frac{v_r^2}{r^2} + \left(\frac{\partial v_z}{\partial z} \right)^2 \right\} + \left(\frac{\partial v_z}{\partial r} + \frac{\partial v_r}{\partial z} \right)^2$$

The equations of continuity, momentum and energy are summarized in Tables 6.1, 6.2 and 6.3, respectively. The vectorial forms in Tables 6.1 and 6.3 are provided for scholars of heat transfer that would like to go to three dimensional applications. The equations in cylindrical coordinates may be obtained from the rectangular coordinate equations by the use of the appropriate transformations.

Example 6.1

Problem: Derive the continuity equation in rectangular coordinates for a three-dimensional flow having velocity components u, v and w in the primary directions.

Solution
Consider a differential volume element $\Delta x\Delta y\Delta z$.   The conservation of mass may be stated as

$$\left(\begin{array}{c} \text{Net rate of mass flow entering} \\ \text{volume element in x direction} \end{array} \right) + \left(\begin{array}{c} \text{Net rate of mass flow entering} \\ \text{volume element in y direction} \end{array} \right)$$

$$+ \left(\begin{array}{c} \text{Net rate of mass flow entering} \\ \text{volume element in z direction} \end{array} \right) = \left(\begin{array}{c} \text{Net rate of increase of mass} \\ \text{within the volume element} \end{array} \right).$$

(i)

The net rate of mass flow entering the volume element in the x direction is

$$-\frac{\partial(\rho u)}{\partial x}\Delta x \Delta y \Delta z \, .$$

Similarly, the net rate of mass flow entering the volume element in the y direction is

$$-\frac{\partial(\rho v)}{\partial y}\Delta x \Delta y \Delta z \, .$$

Similarly, the net rate of mass flow entering the volume element in the z direction is

$$-\frac{\partial(\rho w)}{\partial z}\Delta x \Delta y \Delta z \, .$$

The net rate of increase of mass within the volume element is

$$\frac{\partial \rho}{\partial t}\Delta x \Delta y \Delta z \, .$$

Substituting in Eq. (i),

$$\frac{\partial(\rho u)}{\partial x}+\frac{\partial(\rho v)}{\partial y}+\frac{\partial(\rho w)}{\partial z}=-\frac{\partial \rho}{\partial t} \, .$$

Written vectorially, the continuity equation is $\frac{\partial \rho}{\partial t}+\nabla.(\rho \vec{V})=0$.

6.5   Dimensionless Numbers

The momentum and energy equations are very difficult to solve except for simple cases. For many cases of practical interest, the convective heat transfer is studied experimentally and the results are presented in the form of empirical equations that relate dimensionless numbers.

The following discussion is restricted to two-dimensional, steady, incompressible, constant-property flow. For simplicity, the body forces are neglected. The effects of body forces are considered in the chapter on natural convection. To nondimensionalize the appropriately reduced form of the governing equations from Tables 6.1-6.3, we select a characteristic length L, a reference velocity U_∞, a reference temperature

T_∞, a reference temperature difference ΔT, and define the following dimensionless variables:

$$X = \frac{x}{L}, \qquad Y = \frac{y}{L}, \qquad P = \frac{p}{\rho U_\infty^2} \qquad\qquad (6.43)$$

$$U = \frac{u}{U_\infty}, \qquad V = \frac{v}{U_\infty}, \qquad \theta = \frac{T - T_\infty}{\Delta T}. \qquad\qquad (6.44)$$

The quantity with value twice the dynamic head has been used to make the pressure dimensionless. The dimensionless continuity, x-momentum, y-momentum and energy equations are as follows:

$$\frac{\partial U}{\partial X} + \frac{\partial V}{\partial Y} = 0 \qquad\qquad (6.45)$$

$$U \frac{\partial U}{\partial X} + V \frac{\partial U}{\partial Y} = -\frac{\partial P}{\partial X} + \frac{1}{Re}\left(\frac{\partial^2 U}{\partial X^2} + \frac{\partial^2 U}{\partial Y^2} \right) \qquad\qquad (6.46)$$

$$U \frac{\partial V}{\partial Y} + V \frac{\partial V}{\partial Y} = -\frac{\partial P}{\partial Y} + \frac{1}{Re}\left(\frac{\partial^2 V}{\partial X^2} + \frac{\partial^2 V}{\partial Y^2} \right) \qquad\qquad (6.47)$$

$$U \frac{\partial \theta}{\partial X} + V \frac{\partial \theta}{\partial Y} = \frac{1}{Re\,Pr}\left(\frac{\partial^2 \theta}{\partial X^2} + \frac{\partial^2 \theta}{\partial Y^2} \right)$$

$$+ \frac{E}{Re}\{2\left(\frac{\partial U}{\partial X} \right)^2 + 2\left(\frac{\partial V}{\partial Y} \right)^2 + \left(\frac{\partial V}{\partial X} + \frac{\partial U}{\partial Y} \right)^2\}$$

$$(6.48)$$

where the dimensionless numbers are defined as

$$E \equiv \frac{U_\infty^2}{C_p \Delta T} = \text{Eckert number} \qquad\qquad (6.49)$$

$$Pr \equiv \frac{C_p \mu}{k} = \frac{v}{\alpha} = \text{Prandtl number} \qquad\qquad (6.50)$$

$$\mathrm{Re} \equiv \frac{\rho U_\infty L}{\mu} = \text{Reynolds number} \tag{6.51}$$

$$Pe \equiv \frac{\rho C_p U_\infty L}{k} = \Pr \mathrm{Re} = \text{Peclet number} \tag{6.52}$$

The Eckert number may be considered as the comparison of the temperature rise caused by the dynamic pressure U_∞^2 / C_P, with the reference temperature difference T. The Prandtl number is the ratio of molecular diffusivity of momentum, υ, to the molecular diffusivity of heat, α. The Reynolds number is the comparison of the inertial force (U_∞^2 / L) to the viscous force $(\upsilon U_\infty^2 / L^2)$. The Peclet number is the comparison of energy transferred by convection ($\rho C_p U_\infty A \Delta T$) to that transferred by conduction (kAΔT/L). It is used to compare the relative size of the term in conduction to that in convection in the governing equations with the objective of simplifying the governing equations. The heat transfer in forced convection depends on the three dimensionless groups, E, Pr and Re. For gases, Pr is of the order of unity. For liquids, Pr ranges from about 10 to 1000. For liquid metals, Pr ranges from about 0.003 to about 0.03.

The heat transfer between the fluid and the wall surface of a solid is given by Newton's law of cooling as

$$q \Delta T \tag{6.53}$$

where h is the heat transfer coefficient and ΔT is the difference between the wall surface and the mean fluid temperatures. For flow over solid bodies, the main stream temperature T is taken as the mean fluid temperature. For flow inside pipes, a bulk fluid temperature as defined in a later chapter is taken as the mean fluid temperature. If the main flow is in the x direction, the heat flux q is related to the temperature gradient by

$$q = -k \frac{\partial T}{\partial y}\bigg|_{wall}. \tag{6.54}$$

From Eqs. (6.59) and (6.60),

$$h\Delta T = -k\frac{\partial T}{\partial y}\bigg|_{y=0}. \tag{6.55}$$

In dimensionless form,

$$Nu \equiv \frac{hL}{k} = -\frac{\partial \theta}{\partial Y}\bigg|_{Y=0} \tag{6.56}$$

where the dimensionless quantities are $Y = y/L$ and $\theta = \dfrac{T - T_\infty}{\Delta T}$. The dimensionless number Nu is called the Nusselt number and compares the convective heat transfer coefficient to the conductive coefficient. It is clear that the Nusselt number depends on the same groups as the temperature distribution. Hence, the Nusselt number is a function of the Eckert, Prandtl and Reynolds numbers, and the following functional relationship may be written:

$$Nu = Nu(Re,Pr,E). \tag{6.57}$$

The Eckert number only enters the problem when the viscous dissipation term in the energy equation is significant. For moderate velocities, the viscous dissipation term may be neglected. Under such conditions, the forced convection is characterized by

$$Nu = Nu(Re,Pr). \tag{6.58}$$

Hence, in experimental heat transfer, the number of variables to be studied is significantly reduced. The Nusselt number or heat transfer coefficient is correlated to only two dimensionless numbers.

6.6   The Boundary Layer Equations

Typically, the effects of the viscous forces originate at the solid boundary of the body of fluid. The fluid contained in the region of substantial velocity change is called the hydrodynamic boundary layer, Prandtl, 1904 [2]. Similarly, if the fluid and the solid are at different temperatures, the region of substantial temperature change in the fluid is called the thermal boundary layer. The hydrodynamic boundary layer is caused by viscosity of the fluid, and it shows the resistance to flow by the fluid. The thermal boundary layer is caused by the thermal conductivity of the fluid, and it shows the resistance to heat transfer by the fluid.

The boundary layer equations may be obtained from the equations provided in Tables 6.1-6.3, with simplification and by an order-of-magnitude study of each term in the equations. It is assumed that the main flow is in the x direction. The terms that are too small are neglected. Consider the momentum and energy equations for the two-dimensional, steady flow of an incompressible fluid with constant properties. The dimensionless equations are given by Eqs. (6.46) to (6.48). The principal assumption made in the boundary layer is that the hydrodynamic boundary layer thickness δ and the thermal boundary layer thickness δ_t are small compared to a characteristic dimension L of the body. In mathematical terms,

$$\Delta \equiv \frac{\delta}{L} \ll 1 \quad \text{and} \quad \Delta_t \equiv \frac{\delta_t}{L} \ll 1. \tag{6.59}$$

The Reynolds number is assumed very large, and of the order of $1/\Delta^2$ and the Peclet number is of the order of $1/\Delta_t^2$. All other quantities in the equations can be measured in units of 1, Δ, Δ_t. The variables U, X and θ are assumed to be order unity, and Y is of order Δ or Δ_t.

Consider the continuity equation Eq. (6.45). In this equation, the two terms must be of the same order of magnitude. Since U and X are of the order unity, the derivative $\partial U/\partial X$ is of the order unity, and $\partial V/\partial Y$ must be of the same order. Since Y is assumed to be or order Δ, V must be of the order Δ also. The dimensionless continuity, x-momentum, y-momentum and energy equations are now written, and the order of magnitude of each term is written beneath in units of 1, Δ and Δ_t.

$$\frac{\partial U}{\partial X} + \frac{\partial V}{\partial Y} = 0 \tag{6.60}$$

$$\frac{1}{1} \qquad \frac{\Delta}{\Delta}$$

$$U\frac{\partial U}{\partial X}+V\frac{\partial U}{\partial Y}=-\frac{\partial P}{\partial X}+\frac{1}{Re}\left(\frac{\partial^2 U}{\partial X^2}+\frac{\partial^2 U}{\partial Y^2}\right)$$ (6.61)

$$1\frac{1}{1} \quad \Delta\frac{1}{\Delta} \quad \quad \Delta^2\left(\frac{1}{\Delta} \quad \frac{1}{\Delta^2}\right)$$

$$U\frac{\partial V}{\partial X}+V\frac{\partial V}{\partial Y}=-\frac{\partial P}{\partial Y}+\frac{1}{Re}\left(\frac{\partial^2 V}{\partial X^2}+\frac{\partial^2 V}{\partial Y^2}\right)$$ (6.62)

$$1\frac{\Delta}{1} \quad \Delta\frac{\Delta}{\Delta} \quad \quad \Delta^2\left(\frac{\Delta}{1} \quad \frac{\Delta}{\Delta^2}\right)$$

$$U\frac{\partial\theta}{\partial X}+V\frac{\partial\theta}{\partial Y}=\frac{1}{Re\,Pr}\left(\frac{\partial^2\theta}{\partial X^2}+\frac{\partial^2\theta}{\partial Y^2}\right)$$

$$1\frac{1}{1} \quad \Delta\frac{1}{\Delta} \quad \quad \Delta_t^2\left(\frac{1}{1} \quad \frac{1}{\Delta_t^2}\right)$$

$$+\frac{E}{Re}\{2\left(\frac{\partial U}{\partial X}\right)^2+2\left(\frac{\partial V}{\partial Y}\right)^2+\left(\frac{\partial V}{\partial X}+\frac{\partial U}{\partial Y}\right)^2\}$$

$$\Delta^2\{\frac{1}{1} \quad \quad \frac{\Delta^2}{\Delta^2} \quad \quad \left(\frac{\Delta}{1},\frac{1}{\Delta}\right)^2\}$$

(6.63)

    The order of magnitude exercise leads to the deductions detailed below. The continuity equation is unchanged. In Eq. (6.61), the term $\partial^2 U/\partial X^2$ can be neglected in comparison with the term $\partial^2 U/\partial Y^2$. In Eq. (6.62), the pressure-gradient term has to be of the order Δ since all the other terms are of that order. This means that $\partial P/\partial Y$ is very small and the pressure P across the boundary layer is approximately constant. Hence, the y-momentum equation is not necessary in the analysis. In Eq. (6.63), the term $\partial^2\theta/\partial X^2$ is very small in comparison with the term

$\partial^2\theta / \partial Y^2$. In the viscous dissipation term, all other terms within the parentheses are to be neglected in comparison with $\partial U / \partial Y$. The term $(E/Re)(\partial U / \partial Y)^2$ is of order of unity if the Eckert number is selected to be of the order of unity. The dimensionless boundary-layer equations for two-dimensional, steady flow of an incompressible, constant-property flow are the continuity, x-momentum and the energy equations, which are

$$\frac{\partial U}{\partial X} + \frac{\partial V}{\partial Y} = 0 \tag{6.64}$$

$$U\frac{\partial U}{\partial X} + V\frac{\partial U}{\partial Y} = -\frac{dp}{dX} + \frac{1}{Re}\frac{\partial^2 U}{\partial Y^2} \tag{6.65}$$

$$U\frac{\partial \theta}{\partial X} + V\frac{\partial \theta}{\partial Y} = \frac{1}{Re\,Pr}\frac{\partial^2 \theta}{\partial Y^2} + \frac{E}{Re}\left(\frac{\partial U}{\partial Y}\right)^2. \tag{6.66}$$

PROBLEMS

6.1.   Write the boundary layer equations in dimensional form.

6.2.   Consider a cylindrical elemental control volume of dimensions Δr, $r\Delta\theta$, and Δz in the r, θ and z directions, respectively. Derive the continuity equation in cylindrical coordinates.

6.3.   Consider a spherical elemental control volume of dimensions Δr, $r\sin\varphi\Delta\varphi$, and $r\Delta\theta$ in the r, φ and θ directions, respectively. Derive the continuity equation in spherical coordinates.

6.4.   Take into consideration two-dimensional, rectilinear, steady, incompressible, constant-property, laminar boundary layer flow in the x direction along a flat plate. Assume that viscous energy dissipation may be neglected. Write the continuity, momentum and energy equations.

6.5.   In Prob. 6.4, consider the laminar flow to be along a curved body with the x direction measured along the curved surface and the y direction perpendicular to the surface.

6.6.   In the derivation of the boundary layer equations, two of the assumptions made are as follows:

$$\frac{1}{Re} \approx \Delta^2 \quad \text{and} \quad \frac{1}{Re\,Pr} \approx \Delta_t^2$$

Use these assumptions to find a relation between the hydrodynamic boundary layer thickness δ and the thermal boundary layer thickness δ_t. For gases, Pr is of the order of unity. For liquids, Pr ranges from about 10 to 1000. For liquid metals, Pr ranges from about 0.003 to about 0.03. Deduce the relative thicknesses of δ and δ_t for gases, liquids and liquid metals.

6.7.   From the momentum equations for steady, two-dimensional, incompressible, Newtonian fluid with constant properties in rectangular coordinates, obtain the x-momentum equation for the parallel flow (i.e., v = 0). Obtain the corresponding energy equation.

6.8.   From the momentum equations for steady, two-dimensional, incompressible, Newtonian fluid with constant properties in cylindrical coordinates, obtain the z-momentum equation for the parallel flow (i.e., $v_r = 0$). Obtain the corresponding energy equation.

REFERENCES

1.   H Schlichting. Boundary Layer Theory. 7th ed. New York: McGraw-Hill, 1979.

2.   L Prandtl. Uber Flussigkeitsbewegung bei sehr kleiner Reibung. Proc 3rd Int Math Congr, Heidelberg, 484-491, Teuber, Leipzig, 1904; also in English as: Motion of Fluids with Very Little Viscosity. NACA TM 452, 1928.

Dimensionless Convection Numbers

Dimensionless numbers help in convection heat transfer engineering
Used to compare relative values in the practice of engineering
In convection, there is the Eckert number and the Prandtl number,
There is also the Reynolds number, Peclet number and Nusselt number.

Eckert compares dynamic-pressure-caused temperature difference
To the selected reference temperature difference
Prandtl compares the momentum molecular diffusivity
To the value of the thermal molecular diffusivity.

Reynolds compares the inertia forces to the viscous forces,
Peclet compares convection energy to conduction energy,
Nusselt number compares the convection heat transfer coefficient
To the magnitude of the conductive heat coefficient.

K.V. Wong

7

External Forced Convection

The boundary layer problem is difficult to solve exactly. There are several approximate methods to solve the problem. This chapter looks at external forced convection, that is, flow outside and around a solid body like a plate. The next chapter discusses flows inside a solid structure such as a pipe, or between two plates.

7.1 Momentum Integral Method of Analysis

Several approximate methods exist for solving the boundary layer equations. The momentum-integral method of analysis is an important method. The principal steps of the method are listed below.

(1)   Integrate x-momentum equation with respect to y over the boundary layer thickness $\delta(x)$. Eliminate velocity component $v(x,y)$ in the equation by means of the continuity equation, resulting in the momentum integral equation.

(2)   A polynomial profile is chosen for the velocity component $u(x,y)$ over the boundary layer $0 \leq y \leq \delta$. Use the boundary conditions to express
$U(x,y) = f[y, \delta(x)]$ in $0 \leq y \leq \delta$.

(3)   Substitute $u(x,y)$ into the momentum integral equation derived in Step 1, and integrate with respect to y. The ordinary differential equation for $\delta(x)$ is obtained; solve for $\delta(x)$.

(4)   Obtain $u(x,y)$ from Step (2), knowing $\delta(x)$.

The continuity equation for the boundary layer in two-dimensional rectilinear coordinates is

$$\frac{\partial u}{\partial x} + \frac{\partial v}{\partial y} = 0. \tag{7.1}$$

The x-momentum equation for the boundary layer in two-dimensional rectilinear coordinates is

149

$$u \frac{\partial u}{\partial x} + v \frac{\partial u}{\partial y} = v \frac{\partial^2 u}{\partial y^2}. \qquad (7.2)$$

Step 1:

$$\int_0^{\delta(x)} u \frac{\partial u}{\partial x} dy + \int_0^{\delta(x)} v \frac{\partial u}{\partial y} dy = v \left(\frac{\partial u}{\partial y} \Big|_{y=\delta} - \frac{\partial u}{\partial y} \Big|_{y=0} \right) = -v \frac{\partial u}{\partial y} \Big|_{y=0}$$

$$(7.3)$$

since $\dfrac{\partial u}{\partial y}\Big|_{y=\delta} = 0$ by the boundary layer concept.   The velocity component v is eliminated from Eq. (7.3) by using Eq. (7.1),

$$\frac{d}{dx} \left[\int_0^{\delta} u(u_\infty - u) dy \right] = v \frac{\partial u}{\partial y} \Big|_{y=0} \quad \text{in} \quad 0 \le y \le 0 \qquad (7.4)$$

where $u = u(x, y)$ and $\delta = \delta(x)$.

Equation (7.4) is called the momentum integral equation.

Step 2:
        In the present analysis, we choose a cubic polynomial for the velocity.

$$u(x, y) = a_o + a_1 y + a_2 y^2 + a_3 y^3 \qquad (7.5)$$

The boundary conditions are as follows:

$$u\big|_{y=0} = 0 \qquad\qquad\qquad u\big|_{y=\delta} = u_\infty$$

$$\frac{\partial u}{\partial y}\Big|_{y=\delta} = 0 \qquad \text{and} \qquad \frac{\partial^2 u}{\partial y^2}\Big|_{y=0} = 0. \quad (7.6)$$

The first two relations are the boundary conditions of the problem, the third one results from the boundary layer concept. The last one is the derived condition which is obtained by evaluating the x-momentum equation, Eq. (7.2), at y = 0, where u = v = 0. The solution to Eq. (7.5) gives

$$\frac{u(x,y)}{u_\infty} = \frac{3}{2}\left(\frac{y}{\delta}\right) - \frac{1}{2}\left(\frac{y}{\delta}\right)^3 \tag{7.7}$$

Step 3:

The velocity profile, Eq. (7.7), is substituted into the momentum integral equation, Eq. (7.4),

$$u_\infty^2 \frac{d}{dx}\{\int_0^\delta \left[\frac{3}{2}\frac{y}{\delta} - \frac{1}{2}\left(\frac{y}{\delta}\right)^3\right]\left[1 - \frac{3}{2}\frac{y}{\delta} + \frac{1}{2}\left(\frac{y}{\delta}\right)^3\right]dy\} = \nu u_\infty \frac{3}{2\delta} \tag{7.8}$$

$$\frac{d}{dx}\left[\frac{39}{280}\delta(x)\right] = \frac{3\nu}{2u_\infty\delta(x)}$$

$$\delta d\delta = \frac{140}{13}\frac{\nu}{u_\infty}dx \quad \text{with boundary condition } \delta(x) = 0 \text{ at } x = 0.$$
$$\tag{7.9}$$

Integrating, $\quad \delta^2(x) = \frac{280}{13}\frac{\nu x}{U_\infty}$ $\tag{7.10}$

$$\delta(x) = 4.64\sqrt{\frac{\nu x}{U_\infty}}$$

$$\frac{\delta(x)}{x} = \frac{4.64}{Re_x^{0.5}} \quad \text{where } Re_x = \frac{u_\infty x}{\nu}. \tag{7.11}$$

Step 4:

$$\frac{u}{u_\infty} = \frac{3}{2}\left(\frac{4.64}{Re_x^{0.5}}\right) - \frac{1}{2}\left(\frac{4.64}{Re_x^{0.5}}\right)^3. \tag{7.12}$$

7.1.1 Drag Coefficient

The local drag force τ_x per unit area exerted by the fluid flowing over a flat plate is related to a local-drag coefficient c_x as

$$\tau_x = c_x \frac{\rho u_\infty^2}{2g} \qquad (^{lb_f}\!\big/\!_{ft^2} \text{ or } ^{N}\!\big/\!_{m^2}). \qquad (7.13)$$

τ_x is the local shear stress.

$$\tau_x = \frac{\mu}{g} \frac{\partial u}{\partial y}\bigg|_{y=0} \qquad (7.14)$$

$$c_x = \frac{2v}{u_\infty^2} \frac{\partial u}{\partial y}\bigg|_{y=0} \qquad (7.15)$$

Since $\dfrac{\partial u}{\partial y}\bigg|_{y=0} = \dfrac{3u_\infty}{2\delta(x)}$,

$$c_x = \frac{3v}{u_\infty} \sqrt{\frac{13}{280} \frac{u_\infty}{vx}} = \sqrt{\frac{117}{280} \frac{v}{u_\infty x}} = \frac{0.648}{Re_x^{0.5}}. \qquad (7.16)$$

The mean value of the drag coefficient $c_{m,L}$ over the length $x = 0$ to L is defined as

$$c_{m,L} = \frac{1}{L} \int_{x=0}^{L} c_x dx. \qquad (7.17)$$

It can be shown that $c_{m,L} = 2c_x|_{x=L}$ \qquad (7.18)

Example 7.1

Problem: For flow along a flat plate subjected to the three conditions stated below, find the second-degree-polynomial representation of the velocity profile.

$$u\big|_{y=0} = 0 \quad u\big|_{y=\delta} = u_\infty \quad \text{and} \quad \frac{\partial u}{\partial y}\bigg|_{y=\delta} = 0.$$

Solution

$$\frac{u(x,y)}{u_\infty} = a + b\left(\frac{y}{\delta}\right) + c\left(\frac{y}{\delta}\right)^2$$

Apply $u = 0$ at $y = 0$: $a = 0$
Apply $u = u_\infty$ at $y = \delta$: $1 = b + c$
Apply $\dfrac{\partial u}{\partial y} = 0$ at $y = \delta$: $0 = b + 2c$

Simultaneously solving for the unknowns: $b = 2$, $c = -1$.

Therefore, $\dfrac{u(x,y)}{u_\infty} = 2\left(\dfrac{y}{\delta}\right) - \left(\dfrac{y}{\delta}\right)^2.$

Example 7.2

Problem: Find the fourth-degree-polynomial representation of the velocity profile for flow along a flat plate.

Solution

$$\frac{u(x,y)}{u_\infty} = a_o + a_1\left(\frac{y}{\delta}\right) + a_2\left(\frac{y}{\delta}\right)^2 + a_3\left(\frac{y}{\delta}\right)^3 + a_4\left(\frac{y}{\delta}\right)^4$$

Apply $u = 0$ at $y = 0$: $a_o = 0$

Apply $\dfrac{\partial^2 u}{\partial y^2} = 0$ at $y = 0$: $a_2 = 0$

Apply $u = u_\infty$ at $y = \delta$: $1 = a_1 + a_3 + a_4$

Apply $\dfrac{\partial u}{\partial y} = 0$ at $y = \delta$: $0 = a_1 + 3a_3 + 4a_4$

Apply $\dfrac{\partial^2 u}{\partial y^2} = 0$ at $y = \delta$: $0 = 6a_3 + 12a_4$

Simultaneously solving for the unknowns: $a_1 = 2$, $a_3 = -2$, $a_4 = 1$.

Hence, $\dfrac{u(x,y)}{u_\infty} = 2\left(\dfrac{y}{\delta}\right) - 2\left(\dfrac{y}{\delta}\right)^3 + \left(\dfrac{y}{\delta}\right)^4 .$

Example 7.3

Problem: Find the third-degree polynomial representation of the velocity profile for flow along a flat plate.

Solution

$$\dfrac{u(x,y)}{u_\infty} = a_o + a_1\left(\dfrac{y}{\delta}\right) + a_2\left(\dfrac{y}{\delta}\right)^2 + a_3\left(\dfrac{y}{\delta}\right)^3$$

Apply $u = 0$ at $y = 0$: $a_0 = 0$

Apply $\dfrac{\partial^2 u}{\partial y^2} = 0$ at $y = 0$: $a_2 = 0$

Apply $u = u_\infty$ at $y = \delta$: $1 = a_1 + a_3$

Apply $\dfrac{\partial u}{\partial y} = 0$ at $y = \delta$: $0 = a_1 + 3a_3$

Simultaneously solving for the unknowns: $a_1 = 1.5$, $a_3 = -0.5$.

Hence, $\dfrac{u(x,y)}{u_\infty} = \dfrac{3}{2}\left(\dfrac{y}{\delta}\right) - \dfrac{1}{2}\left(\dfrac{y}{\delta}\right)^3 .$

7.2 Integral Method of Analysis for Energy Equation

The integral method of analysis for the energy equation is an important method. The principal steps of the method are listed below.

(1)   Integrate energy equation with respect to y over a distance that exceeds both the momentum boundary layer thickness $\delta(x)$ and the thermal boundary layer thickness $\delta_t(x)$. Eliminate velocity component $v(x,y)$ in the equation by means of the continuity equation, resulting in the energy integral equation.

(2)   Polynomial approximations are selected to represent both the temperature distribution θ(x,y) and the velocity component u(x,y).

(3)   Substitute the velocity and temperature profiles determined in Step 2 into the energy integral equation derived in Step 1, and integrate with respect to y.

(4)   Knowing δ_t, the temperature distribution in the boundary layer is determined from Step 2.

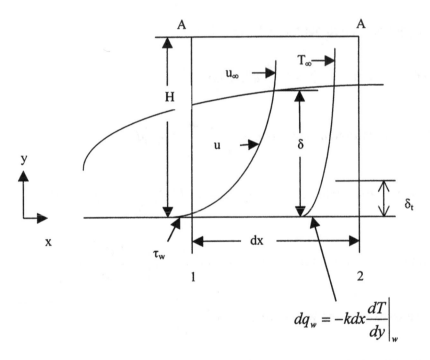

Figure 7.1   Control volume for the integral energy analysis of laminar boundary layer.

Consider the control volume bounded by the planes 1, 2, A-A, and the wall as shown in the figure. Assume that the thermal boundary layer is thinner than the hydrodynamic boundary layer, as shown. Let

T_w = wall temperature

T_∞ = free-stream temperature

dq_w = heat given up to the fluid over the length dx

Energy conservation for the control volume gives

energy convected in +   viscous work within   +   Heat transfer a
                          element   wall

= energy convected out

Energy convected in through plane 1 is $\rho C_P \int_0^H uTdy$.

Energy convected out through plane 2 is

$$\rho C_P \int_0^H uTdy + \frac{d}{dx}\left(\rho C_P \int_0^H uTdy \right)dx .$$

Mass-flow through plane A-A is $\dfrac{d}{dx}\left(\int_0^H \rho u dy \right)dx$,

and this carries with it an energy equal to $C_p T_\infty \dfrac{d}{dx}\left(\int_0^H \rho u dy \right)dx$.

Net viscous work done within the element is $\mu\left[\int_0^{H'} \left(\dfrac{du}{dy}\right)^2 dy \right]dx$.

Heat transfer at the wall is $dq_w = -kdx\dfrac{\partial T}{\partial y}\Big|_w$.

Combining these quantities and collecting terms,

$$\frac{d}{dx}\left[\int_0^H (T_\infty - T)u dy \right] + \frac{\mu}{\rho C_P}\left[\int_0^H \left(\frac{du}{dy}\right)^2 dy \right] = \alpha\frac{\partial T}{\partial y}\Big|_w . \qquad (7.19)$$

This is the integral energy equation of the boundary layer for constant properties.

7.3 Hydrodynamic and Thermal Boundary Layers on a Flat Plate, Where Heating Starts at $x = x_o$.

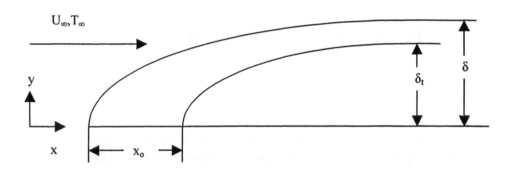

Figure 7.2 Figure illustrating hydrodynamic and thermal boundary layers on a flat plate where heating starts at $x = x_o$.

From Example 7.3, we found that the cubic representation of the velocity profile for flow along a flat plate is

$$\frac{u(x,y)}{u_\infty} = \frac{3}{2}\left(\frac{y}{\delta}\right) - \frac{1}{2}\left(\frac{y}{\delta}\right)^3.$$

A cubic representation of the temperature profile for the same flow is

$$\frac{\theta}{\theta_w} = a_o + a_1\left(\frac{y}{\delta_t}\right) + a_2\left(\frac{y}{\delta_t}\right)^2 + a_3\left(\frac{y}{\delta_t}\right)^3$$

Apply $\dfrac{\theta}{\theta_w} = 1$ at $y = 0$: $a_o = 1$

Apply $\dfrac{\partial^2\theta}{\partial y^2} = 0$ at $y = 0$: $a_2 = 0$

Apply $\dfrac{\theta}{\theta_w} = 0$ at $y = \delta_t$: $0 = 1 + a_1 + a_3$

Apply $\dfrac{\partial \theta}{\partial y} = 0$ at $y = \delta_t$: $0 = a_1 + 3a_3$

Simultaneously solving for the unknowns: $a_o = 1$, $a_1 = -1.5$, $a_3 = 0.5$.

Hence, $\qquad\qquad \dfrac{\theta}{\theta_w} = 1 - \dfrac{3}{2}\left(\dfrac{y}{\delta_t}\right) + \dfrac{1}{2}\left(\dfrac{y}{\delta_t}\right)^3 .$

Inserting the cubic representations of the temperature profile and the velocity profile in Eq. (7.19), and neglecting viscous dissipation,

$$\dfrac{d}{dx}\left[\int_0^H (T_\infty - T)u\,dy\right] = \dfrac{d}{dx}\left[\int_0^H (\theta_\infty - \theta)u\,dy\right]$$

$$= \theta_\infty u_\infty \dfrac{d}{dx}\left\{\int_0^H \left[1 - \dfrac{3}{2}\dfrac{y}{\delta_t} + \dfrac{1}{2}\left(\dfrac{y}{\delta_t}\right)^3\right]\left[\dfrac{3}{2}\dfrac{y}{\delta} - \dfrac{1}{2}\left(\dfrac{y}{\delta}\right)^3\right]dy\right\}$$

$$= \alpha\dfrac{\partial T}{\partial y}\bigg|_{y=0} = \dfrac{3\alpha\theta_\infty}{2\delta_t}. \qquad\qquad (7.20)$$

For the case above, the thermal boundary layer is thinner than the hydrodynamic boundary layer.

$$\dfrac{3\alpha\theta_\infty}{2\delta_t}$$

$$= \theta_\infty u_\infty \dfrac{d}{dx}\left\{\int_0^{\delta_t}\left[\dfrac{3}{2\delta}y - \dfrac{9}{4\delta\delta_t}y^2 + \dfrac{3}{4\delta\delta_t^3}y^4 - \dfrac{1}{2\delta^3}y^3 + \dfrac{3}{4\delta_t\delta^3}y^4 - \dfrac{1}{4\delta_t^3\delta^3}y^6\right]dy\right\}$$

$$\dfrac{3\alpha\theta_\infty}{2\delta_t} = \theta_\infty u_\infty \dfrac{d}{dx}\left[\dfrac{3}{4\delta}y^2 - \dfrac{3}{4\delta_t\delta}y^3 + \dfrac{3}{20\delta\delta_t^3}y^5 - \dfrac{1}{8\delta^3}y^4 + \dfrac{3}{20\delta_t\delta^3}y^5 - \dfrac{1}{28\delta_t^3\delta^3}y^7\right]_0^{\delta_t}$$

$$\frac{3\alpha\theta_\infty}{2\delta_t} = \theta_\infty u_\infty \frac{d}{dx}\left[\delta\left(-\frac{3}{20}\xi^2 - \frac{3}{280}\xi^4\right)\right] \quad \text{where } \xi = \frac{\delta_t}{\delta}.$$

Since $\delta_t < \delta, \xi < 1$, and term in ξ^4 is small,

$$\frac{3}{20}\theta_\infty u_\infty \frac{d}{dx}\left(\delta\xi^2\right) = \frac{3}{2}\frac{\alpha\theta_\infty}{\xi\delta} \tag{7.21}$$

Performing the differentiation gives

$$\frac{1}{10}u_\infty\left(2\delta\xi\frac{d\xi}{dx} + \xi^2\frac{d\delta}{dx}\right) = \frac{\alpha}{\delta\xi}$$

$$\frac{1}{10}u_\infty\left(2\delta^2\xi^2\frac{d\xi}{dx} + \xi^3\delta\frac{d\delta}{dx}\right) = \alpha.$$

But $\quad \delta d\delta = \frac{140}{13}\frac{\upsilon}{u_\infty}dx,$

$$\delta^2 = \frac{280}{13}\frac{\upsilon x}{u_\infty}.$$

Hence, $\quad \xi^3 + 4x\xi^2\dfrac{d\xi}{dx} = \dfrac{13}{14}\dfrac{\alpha}{\upsilon} \tag{7.22}$

Since $\quad \xi^2\dfrac{d\xi}{dx} = \dfrac{1}{3}\dfrac{d}{dx}\left(\xi^3\right)$

$$\left(\xi^3\right) + \frac{4}{3}x\frac{d}{dx}\left(\xi^3\right) = \frac{13}{14}\frac{\alpha}{\upsilon}.$$

It is a first-order linear differential equation in ξ^3, with the solution

$$\xi^3 = Cx^{-\frac{3}{4}} + \frac{13}{14}\frac{\alpha}{\upsilon}.$$

The boundary conditions are as follows:

$$\delta_t = 0 \text{ at } x = x_0$$
$$\xi = 0 \text{ at } x = x_0.$$

Hence, $C = -\dfrac{13}{14}\dfrac{\alpha}{\upsilon}x_o^{\frac{3}{4}}.$

Thus, $\xi = \dfrac{\delta_t}{\delta} = \dfrac{1}{1.026}\operatorname{Pr}^{-\frac{1}{3}}\left[1-\left(\dfrac{x_o}{x}\right)^{\frac{3}{4}}\right]^{\frac{1}{3}}$ where $\operatorname{Pr} = \dfrac{\upsilon}{\alpha}.$

$$(7.23)$$

When the plate is heated over the entire length, $x_o = 0$. Under this condition,

$$\frac{\delta_t}{\delta} = \xi = \frac{1}{1.026}\operatorname{Pr}^{-\frac{1}{3}}.$$

The heat transfer coefficient, $h = \dfrac{-k\left(\dfrac{\partial T}{\partial y}\right)_w}{T_w - T_\infty} = \dfrac{3}{2}\dfrac{k}{\delta_t} = \dfrac{3}{2}\dfrac{k}{\xi\delta}.$

Substituting $\dfrac{\delta}{x} = 4.64\left(\dfrac{\upsilon}{u_\infty x}\right)^{\frac{1}{2}}$ and using Eq. (7.23), we get

$$h_x = 0.332 k \operatorname{Pr}^{\frac{1}{3}}\left(\frac{u_\infty}{\upsilon x}\right)^{\frac{1}{2}}\left[1-\left(\frac{x_o}{x}\right)^{\frac{3}{4}}\right]^{-\frac{1}{3}}.$$

Multiplying both sides by x/k,

$$Nu_x = \frac{h_x x}{k} = 0.332 \, \mathrm{Pr}^{\frac{1}{3}} \, \mathrm{Re}_x^{\frac{1}{2}} \left[1 - \left(\frac{x_o}{x} \right)^{\frac{3}{4}} \right]^{\frac{1}{3}}. \qquad (7.24)$$

For a plate heated over its entire length, $x_o = 0$ and

$$Nu_x = 0.332 \, \mathrm{Pr}^{\frac{1}{3}} \, \mathrm{Re}_x^{\frac{1}{2}}.$$

When $x_o = 0$, $\quad \bar{h} = \dfrac{\int_0^L h_x dx}{\int_0^L dx} = 2h_{x=L}, \quad \overline{Nu_L} = \dfrac{\bar{h}L}{k} = 2Nu_{x=L}.$ The film

temperature is $T_f = \dfrac{T_w + T_\infty}{2}$. Evaluate the properties at this mean

temperature.

Example 7.4

Problem: Using linear profiles for the velocity and the temperature within the boundary layer, obtain an expression for the velocity boundary layer thickness in terms of the local Reynolds number. Hence, derive an expression for the local Nusselt number.

Solution

Assume that the linear velocity profile is $u = u_\infty \dfrac{y}{\delta}$. $\qquad$ (i)

Assume that the linear velocity profile is $\theta = \dfrac{y}{\delta_t}$. $\qquad$ (ii)

The Nusselt number is $Nu_x =$

$$\frac{hx}{k} = \frac{x}{k} \cdot k \frac{\partial T}{\partial y}\bigg|_w \bigg/ (T_\infty - T_w) = x \frac{\partial \theta}{\partial y}\bigg|_w = \frac{x}{\delta_t}. \qquad (iii)$$

From Eq. (7.4) and Eq. (i) above,

LHS of Eq. (7.4)

$$= \frac{d}{dx}\left[\int_0^\delta (u_\infty - u)u\,dy\right] = \frac{d}{dx}\left[\int_0^\delta u_\infty^2\left(1 - \frac{u}{u_\infty}\right)\frac{u}{u_\infty}\,dy\right] = \frac{1}{6}u_\infty^2\frac{d\delta}{dx}.$$

RHS of Eq. (7.4) $\left.\dfrac{\partial u}{\partial y}\right|_w = u_\infty\dfrac{d}{dy}\left(\dfrac{y}{\delta}\right) = \dfrac{u_\infty}{\delta}.$

Hence, $\delta d\delta = \dfrac{6\upsilon}{\rho u_\infty}\,dx.$

Integrating, $\delta^2 = \dfrac{12\upsilon}{u_\infty}x$ and $\delta = \dfrac{3.464x}{\sqrt{Re_x}}.$ (iv)

From the energy integral equation, Eq. (7.19) and Eq. (ii) above,

$$\frac{d}{dx}\left[\int_0^{\delta_t} u_\infty \frac{y}{\delta}\left(1 - \frac{y}{\delta_t}\right)dy\right] = \frac{\alpha}{\delta_t}$$

$$u_\infty \frac{d}{dx}\left[\frac{\delta_t^2}{\delta.2} - \frac{1}{3}\frac{\delta_t^3}{\delta_t.\delta}\right] = \frac{\alpha}{\delta_t}.$$

Hence, $\dfrac{u_\infty}{6}\dfrac{d}{dx}\left(\dfrac{\delta_t^2}{\delta}\right) = \dfrac{\alpha}{\delta_t}.$ (v)

Let $\xi = \dfrac{\delta_t}{\delta}$

$$\frac{u_\infty}{6}\left(2\xi\frac{d\xi}{dx}\delta + \xi^2\frac{d\delta}{dx}\right) = \frac{\alpha}{\delta\xi}$$ (vi)

$$\frac{u_\infty}{6}\left(\frac{24\upsilon}{u_\infty}x\xi^2\frac{d\xi}{dx} + \frac{6\upsilon}{u_\infty}\xi^3\right) = \alpha.$$

Hence, $\qquad 4x\xi^2\dfrac{d\xi}{dx}+\xi^3=\dfrac{\alpha}{\upsilon}.$ (vii)

Let $\xi^3=\eta.$ $\qquad \dfrac{4}{3}x\dfrac{d\eta}{dx}+\eta=\dfrac{\alpha}{\upsilon}.$ (viii)

The solution to Eq. (viii) is $\eta x^{\frac{3}{4}}=\dfrac{\alpha}{\upsilon}x^{\frac{3}{4}}+c.$

If $x=x_o,$ $\xi=0$ (that is, the heating starts at x_o from the leading edge),

then $c=-\dfrac{\alpha}{\upsilon}x_o^{\frac{3}{4}}.$

Hence, $\qquad \xi=\dfrac{\delta_t}{\delta}=\left\{\dfrac{\left[1-\left(\dfrac{x_o}{x}\right)^{\frac{3}{4}}\right]^{\frac{1}{3}}}{\text{Pr}}\right\}.$ (ix)

If $x_o=0,$ then $\xi=\dfrac{\delta_t}{\delta}=\text{Pr}^{-\frac{1}{3}}$ is a constant.

From Eqs. (iii), (iv) and (ix), we get

$$Nu_x = \frac{x}{\delta_t}$$

$$= x \cdot \left(\frac{12\upsilon}{u_\infty} x\right)^{-\frac{1}{2}} \left[\left[1 - \left(\frac{x_o}{x}\right)^{\frac{3}{4}}\right]\right]^{-\frac{1}{3}} Pr^{\frac{1}{3}} = 0.28867 \, Re_x^{\frac{1}{2}} \, Pr^{\frac{1}{3}} \left[1 - \left(\frac{x_o}{x}\right)^{\frac{3}{4}}\right]^{-\frac{1}{3}}.$$

$$(x)$$

If $x_o = 0$, then $Nu_x = 0.2887 \, Re_x^{\frac{1}{2}} \, Pr^{\frac{1}{3}}$.

Example 7.5

Problem: (a) Derive the boundary layer thickness and local drag coefficient, assuming the velocity profile is a quartic polynomial. (b) Derive the convective heat transfer Nusselt number, assuming the temperature profile is a quartic polynomial.

Solution

   (a) Assume that the dimensionless velocity profile is a quartic polynomial.

$$u = \frac{u'}{u_\infty} = a_o + a_1\left(\frac{y}{\delta}\right) + a_2\left(\frac{y}{\delta}\right)^2 + a_3\left(\frac{y}{\delta}\right)^3 + a_4\left(\frac{y}{\delta}\right)^4 \qquad \text{(i)}$$

$$\frac{\partial u}{\partial y} = a_1\left(\frac{1}{\delta}\right) + 2a_2\left(\frac{y}{\delta^2}\right) + 3a_3\left(\frac{y^2}{\delta^3}\right) + 4a_4\left(\frac{y^3}{\delta^4}\right)$$

$$\frac{\partial^2 u}{\partial y^2} = 2a_2\left(\frac{1}{\delta^2}\right) + 6a_3\left(\frac{y}{\delta^3}\right) + 12a_4\left(\frac{y^2}{\delta^4}\right)$$

Boundary conditions: At y = 0, u = 0, $\dfrac{\partial^2 u}{\partial y^2} = 0$

   At y = δ,   u = 1, $\dfrac{\partial u}{\partial y} = 0, \quad \dfrac{\partial^2 u}{\partial y^2} = 0,$

From B.C. at y = 0,   u = 0: → $a_o = 0$.

$$\frac{\partial^2 u}{\partial y^2} = 0 \; : \; \rightarrow \quad a_2 = 0.$$

At $y = \delta$,

$$\frac{\partial^2 u}{\partial y^2} = 0 : \qquad 6a_3 + 12a_4 = 0$$

$$\frac{\partial u}{\partial y} = 0, : \qquad a_1 + 3a_3 + 4a_4 = 0$$

$$u = 1: \qquad a_1 + a_3 + a_4 = 1.$$

Hence, $a_1 = 2$, $a_3 = -2$, $a_4 = 1$. The dimensionless velocity profile is

$$u = 2\left(\frac{y}{\delta}\right) - 2\left(\frac{y}{\delta}\right)^3 + \left(\frac{y}{\delta}\right)^4 \tag{ii}$$

From the momentum integral equation,

$$\text{LHS} = \frac{d}{dx}\left[\int_0^\delta u(u_\infty - u)dy\right] = u_\infty^2 \frac{d}{dx}\left[\int_0^\delta \frac{u}{u_\infty}\left(1 - \frac{u}{u_\infty}\right)dy\right]$$

$$= u_\infty^2 \frac{d}{dx}\left[\int_0^\delta \left\{2\left(\frac{y}{\delta}\right) - 2\left(\frac{y}{\delta}\right)^3 + \left(\frac{y}{\delta}\right)^4\right\}\left\{1 - 2\left(\frac{y}{\delta}\right) + 2\left(\frac{y}{\delta}\right)^3 - \left(\frac{y}{\delta}\right)^4\right\}dy\right]$$

$$= u_\infty^2 \frac{d}{dx}\left[\int_0^\delta \{2\left(\frac{y}{\delta}\right) - 2\left(\frac{y}{\delta}\right)^3 + \left(\frac{y}{\delta}\right)^4 - 4\left(\frac{y}{\delta}\right)^2 + 4\left(\frac{y}{\delta}\right)^4 - 2\left(\frac{y}{\delta}\right)^5 + 4\left(\frac{y}{\delta}\right)^4\right.$$

$$\left. - 4\left(\frac{y}{\delta}\right)^6 + 2\left(\frac{y}{\delta}\right)^7 - 2\left(\frac{y}{\delta}\right)^5 + 2\left(\frac{y}{\delta}\right)^7 - \left(\frac{y}{\delta}\right)^8\}dy\right]$$

$$= u_\infty^2 \frac{d}{dx} \{ \frac{2}{\delta} \frac{1}{2} \delta^2 - \frac{4}{\delta^2} \frac{\delta^3}{3} - \frac{2}{\delta^3} \frac{\delta^4}{4} + \frac{9}{\delta^4} \frac{\delta^5}{5} - \frac{4}{\delta^5} \frac{\delta^6}{6} - \frac{4}{\delta^6} \frac{\delta^7}{7} + \frac{4}{\delta^7} \frac{\delta^8}{8} - \frac{1}{\delta^8} \frac{\delta^9}{9} \}$$

$$= \frac{37}{315} u_\infty^2 \frac{d\delta}{dx} \qquad\qquad\qquad\qquad \text{(iii)}$$

RHS of the momentum integral equation is

$$= \upsilon \frac{\partial u}{\partial y}\bigg|_{y=0} = \upsilon \left[\frac{2}{\delta} - \frac{2}{3} \frac{y^2}{\delta^3} + \frac{1}{4} \frac{y^3}{\delta^4} \right]_{y=0} u_\infty = \upsilon u_\infty \frac{2}{\delta}. \qquad \text{(iv)}$$

Putting Eq. (iii) equal to Eq. (iv),

$$\frac{37}{315} u_\infty^2 \frac{d\delta}{dx} = \upsilon u_\infty \frac{2}{\delta}$$

$$\delta d\delta = \frac{630}{37} \frac{\upsilon}{u_\infty} dx.$$

Integrating, $\qquad \frac{1}{2} \delta^2 (x) = \frac{630}{37} \frac{\upsilon}{u_\infty} x.$

Hence, $\qquad \dfrac{\delta(x)}{x} = \dfrac{5.836}{\sqrt{Re_x}} \qquad\qquad\qquad\qquad \text{(v)}$

shear stress at the wall, $\tau = \mu \dfrac{\partial u}{\partial y}\bigg|_{y=0} = \mu. \dfrac{2}{\delta} u_\infty = \dfrac{2\mu\sqrt{Re_x}}{5.836x} u_\infty$ (vi)

the drag coefficient, $\qquad C_x = \dfrac{\tau}{0.5\rho u_\infty^2} = \dfrac{4}{5.836} \dfrac{\sqrt{\upsilon}}{\sqrt{u_\infty x}} = \dfrac{0.685}{\sqrt{Re_x}}.$

$$\text{(vii)}$$

(a)   Assume that the dimensionless temperature profile is a quartic polynomial.

$$\theta = \frac{T - T_w}{T_\infty - T_w} = a_o + a_1\left(\frac{y}{\delta_t}\right) + a_2\left(\frac{y}{\delta_t}\right)^2 + a_3\left(\frac{y}{\delta_t}\right)^3 + a_4\left(\frac{y}{\delta_t}\right)^4$$

$$\frac{\partial\theta}{\partial y} = a_1\left(\frac{1}{\delta_t}\right) + 2a_2\left(\frac{y}{\delta_t^2}\right) + 3a_3\left(\frac{y^2}{\delta_t^3}\right) + 4a_4\left(\frac{y^3}{\delta_t^4}\right)$$

$$\frac{\partial^2\theta}{\partial y^2} = 2a_2\left(\frac{1}{\delta_t}\right) + 6a_3\left(\frac{y^2}{\delta_t^2}\right) + 12a_4\left(\frac{y^3}{\delta_t^3}\right)$$

Boundary conditions: At $y = 0$, $\theta = 0$, $\dfrac{\partial^2\theta}{\partial y^2} = 0$

At $y = \delta_t$, $\qquad \theta = 1$, $\qquad \dfrac{\partial\theta}{\partial y} = 0$, $\qquad \dfrac{\partial^2\theta}{\partial y^2} = 0$,

From B.C. at $y = 0$, $\qquad \theta = 0 : \rightarrow \qquad a_o = 0$,

$$\frac{\partial^2\theta}{\partial y^2} = 0 : \rightarrow \quad a_2 = 0,$$

At $y = \delta_t$,

$\dfrac{\partial^2\theta}{\partial y^2} = 0 : \qquad 6a_3 + 12a_4 = 0$

$\dfrac{\partial\theta}{\partial y} = 0 : \qquad a_1 + 3a_3 + 4a_4 = 0$

$\theta = 1 : \qquad a_1 + a_3 + a_4 = 1$

Hence, $a_1 = 2$, $a_3 = -2$, $a_4 = 1$. The dimensionless temperature profile is

$$\theta = 2\left(\frac{y}{\delta_t}\right) - 2\left(\frac{y}{\delta_t}\right)^3 + \left(\frac{y}{\delta_t}\right)^4. \qquad \text{(viii)}$$

Substituting the velocity and temperature profiles into the energy integral equation,

$$\text{LHS} = \frac{d}{dx}\left[\int_0^{\delta_t} u(1-\theta)dy\right] = u_\infty \frac{d}{dx}\left[\int_0^{\delta_t} \frac{u}{u_\infty}(1-\theta)dy\right]$$

$$= u_\infty \frac{d}{dx}\left[\int_0^{\delta_t}\left(\frac{2}{\delta}y - \frac{2}{\delta^3}y^3 + \frac{1}{\delta^4}y^4\right)\left(1 - \frac{2}{\delta_t}y + \frac{2}{\delta_t^3}y^3 - \frac{4}{\delta_t^4}y^4\right)dy\right]$$

$$= u_\infty \frac{d}{dx}\left[\int_0^{\delta_t}\left(\begin{array}{l}(\frac{2}{\delta}y - \frac{2}{\delta^3}y^3 + \frac{1}{\delta^4}y^4 - \frac{4}{\delta\delta_t}y^2 + \frac{4}{\delta^3\delta_t}y^4 - \frac{2}{\delta^4\delta_t}y^5 + \frac{4}{\delta\delta_t^3}y^4 \\ - \frac{4}{\delta^3\delta_t^3}y^6 + \frac{2}{\delta^4\delta_t^3}y^7 - \frac{2}{\delta\delta_t^4}y^5 + \frac{2}{\delta^3\delta_t^4}y^7 - \frac{1}{\delta^4\delta_t^4}y^8)\end{array}\right)dy\right]$$

$$= u_\infty \frac{d}{dx}\left[\frac{\delta_t^2}{\delta} - \frac{\delta_t^4}{2\delta^3} + \frac{\delta_t^5}{5\delta^4} - \frac{4\delta_t^2}{3\delta} + \frac{4\delta_t^4}{5\delta^3} - \frac{\delta_t^5}{3\delta^4}\right.$$
$$\left. + \frac{4\delta_t^2}{5\delta} - \frac{\delta_t^4}{7\delta^3} + \frac{\delta_t^5}{4\delta^4} - \frac{\delta_t^2}{3\delta} + \frac{\delta_t^4}{4\delta^3} - \frac{\delta_t^5}{9\delta^4}\right]$$

$$= u_\infty \frac{d}{dx}\delta\left[\frac{2}{15}\left(\frac{\delta_t}{\delta}\right)^2 + \frac{41}{140}\left(\frac{\delta_t}{\delta}\right)^4 + \frac{1}{180}\left(\frac{\delta_t}{\delta}\right)^5\right] \qquad \text{(ix)}$$

Let $\xi = \dfrac{\delta_t}{\delta}$ and $\delta_t < \delta$. Hence, the fourth and fifth power terms in Eqn. (ix) are smaller than the second power term. Neglecting these higher power terms, and substituting the LHS into the energy integral equation,

$$u_\infty \frac{d}{dx}\left[\delta\xi^2 \frac{2}{15}\right] = \frac{2\alpha}{\delta_t} \qquad \text{(x)}$$

$$\frac{d}{dx}\left[\alpha\xi^2\right] = \frac{15\alpha}{\delta_t u_\infty}$$

$$\delta\xi\frac{d}{dx}\left[\delta\xi^2\right] = \frac{15\alpha}{u_\infty}$$

$$\xi^3\delta\frac{d\delta}{dx} + 2\delta^2\xi^2\frac{d\xi}{dx} = \frac{15\alpha}{u_\infty}$$

$$\frac{2}{3}\delta^2\frac{d\xi^3}{dx} + \xi^3\delta\frac{d\delta}{dx} = \frac{15\alpha}{u_\infty}. \qquad \text{(xi)}$$

Previously in part (a) above it had been derived that

$$\delta^2(x) = \frac{1260}{37}\frac{\upsilon}{u_\infty}x \qquad \text{or} \qquad \delta\frac{d\delta}{dx} = \frac{630}{37}\frac{\upsilon}{u_\infty}. \qquad \text{(xii)}$$

Substituting Eq. (xii) into Eq. (xi),

$$\frac{2}{3}\frac{1260}{37}\frac{\upsilon x}{u_\infty}\frac{d\xi^3}{dx} + \xi^3\frac{630}{37}\frac{\upsilon x}{u_\infty} = \frac{15\alpha}{u_\infty}$$

$$x\frac{d\xi^3}{dx} + \frac{3}{4}\xi^3 = \frac{37\alpha}{56\upsilon} \qquad \text{(xiii)}$$

The general solution to Eq. (xiii) is

$$\xi^3 = cx^{-\frac{3}{4}} + \frac{37}{42}\frac{a}{\upsilon}. \qquad \text{(xiv)}$$

When x = x$_0$, $\xi = 0$, hence $c = -\frac{37}{42}\frac{a}{\upsilon}x_0^{\frac{3}{4}}$.

So,
$$\xi^3 = \frac{37}{42\,\mathrm{Pr}}\left[1-\left(\frac{x_o}{x}\right)^{\frac{3}{4}}\right] \tag{xv}$$

When $x_0 = 0$, $\xi(x) = \dfrac{\delta_t(x)}{\delta(x)} = 0.959\,\mathrm{Pr}^{-\frac{1}{3}}$ (xvi)

$$\delta_t(x) = \xi(x)\delta(x) = 5.597x\,\mathrm{Re}_x^{-\frac{1}{2}}\,\mathrm{Pr}^{-\frac{1}{3}} \tag{xvii}$$

The heat transfer coefficient $h(x) = k\dfrac{\partial\theta}{\partial y}\bigg|_{y=0} = \dfrac{2}{\delta_t}k$.

The Nusselt number is $Nu(x) = \dfrac{h(x)x}{k} = \dfrac{2k}{\delta_t}\cdot\dfrac{x}{k} = 0.357\,\mathrm{Re}_x^{\frac{1}{2}}\,\mathrm{Pr}^{\frac{1}{3}}$

$$\tag{xviii}$$

Example 7.6

Problem: (a) Derive the boundary layer thickness and local drag coefficient, assuming the velocity profile is a polynomial of the fifth degree. (b)   Derive the convective heat transfer Nusselt number, assuming the temperature profile is a polynomial of the fifth degree.

Solution
(a)   Assume that the velocity profile is a polynomial of the fifth degree.

$$u = \frac{u'}{u_\infty} = a_o + a_1\left(\frac{y}{\delta}\right) + a_2\left(\frac{y}{\delta}\right)^2 + a_3\left(\frac{y}{\delta}\right)^3 + a_4\left(\frac{y}{\delta}\right)^4 + a_5\left(\frac{y}{\delta}\right)^5$$

$$\tag{i}$$

$$\frac{\partial u}{\partial y} = a_1\left(\frac{1}{\delta}\right) + 2a_2\left(\frac{y}{\delta^2}\right) + 3a_3\left(\frac{y^2}{\delta^3}\right) + 4a_4\left(\frac{y^3}{\delta^4}\right) + 5a_5\left(\frac{y^4}{\delta^5}\right)$$

$$\frac{\partial^2 u}{\partial y^2} = 2a_2\left(\frac{1}{\delta^2}\right) + 6a_3\left(\frac{y}{\delta^3}\right) + 12a_4\left(\frac{y^2}{\delta^4}\right) + 20a_5\left(\frac{y^3}{\delta^5}\right)$$

$$\frac{\partial^4 u}{\partial y^4} = 24a_4\left(\frac{1}{\delta^4}\right) + 120a_5\left(\frac{y}{\delta^5}\right)$$

Boundary conditions: At y = 0, u = 0, $\quad \dfrac{\partial^2 u}{\partial y^2} = 0$

At y = δ, $\qquad$ u = 1, $\qquad\qquad \dfrac{\partial u}{\partial y} = 0, \quad \dfrac{\partial^2 u}{\partial y^2} = 0, \quad \dfrac{\partial^4 u}{\partial y^4} = 0.$

From B.C. at y = 0, $\qquad$ u = 0: $\quad \rightarrow \qquad$ a$_o$ = 0.

$$\frac{\partial^2 u}{\partial y^2} = 0 : \rightarrow \qquad a_2 = 0.$$

At y = δ,

$\dfrac{\partial^4 u}{\partial y^4} = 0:$ $\qquad$ 24a$_4$ + 120a$_5$ = 0 $\qquad$ i.e. a$_4$ = -5a$_5$

$\dfrac{\partial^2 u}{\partial y^2} = 0:$ $\qquad$ 6a$_3$ + 12a$_4$ +20a$_5$ = 0 $\quad$ i.e. $\quad$ a$_3$ = 20/3 a$_5$

$\dfrac{\partial u}{\partial y} = 0,:$ $\qquad$ a$_1$ +3a$_3$ + 4a$_4$ +5a$_5$ = 0 $\quad$ i.e. $\quad$ a$_1$ = -5a$_5$

$\quad$ u = 1: $\qquad$ a$_1$ + a$_3$ + a$_4$ + a$_5$ = 1 $\qquad$ i.e. a$_5$ = -3/7

Hence, a$_1$ = 15/7, a$_3$ = -20/7, a$_4$ = 15/7, a$_5$ = -3/7. The dimensionless velocity profile is

$$u = \frac{u'}{u_\infty} = \frac{15}{7}\left(\frac{y}{\delta}\right) - \frac{20}{7}\left(\frac{y}{\delta}\right)^3 + \frac{15}{7}\left(\frac{y}{\delta}\right)^4 - \frac{3}{7}\left(\frac{y}{\delta}\right)^5. \quad \text{(ii)}$$

From the momentum integral equation,

$$\text{LHS} = \frac{d}{dx}\left[\int_0^\delta u(u_\infty - u)dy\right] = u_\infty^2 \frac{d}{dx}\left[\int_0^\delta \frac{u}{u_\infty}\left(1 - \frac{u}{u_\infty}\right)dy\right]$$

$$= u_\infty^2 \frac{d}{dx}\left[\int_0^\delta \left\{\frac{15}{7}\left(\frac{y}{\delta}\right) - \frac{20}{7}\left(\frac{y}{\delta}\right)^3 + \frac{15}{7}\left(\frac{y}{\delta}\right)^4 - \frac{3}{7}\left(\frac{y}{\delta}\right)^5\right\}\left\{1 - \frac{15}{7}\left(\frac{y}{\delta}\right) + \frac{20}{7}\left(\frac{y}{\delta}\right)^3 - \frac{15}{7}\left(\frac{y}{\delta}\right)^4 + \frac{3}{7}\left(\frac{y}{\delta}\right)^5\right\}dy\right]$$

$$= u_\infty^2 \frac{d}{dx}\left[\int_0^\delta \left\{\frac{15}{7}\left(\frac{y}{\delta}\right) - \frac{20}{7}\left(\frac{y}{\delta}\right)^3 + \frac{15}{7}\left(\frac{y}{\delta}\right)^4 - \frac{3}{7}\left(\frac{y}{\delta}\right)^5 - \frac{225}{49}\left(\frac{y}{\delta}\right)^2 + \frac{300}{49}\left(\frac{y}{\delta}\right)^4\right.\right.$$

$$- \frac{225}{49}\left(\frac{y}{\delta}\right)^5 + \frac{45}{49}\left(\frac{y}{\delta}\right)^6 + \frac{300}{49}\left(\frac{y}{\delta}\right)^4$$

$$- \frac{400}{49}\left(\frac{y}{\delta}\right)^6 + \frac{300}{49}\left(\frac{y}{\delta}\right)^7 - \frac{60}{40}\left(\frac{y}{\delta}\right)^5 - \frac{225}{49}\left(\frac{y}{\delta}\right)^5 + \frac{300}{49}\left(\frac{y}{\delta}\right)^7 - \frac{225}{49}\left(\frac{y}{\delta}\right)^8$$

$$\left.\left.+ \frac{45}{49}\left(\frac{y}{\delta}\right)^9 + \frac{45}{49}\left(\frac{y}{\delta}\right)^6 - \frac{60}{49}\left(\frac{y}{\delta}\right)^8 + \frac{45}{49}\left(\frac{y}{\delta}\right)^9 - \frac{9}{49}\left(\frac{y}{\delta}\right)^{10}\right\}dy\right]$$

$$= 0.113526\,u_\infty^2\,\frac{d\delta}{dx} \quad \text{(iii)}$$

RHS of the momentum integral equation is

$$= \upsilon\frac{\partial u}{\partial y}\bigg|_{y=0} = \upsilon a_1 = \upsilon u_\infty \frac{2.14286}{\delta} \quad \text{(iv)}$$

Putting Eq. (iii) equal to Eq. (iv),

$$0.113526u_\infty^2 \frac{d\delta}{dx} = \upsilon u_\infty \frac{2.14286}{\delta}$$

$$\delta d\delta = 18.875475 \frac{\upsilon}{u_\infty} dx.$$

Integrating, $\quad \dfrac{1}{2}\delta^2(x) = 18.875475 \dfrac{\upsilon}{u_\infty} x.$

Hence, $\quad \dfrac{\delta(x)}{x} = \dfrac{6.144}{\sqrt{Re_x}}.$ $\qquad\qquad$ (v)

Shear stress at the wall, $\tau = \mu \left.\dfrac{\partial u}{\partial y}\right|_{y=0} = \mu \dfrac{2.14286}{\delta} u_\infty = \dfrac{\mu\sqrt{Re_x}}{2.8672} u_\infty.$

$$\text{(vi)}$$

The drag coefficient, $\quad C_x = \dfrac{\tau}{0.5\rho u_\infty^2} = \dfrac{2}{2.8672}\dfrac{\sqrt{\upsilon}}{\sqrt{u_\infty x}} = \dfrac{0.6975}{\sqrt{Re_x}}.$ $\quad$ (vii)

(b) Assume that the dimensionless temperature profile is a polynomial of the fifth degree:

$$\theta = \frac{T - T_w}{T_\infty - T_w} = a_o + a_1\left(\frac{y}{\delta_t}\right) + a_2\left(\frac{y}{\delta_t}\right)^2 + a_3\left(\frac{y}{\delta_t}\right)^3 + a_4\left(\frac{y}{\delta_t}\right)^4 + a_5\left(\frac{y}{\delta_t}\right)^5$$

$$\text{(viii)}$$

$$\frac{\partial\theta}{\partial y} = a_1\left(\frac{1}{\delta_t}\right) + 2a_2\left(\frac{y}{\delta_t^2}\right) + 3a_3\left(\frac{y^2}{\delta_t^3}\right) + 4a_4\left(\frac{y^3}{\delta_t^4}\right) + 5a_5\left(\frac{y^4}{\delta_t^5}\right)$$

$$\frac{\partial^2\theta}{\partial y^2} = 2a_2\left(\frac{1}{\delta_t}\right) + 6a_3\left(\frac{y}{\delta_t^3}\right) + 12a_4\left(\frac{y^2}{\delta_t^4}\right) + 20a_5\left(\frac{y^3}{\delta_t^5}\right)$$

$$\frac{\partial^3 \theta}{\partial y^3} = 6a_3 \left(\frac{1}{\delta_t^3} \right) + 24a_4 \left(\frac{y}{\delta_t^4} \right) + 60a_5 \left(\frac{y^2}{\delta_t^5} \right)$$

$$\frac{\partial^4 \theta}{\partial y^4} = 24a_4 \left(\frac{1}{\delta_t^4} \right) + 120a_5 \left(\frac{y}{\delta_t^5} \right)$$

Boundary conditions: At $y = 0$, $\theta = \dfrac{\theta'}{\theta_w'} = 1$, $\dfrac{\partial^2 \theta}{\partial y^2} = 0$ (ix)

At $y = \delta_t$, $\theta = 0$, $\dfrac{\partial \theta}{\partial y} = 0$, $\dfrac{\partial^2 \theta}{\partial y^2} = 0$, $\dfrac{\partial^4 \theta}{\partial y^4} = 0$

(x)

From B.C. at $y = 0$, $a_2 = 0$, $a_o = 1$

At $y = \delta_t$, $\dfrac{\partial^4 \theta}{\partial y^4} = 0$: $a_4 = -5a_5$

$\dfrac{\partial^2 \theta}{\partial y^2} = 0$: $6a_3 - 40a_5 = 0$ or $a_3 = 20/3 \ a_5$

$\dfrac{\partial \theta}{\partial y} = 0$: $a_1 + 20a_5 - 20 \ a_5 + 5a_5 = 0$ or $a_1 = -5a_5$

$\theta = 1$: $7/3 \ a_5 = -1$ or $a_5 = -3/7$.

Hence, $a_1 = -15/7$, $a_3 = 20/7$, $a_4 = -15/7$, $a_5 = 3/7$. The dimensionless temperature profile is

$$\theta = \frac{15}{7} \left(\frac{y}{\delta_t} \right) - \frac{20}{7} \left(\frac{y}{\delta_t} \right)^2 + \frac{15}{7} \left(\frac{y}{\delta_t} \right)^3 - \frac{3}{7} \left(\frac{y}{\delta_t} \right)^5 \qquad \text{(xi)}$$

Substituting the velocity and temperature profiles into the energy integral equation,

$$\text{LHS} = \frac{d}{dx}\left[\int_{\delta_i}^{\delta} u(1-\theta)dy\right] = u_\infty \frac{d}{dx}\left[\int_{\delta_i}^{\delta} \frac{u}{u_\infty}(1-\theta)dy\right]$$

$$=$$

$$u_\infty \frac{d}{dx}\left[\int_{\delta_i}^{\delta}\left(\frac{15}{7\delta}y - \frac{20}{7\delta^3}y^3 + \frac{15}{7\delta^4}y^4 - \frac{3}{7\delta^5}y^5\right)\left(1 - \frac{15}{7\delta_i}y + \frac{20}{7\delta_i^3}y^3 - \frac{15}{7\delta_i^4}y^4 + \frac{3}{7\delta_i^5}y^5\right)dy\right]$$

$$= u_\infty \frac{d}{dx}[\int_{\delta_i}^{\delta}(\frac{15}{7\delta}y - \frac{20}{7\delta^3}y^3 + \frac{15}{7\delta^4}y^4 - \frac{3}{7\delta^5}y^5$$

$$-\frac{225}{49\delta\delta_i}y^2 + \frac{300}{49\delta^3\delta_i}y^4 - \frac{225}{49\delta^4\delta_i}y^5 + \frac{45}{49\delta^3\delta_i}y^6$$

$$+\frac{300}{49\delta\delta_i^3}y^4 - \frac{400}{49\delta^3\delta_i^3}y^6 + \frac{300}{49\delta^4\delta_i^3}y^7 - \frac{60}{49\delta^5\delta_i^3}y^8$$

$$-\frac{225}{49\delta\delta_i^4}y^5 + \frac{300}{49\delta^3\delta_i^4}y^7 - \frac{225}{49\delta^4\delta_i^4}y^8 + \frac{45}{49\delta^5\delta_i^4}y^9$$

$$+\frac{45}{49\delta\delta_i^5}y^6 - \frac{60}{49\delta^3\delta_i^5}y^8 + \frac{45}{49\delta^4\delta_i^5}y^9 - \frac{9}{49\delta^5\delta_i^5}y^{10})dy]$$

$$= u_\infty \frac{d}{dx}[\frac{15\delta_i^2}{14\delta} - \frac{20\delta_i^4}{25\delta^3} + \frac{15\delta_i^5}{35\delta^4} - \frac{3\delta_i^6}{42\delta^5} - \frac{75\delta_i^2}{49\delta} + \frac{60\delta_i^4}{49\delta^3}$$

$$-\frac{75\delta_i^5}{98\delta^4} + \frac{45\delta_i^6}{343\delta^5} + \frac{60\delta_i^2}{49\delta} - \frac{400\delta_i^4}{343\delta^3} + \frac{75\delta_i^5}{98\delta^4} - \frac{20\delta_i^6}{147\delta^5}$$

$$-\frac{75\delta_i^2}{98\delta} + \frac{75\delta_i^4}{98\delta^3} - \frac{25\delta_i^5}{49\delta^4} + \frac{9\delta_i^6}{98\delta^5} + \frac{45\delta_i^2}{343\delta} - \frac{20\delta_i^4}{147\delta^3} + \frac{9\delta_i^5}{98\delta^4} - \frac{9\delta_i^6}{539\delta^5}]$$

Let $\xi = \frac{\delta_i}{\delta}$ and $\delta_i < \delta$. Neglecting terms in $\xi^4 \, \xi^5$ and higher,

$$\text{LHS} = u_\infty \frac{d}{dx}\{\delta\left[\frac{15}{14} - \frac{75}{49} + \frac{60}{49} - \frac{75}{98} + \frac{45}{343}\right]\left(\frac{\delta_i}{\delta}\right)^2\} = \frac{45}{343}u_\infty \frac{d}{dx}\left(\delta\xi^2\right)$$

The right-hand side of the energy integral equation is

$$\text{RHS} = \alpha \frac{\partial T}{\partial y}\bigg|_{y=0} = \frac{15}{7}\frac{\alpha}{\delta_t}$$

Equating the LHS to the RHS of the energy integral equation,

$$u_\infty \left(2\delta^2 \xi^2 \frac{d\xi}{dx} + \xi^3 \delta \frac{d\delta}{dx} \right) = \frac{49}{3}\alpha. \tag{xii}$$

It has been previously shown that

$$\delta d\delta = 18.875475\frac{\upsilon}{u_\infty}dx \qquad \text{and} \qquad \delta^2 = 37.750949\frac{\upsilon x}{u_\infty}.$$

Hence, $u_\infty \left[2\left(37.750949\frac{\upsilon x}{u_\infty}\xi^2 \frac{d\xi}{dx} + \xi^3 (18.875475)\frac{\upsilon}{u_\infty} \right) \right] = 16.33\alpha$

$$\xi^3 + 4x\xi^2 \frac{d\xi}{dx} = 0.8653\frac{\alpha}{\upsilon}$$

$$\xi^3 + 4x\frac{d\xi^3}{dx} = 0.8653\frac{\alpha}{\upsilon}. \tag{xiii}$$

The general solution to Eq. (xiii) is

$$\xi^3 = cx^{-\frac{3}{4}} + 0.8653\frac{a}{\upsilon} \tag{xiv}$$

When $x = x_0$, $\xi = 0$, hence $c = -0.8653\frac{a}{\upsilon}x_0^{\frac{3}{4}}$.

So, $\xi^3 = 0.8653\,\text{Pr}^{-1}\left[1 - \left(\frac{x_0}{x} \right)^{\frac{3}{4}} \right]$ \hfill (xv)

$$\xi(x) = \frac{\delta_t(x)}{\delta(x)} = 0.9529 \Pr^{-\frac{1}{3}}\left[1 - \left(\frac{x_o}{x}\right)^{\frac{3}{4}}\right]^{\frac{1}{3}} \qquad \text{(xvi)}$$

$$\delta_t(x) = \xi(x)\delta(x) = 5.8546x \operatorname{Re}_x^{-\frac{1}{2}} \Pr^{-\frac{1}{3}}\left[1 - \left(\frac{x_o}{x}\right)^{\frac{3}{4}}\right]^{\frac{1}{3}} \qquad \text{(xvii)}$$

The heat transfer coefficient $\quad h(x) = k\dfrac{\partial \theta}{\partial y}\bigg|_{y=0} = \dfrac{15}{7}\dfrac{k}{\delta_t}.$

The Nusselt number is

$$Nu(x) = \frac{h(x)x}{k} = 0.366 \Pr^{\frac{1}{3}} \operatorname{Re}_x^{\frac{1}{2}}\left[1 - \left(\frac{x_o}{x}\right)^{\frac{3}{4}}\right]^{-\frac{1}{3}} \qquad \text{(xviii)}$$

Example 7.7

Problem: Derive the boundary layer thickness, the local drag coefficient and the average drag coefficient over a distance L, by assuming a sinusoidal profile for the dimensionless velocity,

$$\frac{u}{u_\infty} = \sin\left(\frac{\pi}{2}\cdot\frac{y}{\delta}\right).$$

Solution

The dimensionless velocity is

$$\frac{u}{u_\infty} = \sin\left(\frac{\pi}{2}\cdot\frac{y}{\delta}\right). \qquad \text{(i)}$$

From the momentum integral equation, the left-hand side is

$$\text{LHS} = u_\infty^2 \frac{d}{dx}\left[\int_0^\delta \left(-\sin^2\left\{\frac{\pi}{2}\frac{y}{\delta}\right\} + \sin\left\{\frac{\pi}{2}\frac{y}{\delta}\right\}\right)dy\right]$$

$$= u_\infty^2 \frac{d}{dx} \left[-\frac{y}{2} + \frac{\sin\left(\frac{\pi y}{\delta}\right)}{\left(\frac{2\pi}{\delta}\right)} - \frac{\cos\left(\frac{\pi y}{2\delta}\right)}{\left(\frac{\pi}{\delta}\right)} \right]_0^\delta$$

$$= u_\infty^2 \frac{d}{dx} \left[\frac{\delta(-\pi + 4)}{2\pi} \right].$$

From the momentum integral equation, the right-hand side is

$$\text{RHS} = \upsilon \frac{\partial u}{\partial y}\bigg|_{y=0} = \upsilon u_\infty \left[\frac{\pi}{2\delta} \cos\left(\frac{\pi y}{2\delta}\right) \right]_{y=0} = \frac{\upsilon u_\infty \pi}{2\delta}.$$

Putting both sides of the momentum integral equation together,

$$\delta d\delta = \frac{\upsilon \pi^2}{u_\infty (4 - \pi)} dx.$$

Hence,

$$\delta^2 = \frac{2\upsilon \pi^2 x}{u_\infty (4 - \pi)} + c_i$$

But when $x = 0$, $\delta = 0$, so $c_1 = 0$.

Therefore,

$$\delta = \sqrt{\frac{2\upsilon \pi^2 x}{u_\infty (4 - \pi)}} = \frac{4.795}{(\text{Re}_x)^{\frac{1}{2}}} \quad \text{where} \quad \text{Re}_x = \frac{u_\infty x}{\upsilon}.$$

(ii)

The dimensionless velocity profile is

$$\frac{u}{u_\infty} = \sin\left\{ 0.328(\text{Re}_x)^{\frac{1}{2}} y \right\}.$$

(iii)

The local drag coefficient is given by

$$C_x = \frac{2\upsilon}{u_\infty^2}\frac{\partial u}{\partial y}\bigg|_{y=0} = \frac{2\upsilon}{u_\infty^2}(0.328)\sqrt{\frac{u_\infty x}{\nu}} = 0.656\sqrt{\frac{\upsilon x}{u_\infty}} = \frac{0.656}{(\mathrm{Re}_x)^{\frac{1}{2}}}.$$

(iv)

The average drag coefficient over a distance L is given by

$$C_L = \frac{1}{L}\int_0^L C_x dx = 2C_x\big|_{x=L} = \frac{1.312}{(\mathrm{Re}_L)^{\frac{1}{2}}}.$$

(v)

7.4 Similarity Solution

7.4.1 Laminar Flow Along a Flat Plate

Consider two-dimensional flow, $\dfrac{\partial \rho}{\partial t} = 0, \dfrac{\partial P}{\partial y} = 0;$

Assume that the properties are constant, and $\dfrac{\partial^2 u}{\partial x^2} \ll \dfrac{\partial^2 u}{\partial y^2}.$

The continuity equation for two-dimensional, incompressible flow is

$$\frac{\partial u}{\partial x} + \frac{\partial v}{\partial y} = 0.$$

(7.25)

The significant boundary layer momentum equation is

$$u\frac{\partial u}{\partial x} + v\frac{\partial u}{\partial y} = \upsilon\frac{\partial^2 u}{\partial y^2} - \frac{1}{\rho}\frac{dP}{dx}$$

(7.26)

$$= 0 \text{ for flat plate.}$$

The boundary conditions for the momentum equation are as follows:

At $y = 0$, $u = v = 0$

At $y \to \infty$, $u \to u_\infty$

For moderate velocities (assuming no viscous dissipation), the energy equation is

$$u\frac{\partial T}{\partial x} + v\frac{\partial T}{\partial y} = \alpha\frac{\partial^2 T}{\partial y^2}. \tag{7.27}$$

In other words, Eq. (7.27) is valid when the Eckert number is small.

The boundary conditions for the energy equation are as follows:
At $y = 0$, $T = T_w$
At $y \to \delta_t$, $T \to T_\infty$.

Recall that for incompressible flow, the streamfunction $\psi(x,y)$ is defined by

$$u = \frac{\partial\psi}{\partial y} \quad \text{and} \quad v = -\frac{\partial\psi}{\partial x}, \quad \text{which automatically satisfies the}$$

continuity equation.

With the substitution of the streamfunction, the momentum equation becomes

$$\psi_y\psi_{xy} - \psi_x\psi_{yy} = \upsilon\psi_{yyy}. \tag{7.28}$$

The boundary conditions become as follows:

At $y = 0$, $\psi_x = \psi_y = 0$

As $y \to \infty$, $\psi_y = u_\infty$.

We now examine the solution procedure for this third-order partial differential equation in one dependent variable, ψ. We started with two partial differential equations, coupled together, where there were two dependent variables, u and v. This was achieved by a transformation of u and v to ψ; this procedure is called group transformation of the dependent variables. Now, we will group the independent variables x and y; this is called a similarity variable technique. The similarity variables can be obtained from group theory (not in the scope of the present book). So we shall examine the predetermined similarity variables, that is,

$$\eta = y\sqrt{\frac{u_\infty}{vx}} \qquad \text{and} \qquad f(\eta) = \frac{\psi}{\sqrt{xvu_\infty}} \qquad (7.29)$$

We will now proceed to write the streamfunction equation in terms of f and η.

$$\frac{\partial \psi}{\partial y} = \psi_y \qquad = \frac{\partial}{\partial y}(\sqrt{xvu_\infty}\, f)$$

$$= \frac{\partial}{\partial \eta}(\sqrt{xvu_\infty}\, f)\frac{\partial \eta}{\partial y}$$

$$= \sqrt{xvu_\infty}\, f'\sqrt{\frac{u_\infty}{vx}} = u_\infty f' \qquad (7.30)$$

$$\frac{\partial^2 \psi}{\partial y^2} = \psi_{yy} \qquad = \frac{\partial}{\partial y}(u_\infty f)$$

$$= \frac{\partial}{\partial \eta}(u_\infty f')\frac{\partial \eta}{\partial y} = \frac{u_\infty f''\eta}{y} \qquad (7.31)$$

$$\frac{\partial \psi}{\partial x} = \psi_x \qquad = \frac{\partial}{\partial x}(\sqrt{xvu_\infty}\, f)$$

$$= \frac{\partial}{\partial x}(\sqrt{xvu_\infty})f + \frac{\partial f}{\partial \eta}\frac{\partial \eta}{\partial x}\sqrt{xvu_\infty}$$

$$= \frac{1}{2}(\sqrt{\frac{vu_\infty}{x}})f - f'\frac{y}{2x}\sqrt{\frac{u_\infty}{vx}}\sqrt{xvu_\infty}$$

$$= \frac{1}{2}(\sqrt{\frac{vu_\infty}{x}})f - f'\frac{y}{2x}u_\infty \qquad (7.32)$$

$$\frac{\partial^2 \psi}{\partial x \partial y} = \psi_{xy} \quad = \frac{\partial}{\partial x}(u_\infty f')$$

$$= u_\infty \frac{\partial f'}{\partial \eta}\frac{\partial \eta}{\partial x} = -\frac{u_\infty \eta}{2x}f'' \tag{7.33}$$

$$\frac{\partial^3 \psi}{\partial y^3} = \psi_{yyy} \quad = \frac{\partial}{\partial y}(\frac{u_\infty \eta}{y}f'')$$

$$= u_\infty[\frac{\partial}{\partial y}(y^{-1})\eta f'' + y^{-1}\frac{\partial}{\partial \eta}(\eta f'')\frac{\partial \eta}{\partial y}]$$

$$= u_\infty[-\frac{\eta}{y^2}f'' + y^{-1}(f'' + \eta f''')\frac{\eta}{y}]$$

$$= \frac{u_\infty \eta}{y^2}[\eta f'''] \tag{7.34}$$

Substituting equations (7.30)-(7.34) in equation (7.28),

$$u_\infty f'\left(-\frac{u_\infty \eta}{2x}f''\right) - [\frac{1}{2}\sqrt{\frac{\upsilon u_\infty}{x}}f - \frac{f'yu_\infty}{2x}]\frac{u_\infty f''\eta}{y} = \upsilon\frac{u_\infty \eta^2}{y^2}f'''. \tag{7.35}$$

Thus, $2f''' + ff'' = 0.$ \tag{7.36}

Since $\eta = 0$ when $y = 0$,

$$u = \psi_y = uf' = 0, \qquad \text{that is,} \quad f' = 0.$$

Since at $\eta = 0$, $v = -\psi_x = \frac{1}{2}(\sqrt{\frac{\upsilon u_\infty}{x}})f - f'\frac{y}{2x}u_\infty = 0.$

then $f = 0$

As $y \to \infty$, $\eta \to \infty$, $\quad u = u_\infty f' = u_\infty$, hence $f' = 1.$

The solution by Howarth, 1938 [1], is presented in Table 7.1 and represented in Fig. 7.3.

Figure 7.3 Howarth solution for laminar flow along a flat plate.

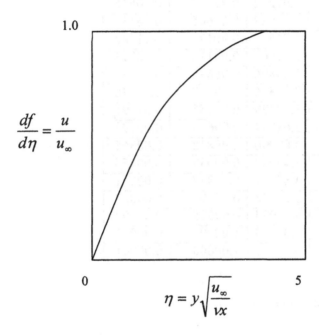

$$\frac{df}{d\eta} = \frac{u}{u_\infty}$$

$$\eta = y\sqrt{\frac{u_\infty}{vx}}$$

Table 7.1   The functions $f(\eta)$, $df/d\eta$ and $d^2f/d\eta^2$ for laminar flow along a flat plate.

$\eta = y\sqrt{\dfrac{u_\infty}{vx}}$	f	$\dfrac{df}{d\eta} = \dfrac{u}{u_\infty}$	$\dfrac{d^2 f}{d\eta^2}$
0	0	0	0.33206
0.2	0.00664	0.06641	0.33199
0.4	0.02656	0.13277	0.33147
0.6	0.05974	0.19894	0.33008
0.8	0.10611	0.26471	0.32739
1.0	0.16557	0.32979	0.32301
1.2	0.23795	0.39378	0.31659
1.4	0.32298	0.45627	0.130787

1.6	0.42032	0.51676	0.29667
1.8	0.52952	0.57477	0.28293
2.0	0.65003	0.62977	0.26675
2.2	0.78120	0.68132	0.24835
2.6	1.07252	0.77246	0.20646
3.0	1.39682	0.84605	0.16136
3.4	1.74696	0.90177	0.11788
3.8	2.11605	0.94112	0.08013
4.2	2.49806	0.96696	0.05052
4.6	2.88826	0.98269	0.02948
5.0	3.28329	0.99155	0.01591
5.4	3.68094	0.99616	0.00793
5.8	4.07990	0.99838	0.00365
6.2	4.47948	0.99937	0.00155
6.6	4.87931	0.99977	0.00061
7.0	5.27926	0.99992	0.00022
7.4	5.67924	0.99998	0.00007
7.8	6.07923	1.00000	0.00002
8.2	6.47923	1.00000	0.00001

7.4.2 Energy Equation

Let the dimensionless temperature $\theta = \dfrac{T - T_\infty}{T_w - T_\infty}$. Equation   (7.27) becomes

$$u \frac{\partial \theta}{\partial x} + v \frac{\partial \theta}{\partial y} = \alpha \frac{\partial^2 \theta}{\partial y^2}. \tag{7.37}$$

Note that

$$\frac{\partial \theta}{\partial x} = \frac{\partial \theta}{\partial \eta} \frac{\partial \eta}{\partial x} = -\frac{\partial \theta}{\partial \eta} \frac{\eta}{2x}$$

$$\frac{\partial \theta}{\partial y} = \frac{\partial \theta}{\partial \eta} \frac{\partial \eta}{\partial y} = \frac{\partial \theta}{\partial \eta} \frac{\eta}{y}$$

$$\frac{\partial^2 \theta}{\partial y^2} = \frac{\partial^2 \theta}{\partial \eta^2} \frac{\eta^2}{y^2}.$$

Substituting in Eq. (7.37),

$$-u_\infty f' \frac{\eta}{2x} \frac{\partial \theta}{\partial \eta} + \{\frac{yu_\infty}{2x} f' - \frac{1}{2}\sqrt{\frac{\upsilon u_\infty}{x}} f\} = \frac{\alpha \eta^2}{y^2} \frac{\partial^2 \theta}{\partial \eta^2}.$$

Simplifying, $\quad \theta'' + \frac{1}{2}\frac{\upsilon}{\alpha} f\theta' = 0$

$$2\theta'' + \mathrm{Pr}\ f\theta' = 0 \tag{7.38}$$

with boundary conditions $\qquad \theta = 1 \quad$ at $\quad \eta = 0$
$$\theta = 0 \quad \text{at} \quad \eta \to \infty.$$

The solution was first given by E. Pohlhausen:

$$\theta(\eta, \mathrm{Pr}) = \frac{\displaystyle\int_{\varsigma=\eta}^{\infty} [f''(\varsigma)]^{\mathrm{Pr}}\, d\varsigma}{\displaystyle\int_{\varsigma=0}^{\infty} [f''(\varsigma)]^{\mathrm{Pr}}\, d\varsigma} \tag{7.39}$$

When $\mathrm{Pr} = 1$, $\quad \theta(\eta) = 1 - f'(\eta) = 1 - \dfrac{u}{u_\infty}.$ The temperature distribution is identical to the velocity distribution.
The temperature gradient at the wall, $[f'(0) = 0.332]$,

$$-\left(\frac{d\theta}{d\eta}\right)_o = a_1(\mathrm{Pr}) = \frac{(0.332)^{\mathrm{Pr}}}{\displaystyle\int_0^\infty [f''(\varsigma)]^{\mathrm{Pr}}\, d\varsigma}.$$

The constant a_1 depends solely on the Prandtl Number, $a_1(\mathrm{Pr})$. This dimensionless coefficient of heat transfer, a1, and the dimensionless adiabatic wall temperature for a flat plate at zero incidence is presented in Table 7.2, and represented in Fig. 7.4.

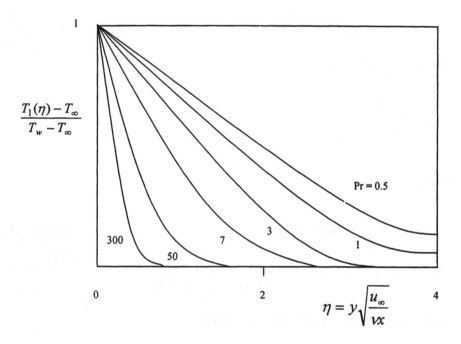

$$\frac{T_1(\eta) - T_\infty}{T_w - T_\infty}$$

$$\eta = y\sqrt{\frac{u_\infty}{vx}}$$

Figure 7.4.   Temperature distribution on a heated flat plate at zero incidence with small velocity plotted for different Prandtl Numbers, Pr.

Table 7.2   Dimensionless coefficient of heat transfer, a_1, and dimensionless adiabatic wall temperature, b, for a flat plate at zero incidence, Schlichting[2].

Pr	0.6	0.7	0.8	0.9	1.0	1.1	7.0	10.0	15.0
a_1	0.276	0.293	0.307	0.320	0.332	0.344	0.645	0.730	0.835
b	0.770	0.835	0.895	0.950	1.000	1.050	2.515	2.965	3.525

PROBLEMS

7.1.   Find the expression for the momentum boundary layer thickness as a function of x, by using the linear velocity profile

$$\frac{u}{U_\infty} = \frac{y}{\delta}.$$

7.2.   Using the linear velocity distribution in Prob. 7.1, obtain an expression for the local-drag coefficient c_x, and the average-drag coefficient c_L over the length $0 \leq x \leq L$.

7.3.   Use the momentum integral method and the energy integral method to arrive at expressions for the momentum boundary layer thickness and the thermal boundary layer thickness, for flow over a flat plate. Assume second-degree polynomials for the velocity profile and the temperature profile.

7.4.   Use the following velocity and temperature profiles for flow over a flat plate:

$$\frac{u}{U_\infty} = \frac{y}{\delta}$$

$$\theta = \frac{T - T_w}{T_\infty - T_w} = \frac{3y}{2\delta_t} - \frac{1}{2}\left(\frac{y}{\delta_t}\right)^3.$$

Obtain the Nu_x vs Re_x and Pr relationship.

7.5.   A flat plate is maintained at a constant temperature of T_w. Liquid metal flows with a velocity of U_∞ and a temperature T_∞ along it. Derive the expressions for the thermal boundary layer thickness $\delta_t(x)$, and the local Nusselt number $Nu_x = hx/k$. Use a linear temperature profile, $T(x,y)$ such that

$$\frac{T(x,y) - T_w}{T_\infty - T_w} = \frac{y}{\delta_t(x)}.$$

7.6.   With the velocity distribution

$$\frac{u}{U_\infty} = \frac{3}{2}\frac{y}{\delta} - \frac{1}{2}\left(\frac{y}{\delta}\right)^3,$$

and the boundary layer thickness expressed as

$$\frac{\delta}{x} = \frac{4.64}{Re_x^{0.5}},$$

find the expression for the y component of velocity, v, as a function of x and y. Hence, deduce the expression for v when y = δ.

7.7. Fluid with Pr ~ 1, flows with a velocity U_∞, and temperature T_∞, along a flat plate maintained at a constant temperature T_w. Use a linear velocity profile and a second-degree polynomial for the temperature distribution. Find the expressions for the thermal boundary layer thickness and the local Nusselt number.

7.8. Retaining the viscous-energy dissipation term in the boundary layer equation, derive the energy integral equation.

7.9. Derive the boundary layer thickness, the local drag coefficient and the average drag coefficient over a distance L by assuming a cosine profile for the dimensionless velocity,

$$\frac{u}{u_\infty} = \cos\left(\frac{\pi}{2} - \frac{\pi y}{2\delta}\right).$$

REFERENCES

1. L Howarth. On the Solution of the Laminar Boundary Layer Equations. Proc R Soc (London), A164:546, 1938.

2. H Schlichting. Boundary Layer Theory. 7th ed. New York: McGraw-Hill, 1979.

Integral Method of Analysis

Choose a polynomial profile for the velocity component, u
Use boundary conditions to express the velocity component, u
Substitute velocity u into momentum integral equation
O.D.E. for boundary layer thickness, solve for delta from equation.

Choose a polynomial profile for the temperature distribution, T
Use boundary conditions to express the temperature distribution, T
Substitute u and T into the energy integral equation
O.D.E. for thermal boundary layer thickness, solve delta T from equation.

K.V. Wong

8

Internal Forced Convection

There are many types of internal forced convection. This chapter examines selected examples. Flows with heat transfer between parallel plates and flows in pipes, tubes and ducts are considered.

8.1 Couette Flow

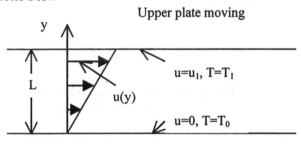

Figure 8.1 Couette Flow

Couette flow is the model for flow between parallel plates. The plates are separated by a distance L, and filled with a fluid with density ρ, viscosity μ and thermal conductivity k, Fig. 8.1. The upper plate moves at a constant velocity u_1 and causes the fluid particles to move in the direction parallel to the plates. The upper and lower plates are kept at uniform temperatures T_1 and T_0 respectively.

A journal and its bearing is one engineering problem that is modeled by the Couette flow. One of the surfaces is stationary while the other is rotating, and the gap between them is filled with a lubricant oil of high viscosity. Since the gap is small compared to the radius of the bearing, the geometry may be treated as two parallel plates. Since the oil is very viscous, the heat generated by viscous energy dissipation may be significant even at moderate flow velocities. The temperature rise in the fluid and the heat transferred through the walls are of interest. In addition, there are many membranes in human and animal bodies. The fluid flow in between these membranes may sometimes be modeled

191

using Couette flow. In case of an inflammation, the velocity profile would change since the gap will change, and hence the stresses will increase. The velocity distribution in the flow is first solved, and then the temperature distribution is derived.

Velocity Distribution

For incompressible flow of a fluid with constant properties, the particles are all moving in the direction parallel to the plates so that the velocity component v normal to the plates must be zero. By putting v = 0 in the continuity equation,

$$\frac{du}{dx} = 0. \tag{8.1}$$

Hence   u = u(y).   The y-momentum equation yields no useful information since v = 0. The x-momentum equation is

$$-\rho\left(u\frac{\partial u}{\partial x} + v\frac{\partial u}{\partial y}\right) = F_x - \frac{\partial P}{\partial x} + \mu\left(\frac{\partial^2 u}{\partial x^2} + \frac{\partial^2 u}{\partial y^2}\right). \tag{8.2}$$

Setting v = 0, and F_x=0 for no body forces, we obtain

$$-\frac{dP}{dx} + \mu\frac{d^2u}{dy^2} = 0. \tag{8.3}$$

Simple shear flow is the characteristic of Couette flow, and no pressure gradient exists in the direction of motion. Since the pressure gradient term is also zero, the governing equation reduces to

$$\frac{d^2u}{dy^2} = 0 \qquad\qquad \text{in } 0 \le y \le L. \tag{8.4}$$

The boundary conditions are the no slip boundary conditions at y = 0, and y = L, that is,

u = 0 at y = 0
u = u_1 at y = L   (8.5)

The solution of Eq. (8.4) with boundary conditions (8.5) gives the velocity distribution as

$$u(y) = \frac{y}{L} u_1 \tag{8.6}$$

Temperature Distribution

We expect the temperature to vary only in the y direction, so T = T(y). The energy conservation equation is in the form

$$\rho C_P \left(u \frac{\partial T}{\partial x} + v \frac{\partial T}{\partial y} \right) = k \left(\frac{\partial^2 T}{\partial x^2} + \frac{\partial^2 T}{\partial y^2} \right) + \frac{\mu}{g_c J} \phi \tag{8.7}$$

where $\quad \phi \equiv 2 \left[\left(\frac{\partial u}{\partial x} \right)^2 + \left(\frac{\partial v}{\partial y} \right)^2 \right] + \left(\frac{\partial v}{\partial x} + \frac{\partial u}{\partial y} \right)^2$ (8.8)

Since u is only a function of y and v= 0, the only term left to describe the viscous dissipation energy is

$$\phi \equiv \left(\frac{du}{dy} \right)^2 \tag{8.9}$$

In Eq. (8.7), v = 0 and T is only a function of y, so

$$k \frac{d^2 T(y)}{dy^2} + \frac{\mu}{g_c J} \left(\frac{du}{dy} \right)^2 = 0. \tag{8.10}$$

Substituting for u = (y/L)u₁, we obtain

$$\frac{d^2 T(y)}{dy^2} = -\frac{\mu u_1^2}{g_c J k L^2} \quad \text{in } 0 \le y \le L. \tag{8.11}$$

The boundary conditions for Eq. (8.11) are taken as temperature equals the upper-plate temperature T_1 at $y = L$, and the lower-plate temperature T_o at $y = 0$, that is,

$$T(y) = T_1 \text{ at } y = L$$
$$T(y) = T_o \text{ at } y = 0 \tag{8.12}$$

The solution of Eq. (8.11) is

$$T(y) = -\frac{1}{2}\frac{\mu u_1^2}{g_c JkL^2} y^2 + C_1 y + C_2. \tag{8.13}$$

Using the boundary conditions, we obtain

$$C_2 = T_o \tag{8.14}$$

$$C_1 = \frac{1}{L}(T_1 - T_o) + \frac{1}{2}\frac{\mu u_1^2}{g_c JkL} \tag{8.15}$$

$$T(y) - T_o = \frac{y}{L}\left\{((T_1 - T_o) + \frac{\mu u_1^2}{2g_c Jk}\left(1 - \frac{y}{L}\right)\right\} \tag{8.16}$$

8.1.1 Case $T_o \neq T_1$

The temperature distribution may be arranged in the form below.

$$\frac{T(y) - T_o}{T_1 - T_o} = \frac{y}{L}\left\{1 + \frac{1}{2}\frac{\mu u_1^2}{g_c Jk(T_1 - T_o)}\left(1 - \frac{y}{L}\right)\right\} \tag{8.17}$$

In dimensionless form, with $\eta = \frac{y}{L}$ and $\theta(\eta) = \frac{T(y) - T_o}{T_1 - T_o}$,

$$\theta(\eta) = \eta\left\{1 + \frac{1}{2}\Pr E(1 - \eta)\right\} \tag{8.18}$$

where $\mathrm{Pr} = \text{Prandtl Number} = \dfrac{C_P \mu}{k}$

and $\quad E = \text{Eckert Number} = \dfrac{u_1^2}{C_P(T_1 - T_o)g_c J}.$

When there is no flow, $u_1 = 0$ and $\mathrm{PrE} = 0$, so that $\theta(\eta) = \eta$. This is the case of pure conduction, and the temperature profile is a straight line.

By definition, the heat flux at the wall is determined from

$$q_{wall} = -k\frac{dT(y)}{dy}\bigg|_{wall}. \tag{8.19}$$

In terms of the dimensionless temperature, this expression is

$$q_{wall} = -k\frac{k(T_1 - T_o)}{L}\frac{d\theta(\eta)}{dy}\bigg|_{wall}. \tag{8.20}$$

The derivative of the temperature is obtained from Eq. (8.18) as

$$\frac{d\theta(\eta)}{d\eta} = 1 + \mathrm{Pr}\, E\left(\frac{1}{2} - \eta\right). \tag{8.21}$$

The heat flux at the upper wall, for instance, is obtained from Eqs.(8.20) and (8.21) by setting $\eta = 1$. Hence,

$$q_{upper\ wall} = -\frac{k(T_1 - T_o)}{L}\left(1 - \frac{1}{2}\mathrm{Pr}\, E\right) \tag{8.22}$$

We will now study the heat flow at the upper wall for the case $T_1 > T_o$ for different values of the parameter PrE, by examining Eq. (8.22). The following cases highlight the main features:

1. PrE = 2. The term $1 - \dfrac{1}{2} \operatorname{Pr} E$ is zero and there is no heat transfer at the upper wall. The derivative of the temperature with respect to η at the upper wall is zero because there is no heat flow.

2. PrE = 0. This corresponds to the pure conduction case, and the temperature profiles is a straight line as stated above.

3. PrE < 2. Both $1 - \dfrac{1}{2} \operatorname{Pr} E$ and $T_1 - T_o$ are positive in Eq. (8.22), so $q_{upper\,wall} < 0$ and the heat flows from the upper wall into the fluid, or in the negative y direction.

4. PrE > 2. The term $1 - \dfrac{1}{2} \operatorname{Pr} E$ is negative, and $T_1 - T_o$ is positive, so Eq. (8.22) states that $q_{upper\,wall} > 0$ or the heat flows from the fluid to the upper wall, or in the positive y direction. The energy generated by viscous dissipation is so large that the lower plate cannot remove it all.

8.1.2 Case $T_0 = T_1$

When the lower plate is at the same temperature as the upper plate, Eq. (8.16) simplifies to

$$T(y) - T_o = \frac{\mu u_1^2}{2g_c Jk} \frac{y}{L}\left(1 - \frac{y}{L}\right). \tag{8.23}$$

From symmetry, the maximum temperature in the fluid occurs at the midpoint between the plates. Putting y = L/2 in Eq. (8.23),

$$T_{max} - T_o = \frac{\mu u_1^2}{8g_c Jk} \tag{8.24}$$

From Eqs. (8.23) and (8.24), the temperature in the fluid is described by

$$\frac{T(\eta) - T_o}{T_{max} - T_o} = 4\eta(1-\eta) \quad \text{where} \quad \eta = \frac{y}{L}. \tag{8.25}$$

The heat flux at the walls is given by Fourier's law, as shown in Eq. (8.19).

8.2   Heat Transfer and Velocity Distribution in Hydrodynamically and Thermally Developed Laminar Flow in Conduits

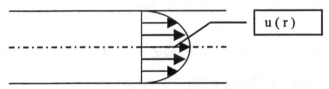

Figure 8.2   Figure for fully developed laminar flow in conduits.

Let us look at an incompressible, constant-property fluid flowing laminarly inside a circular tube in regions away from the inlet where the velocity profile is fully developed. The continuity equation gives

$$\frac{1}{r}\frac{\partial}{\partial r}(rv_r) + \frac{\partial v_z}{\partial z} = 0. \tag{8.26}$$

Since $v_r = 0$, the first term on the left in Eq. (8.26) is also zero. Since $\frac{\partial v_z}{\partial z} = 0$, v_z is a function only of r, and for convenience, we replace v_z by u.

The r-momentum equation is not needed since $v_r = 0$. The z-momentum equation is as follows:

$$\rho\left(v_r\frac{\partial u}{\partial r} + u\frac{\partial u}{\partial z}\right) = F_z - \frac{\partial P}{\partial z} + \mu\{\frac{\partial^2 u}{\partial r^2} + \frac{1}{r}\frac{\partial u}{\partial r} + \frac{\partial^2 u}{\partial z^2}\}. \tag{8.27}$$

Since $v_r = 0$, $F_z = 0$ for no body force, $\frac{\partial u}{\partial z} = 0$ and $\frac{\partial^2 u}{\partial z^2} = 0$,

$$\frac{1}{\mu}\frac{dP}{dz} = \frac{d^2u}{dr^2} + \frac{1}{r}\frac{du}{dr}$$

$$\frac{1}{r}\frac{d}{dr}\left(r\frac{du}{dr}\right) = \frac{1}{\mu}\frac{dP}{dz} \qquad \text{in } 0 \le r \le R. \qquad (8.28)$$

The boundary conditions are that there is no slip at the wall, that is, u = 0 at r = R, and the velocity is finite within the tube. Since there is symmetry about the tube axis, the boundary condition du/dr = 0 at r = 0 is allowable, and it leads to the same result.

Let us consider the case where $\dfrac{dP}{dz} = cons\tan t$. To nondimensionalize the equation, we can define

$$\eta = \frac{r}{R}, \quad u' = \frac{u}{u_m} \qquad (8.29)$$

where u_m is an arbitrary mean velocity. It is acceptable to nondimensionalize with respect to an unknown mean velocity, as long as it is of the correct order of magnitude. Its value may be calculated at a later time. Substituting in Eq. (8.28), we obtain

$$\frac{1}{\eta R^2}\frac{d}{d\eta}\left(\eta u_m \frac{du'}{d\eta}\right) = k. \qquad (8.30)$$

The boundary conditions become

$$u' = 0 \text{ at } \eta = 1 \qquad (8.31a)$$

$$\frac{du'}{d\eta} = 0 \text{ at } \eta = 0. \qquad (8.31b)$$

Integrating Eq. (8.30), we obtain

$$\int \frac{d}{d\eta}\left(\eta u_m \frac{du'}{d\eta}\right)d\eta = \int k\eta R^2 d\eta \qquad (8.32)$$

$$\eta u_m \frac{du'}{d\eta} = \frac{k}{2} \eta^2 R^2 + C_1. \tag{8.33}$$

Using the boundary condition Eq. (8.31b), we find that $C_1 = 0$. Thus,

$$\frac{du'}{d\eta} = \frac{k}{2u_m} \eta R^2 \tag{8.34}$$

Integrating Eq. (8.34), we obtain

$$u' = \frac{k}{4u_m} \eta^2 R^2 + C_2 \tag{8.35}$$

Using the boundary condition Eq. (8.31a), we find that $C2 = -\frac{k}{4u_m} R^2$.

Therefore, the dimensionless velocity distribution may be expressed as

$$u' = -\frac{1}{u_m} \left(\frac{1}{4\mu} \frac{dP}{dz} \right) R^2 \left(1 - \eta^2\right) \tag{8.36}$$

The mean flow velocity is given by

$$u_m = \frac{1}{\pi R^2} \int_0^R 2\pi r u(r) dr = -\frac{R^2}{8\mu} \frac{dP}{dz}. \tag{8.37}$$

Hence, $\dfrac{u(r)}{u_m} = 2\left[1 - \left(\dfrac{r}{R}\right)^2\right].$ $\tag{8.38}$

At the tube axis, $r = 0$, and we expect the maximum velocity to occur because of symmetry.

$$u_o = -\frac{1}{4\mu} \frac{dP}{dz} R^2 \tag{8.39}$$

The velocity distribution may be expressed in terms of this axial velocity maximum as

$$\frac{u(r)}{u_o} = 1 - \left(\frac{r}{R}\right)^2 .$$ (8.40)

It should be noted that the maximum velocity value is twice the value of the mean velocity, that is, $u_o = 2\, u_m$. The friction factor f is defined as

$$f = -\frac{dP/dz}{\left(\frac{1}{2}\rho u_m^2\right)\Big/ D} .$$ (8.41)

The pressure along the tube may be calculated from

$$\int_{P_1}^{P_2} dP = -f \frac{\rho u_m^2}{2D} \int_{z_1}^{z_2} dz .$$ (8.42)

Hence, $P_1 - P_2 = f \frac{1}{D} \frac{\rho u_m^2}{2} (z_2 - z_1).$ (8.43)

8.2.1 Temperature Distribution

For moderate velocities, the viscous dissipation term may be neglected. Under such conditions, the energy conservation equation may be written as

$$\frac{1}{\alpha} u(r) \frac{\partial T}{\partial z} = \frac{\partial^2 T}{\partial r^2} + \frac{1}{r} \frac{\partial T}{\partial r} + \frac{\partial^2 T}{\partial z^2}$$ (8.44)

We define a dimensionless temperature $\theta\,(r)$ such that

$$\theta(r) = \frac{T(r,z) - T_w(z)}{T_m(z) - T_w(z)}$$ (8.45)

where T(r,z) = local temperature in the fluid
          $T_w(z)$ = the tube wall temperature

$T_m(z)$ = mean temperature of the fluid over the cross-sectional area of the tube

$$= \frac{\int_0^R 2\pi r u(r) T(r,z) dr}{\int_0^R 2\pi r u(r) dr}.$$

(8.46)

A fully developed temperature profile is one where θ is not a function of the axial distance z, that is, $\theta = \theta(r)$. Note that the dimensional temperature T is still a function of z; that is why $\dfrac{\partial T}{\partial z}$ is not equal to zero. Differentiating Eq. (8.45) with respect to z,

$$\frac{d\theta(r)}{dz} = \frac{\partial}{\partial z}\left[\frac{T(r,z) - T_w(z)}{T_m(z) - T_w(z)}\right] = 0$$

(8.47)

$$\frac{\partial}{\partial z}\left[\frac{T(r,z) - T_w(z)}{T_m(z) - T_w(z)}\right] = \frac{(T_m - T_w)\dfrac{\partial}{\partial z}(T - T_w) - (T - T_w)\dfrac{\partial}{\partial z}(T_m - T_w)}{(T_m - T_w)^2} = 0$$

(8.48)

When $T_m \neq T_w$,

$$\frac{\partial}{\partial z}(T - T_w) - \frac{T - T_w}{T_m - T_w}\frac{\partial}{\partial z}(T_m - T_w) = 0.$$

(8.49)

For constant heat flux q_w at the wall,

$$q_w = h(T_w - T_m) = \text{constant}$$

(8.50)

For constant heat transfer coefficient h, between the fluid and the wall surface, $T_w - T_m = \text{constant}$. Thus,

$$\frac{d}{dz}(T_w - T_m) = 0$$

(8.51)

or $\dfrac{dT_w}{dz} = \dfrac{dT_m}{dz}$ = constant.   (8.52)

Substituting Eq. (8.52) into Eq. (8.50),

$$\frac{\partial}{\partial z}(T - T_w) = 0 \quad \text{or} \quad \frac{\partial T}{\partial z} = \frac{dT_w}{dz}.$$ (8.53)

From Eqs. (8.52) and (8.53),

$$\frac{\partial T}{\partial z} = \frac{dT_m}{dz} = \text{constant.}$$ (8.54)

This means that the average fluid temperature $T_m(z)$ in the thermally developed region increases linearly with z. Substituting in Eq. (8.44) and noting that $\dfrac{\partial^2 T}{\partial z^2} = 0$ because $\dfrac{\partial T}{\partial z} = 0$, we obtain

$$\frac{1}{\alpha} u(r) \frac{dT_m(z)}{dz} = \frac{1}{r} \frac{\partial}{\partial r}\left(r \frac{\partial T}{\partial r} \right) \quad \text{in } 0 \le r \le R.$$ (8.55)

Substituting for u (r) from Eq. (8.38), we have

$$\frac{\partial}{\partial r}\left(r \frac{\partial T}{\partial r} \right) = Ar\left[1 - \left(\frac{r}{R} \right)^2 \right] \quad \text{in } 0 \le r \le R$$ (8.56)

where $A = \dfrac{2u_m}{\alpha} \dfrac{dT_m(z)}{dz}$ = constant.

The boundary conditions are that at $r = 0$, $\dfrac{\partial T}{\partial r} = 0$ because of symmetry, and that the fluid temperature at the wall is the same as the wall temperature in the thermally fully developed region, that is, $T = T_w(z)$ at $r = R$. Integrating Eq. (8.56) and using the boundary condition at $r = 0$,

$$\frac{\partial T}{\partial r} = A\left(\frac{1}{2}r - \frac{r^3}{4R^2}\right) \tag{8.57}$$

Integrating Eq. (8.57) and using the boundary condition at r = R, the temperature distribution is found as

$$T(r,z) - T_w(z) = -AR^2\{\frac{3}{16} + \frac{1}{16}\left(\frac{r}{R}\right)^4 - \frac{1}{4}\left(\frac{r}{R}\right)^2. \tag{8.58}$$

The mean fluid temperature (or bulk fluid temperature) across the tube, Tm(z), is given by $T_m(z) - T_w(z)$

$$= \frac{\int_0^R 2\pi r u(r)[T(r,z) - T_w(z)]dr}{\int_0^R 2\pi r u(r)dr}$$

$$= \frac{1}{\pi R^2 u_m} \int_0^R 2\pi r u(r)[T(r,z) - T_w(z)]dr$$

$$= -4A \int_0^R r\left(1 - \frac{r^2}{R^2}\right)\left(\frac{3}{16} + \frac{1}{16}\frac{r^4}{R^{\$}} - \frac{1}{4}\frac{r^2}{R^2}\right)dr$$

$$= -\frac{11}{48}\frac{AR^2}{2} \quad \text{where} \quad A = \frac{2u_m}{\alpha}\frac{dT_m(z)}{dz}. \tag{8.59}$$

The wall heat flux is given by

$$k\frac{dT}{dr}\Big|_{r=R} = q_w. \tag{8.60}$$

From Eq. (8.57),
$$\left.\frac{dT}{dr}\right|_{r=R} = \frac{1}{4}AR. \qquad (8.61)$$

Hence, $A = \dfrac{4q_w}{kR}$ \qquad\qquad\qquad\qquad\qquad\qquad\qquad (8.62)

The heat transfer coefficient h between the fluid flow and the wall is given by

$$h[T_m(z) - T_w(z)] = -k\left.\frac{dT}{dr}\right|_{r=R}. \qquad (8.63)$$

Hence, $h = \dfrac{-k}{T_m(z) - T_w(z)} \left.\dfrac{dT}{dr}\right|_{r=R}.$ \qquad\qquad\qquad (8.64)

Substituting Eqs. (8.59) and (8.61) into Eq. (8.64),

$$h = \frac{48}{11}\frac{k}{D} \qquad\qquad \text{where} \quad D = 2R. \qquad (8.65)$$

The Nusselt number is then calculated from

$$Nu = \frac{hD}{k} = \frac{48}{11} = 4.364. \qquad (8.66)$$

In general, the Nusselt number $Nu = \dfrac{hD_e}{k}$, where the equivalent diameter D_e is given by

$$D_e = \frac{4 \times (Flow\ Area)}{Wetted\ Perimeter}. \qquad (8.67)$$

For example, in a full square duct, $D_e = \dfrac{4xa^2}{4a} = a.$

(a)   Constant heat rate

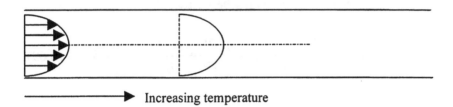

→ Increasing temperature

(b)   Constant surface temperature

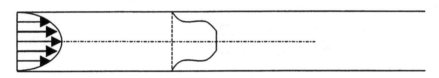

$$T_{wall} < T_{entering\ fluid}$$

Figure 8.3   Fully developed temperature profiles for constant heat rate and constant surface temperature.

8.3   The Circular Tube Thermal-Entry-Length, with Hydrodynamically Fully Developed Laminar Flow

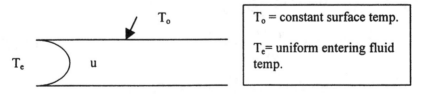

T_o = constant surface temp.

T_e = uniform entering fluid temp.

Figure 8.4   Sketch for the thermal-entry length problem.

For a hydrodynamically fully developed laminar flow, the parabolic velocity profile is applicable. Hence,

$$u' = \frac{u}{u_m} = 2\left(1 - r'^2\right) \quad \text{where } r' = \frac{r}{R}. \tag{8.68}$$

The governing energy equation may be written as

$$\rho C_P \left(v_r \frac{\partial T}{\partial r} + v_z \frac{\partial T}{\partial z} \right) = k \left(\frac{\partial^2 T}{\partial r^2} + \frac{1}{r} \frac{\partial T}{\partial r} + \frac{\partial^2 T}{\partial z^2} \right) + \frac{\mu}{g_c J} \phi \tag{8.69}$$

where $\phi \equiv 2\left[\left(\dfrac{\partial v_r}{\partial r} \right)^2 + \dfrac{v_r^2}{r^2} + \left(\dfrac{\partial v_z}{\partial z} \right)^2 \right] + \left(\dfrac{\partial v_z}{\partial r} + \dfrac{\partial v_r}{\partial z} \right)^2$. For moderate

velocities, $\Phi = 0$. For a hydrodynamically developed flow, $v_r = 0$. Putting $v_z = u$, Eq. (8.69) becomes

$$\frac{u}{\alpha} \frac{\partial T}{\partial z} - \frac{\partial^2 T}{\partial z^2} = \frac{\partial^2 T}{\partial r^2} + \frac{1}{r} \frac{\partial T}{\partial r}. \tag{8.70}$$

Define $\theta = \dfrac{T_o - T}{T_o - T_e}, \; r' = \dfrac{r}{R}, \; u' = \dfrac{u}{u_m}, \; z' = \dfrac{z}{r}. \tag{8.71}$

$$\frac{u' u_m}{\alpha} \cdot \frac{-(T_o - T_e)}{R} \frac{\partial \theta}{\partial z'} - \frac{-(T_o - T_e)}{R^2} \cdot \frac{\partial^2 \theta}{\partial z'^2} = \frac{-(T_o - T_e)}{R^2} \frac{\partial^2 \theta}{\partial r'^2} + \frac{-(T_o - T_e)}{R^2} \cdot \frac{1}{r'} \frac{\partial \theta}{\partial r'}$$

$$\frac{\partial \theta}{\partial r'^2} + \frac{1}{r'} \frac{\partial \theta}{\partial r'} = (\mathrm{Re\,Pr}) \frac{u'}{2} \frac{\partial \theta}{\partial z'} - \frac{\partial^2 \theta}{\partial z'^2}$$

where $D = 2R, \; \mathrm{Re} = \dfrac{u_m D}{\upsilon}, \; \mathrm{Pr} = \dfrac{\upsilon}{\alpha} \tag{8.72}$

$$\frac{\partial^2 \theta}{\partial r'^2} + \frac{1}{r'} \frac{\partial \theta}{\partial r'} = Pe \frac{u'}{2} \frac{\partial \theta}{\partial z'} - \frac{\partial^2 \theta}{\partial z'^2} \quad \text{where } \mathrm{Pe} = \mathrm{RePr}.$$

$$\tag{8.73}$$

For better estimation of the axial conduction term, we define $z^+ = \dfrac{z/R}{Pe}$.

Hence,

$$\frac{\partial^2 \theta}{\partial r'^2} + \frac{1}{r'}\frac{\partial \theta}{\partial r'} = \frac{u'}{2}\frac{\partial \theta}{\partial z^+} - \frac{1}{(Pe)^2}\frac{\partial^2 \theta}{\partial z^{+2}}. \tag{8.74}$$

For Peclet numbers Pe > 100, the last term in Eq. (8.74) may be neglected. Substituting u' from Eq. (8.68) for the cases Pe > 100,

$$\frac{\partial^2 \theta}{\partial r'^2} + \frac{1}{r'}\frac{\partial \theta}{\partial r'} = (1 - r'^2)\frac{\partial \theta}{\partial z^+}. \tag{8.75}$$

8.3.1 Constant Surface Temperature

For a constant surface temperature boundary condition,

$$\theta(0, r') = 1 \quad \text{and} \quad \theta(z^+, 1) = 0 \tag{8.76}$$

Using the method of separation of variables, we assume that

$$\theta = A(r')Z(z^+) \tag{8.77}$$

The resulting equations obtained are

$$Z' + \lambda^2 Z = 0 \tag{8.78}$$

and $\quad A'' + \dfrac{1}{r'}A' + \lambda^2 A(1 - r'^2) = 0 \tag{8.79}$

The solution to Eq. (8.78) is $\quad Z = C\exp(-\lambda^2 z^+)$. Equation (8.79) is of the Sturm-Liouville type, and solutions may be written as $J_n(r')$, cylindrical eigenfunctions or Bessel functions. Therefore, the solution to Eq. (8.79) may be written as

$$\theta(z^+, r') = \sum_{n=0}^{\infty} c_n J_n(r')\exp(-\lambda_n^2 z^+) \tag{8.80}$$

where λ_n's are the eigenvalues and J_n's are the corresponding eigenfunctions. The heat flux at the wall can be calculated from the following formula.

$$\dot{q}_o''(z^+) = -k\left(\frac{\partial T}{\partial r}\right)_{r=R} = k\frac{T_o - T_e}{R}\left(\frac{\partial \theta}{\partial r'}\right)_{r'=1} \tag{8.81}$$

$$\dot{q}_o''(z^+) = \frac{2k}{R}\sum_{n=0}^{\infty} G_n \exp\left(-\lambda_m^2 z^+\right)(T_o - T_e) \quad \text{where } G_n = -\left(\frac{C_n}{2}\right)J_n'(1) \tag{8.82}$$

Table 8.1   Infinite-series solution functions for the circular tube; constant surface temperature; thermal-entry length.

n	λ_n^2	G_n
0	7.312	0.749
1	44.62	0.544
2	113.8	0.463
3	215.2	0.414
4	348.5	0.382

For n > 2, $\lambda_n = 4n + \dfrac{8}{3}$, $G_n = 1.01276\lambda_n^{-\frac{1}{3}}$.

8.3.2 Uniform Heat Flux

For a uniform heat flux boundary condition, we define r', u' and z^+ as before, and the dimensionless temperature as

$$\theta = \frac{T_e - T}{T_e - T_{ref}}, \quad \text{where } T_{ref} = \text{some reference temperature, say}$$

T_{mean}.   (8.83).

The boundary conditions can then be expressed as

$$\theta(0, r') = 0 \quad \text{and} \quad \frac{\partial \theta}{\partial r'}(z^+, 1) = \text{constant}. \tag{8.84}$$

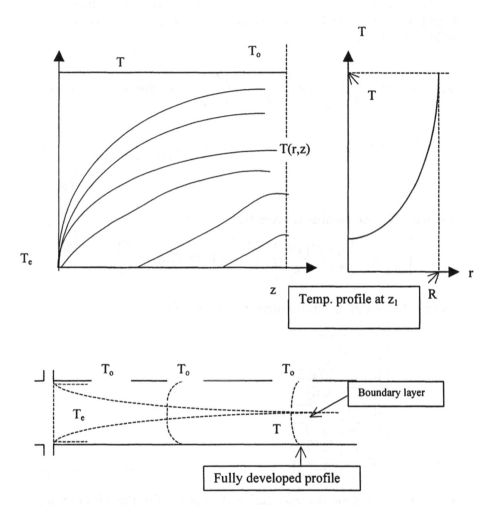

Figure 8.5   Development of the temperature profile in the thermal-entry region of a pipe.

Equation (8.79) is of the Sturm-Liouville type, and with boundary conditions Eq. (8.84), solutions may be written as $R_n(r')$, cylindrical eigenfunctions or Bessel functions. Therefore, the solution to Eq. (8.75) may be written as

$$\theta(z^+, r') = \sum_{n=0}^{\infty} c_n R_n(r') \exp(-\lambda_n^2 z^+) \qquad (8.85)$$

where λ_n's are the eigenvalues and R_n's are the corresponding eigenfunctions. The heat flux at the wall can be calculated from the following formula.

$$\dot{q}_o''(z^+) = -k\left(\frac{\partial T}{\partial r}\right)_{r=R} = k\frac{T_e - T_{ref}}{R}\left(\frac{\partial \theta}{\partial r'}\right)_{r'=1} = \text{constant} \qquad (8.86)$$

$$\dot{q}_o''(z^+) = \frac{2k}{R}\sum_{n=0}^{\infty} G_n \exp\left(-\lambda_m^2 z^+\right)(T_o - T_e) \quad \text{where } G_n = -\left(\frac{C_n}{2}\right)J_n'(1). \qquad (8.87)$$

The local Nusselt number is given by

$$Nu_x = \left[\frac{1}{Nu_\infty} - \frac{1}{2}\sum\frac{\exp\left(-\lambda_n^2 z^+\right)}{A_n \lambda_n^4}\right]^{-1} \quad \text{where } Nu_\infty = \frac{48}{11}.$$

Table 8.2   Values of Nu_x for different values of z^+.

Z^+	Nu_x
0	∞
0.002	12.00
0.004	9.93
0.010	7.49
0.020	6.14
0.100	4.51
∞	4.36

Table 8.3   Infinite-series-solution functions for the circular tube; constant heat rate; thermal-entry length.

n	λ_n^2	A_n
1	25.68	7.630×10^{-3}
2	83.86	2.058×10^{-3}
3	174.2	0.901×10^{-3}
4	296.5	0.487×10^{-3}
5	450.9	0.297×10^{-3}

For larger n,

$$\lambda_n = 4n + \frac{4}{3}; \quad A_n = 0.$$

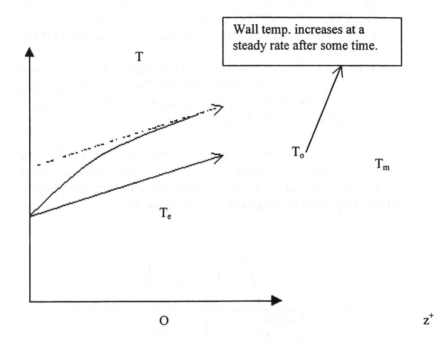

Figure 8.6 Temperature variations in the thermal-entry region of a
tube with constant heat rate per unit of tube length.

8.4   The   Rectangular   Duct   Thermal-Entry   Length,   with
Hydrodynamically Fully Developed Laminar Flow

The thermal entrance region in a hydrodynamically fully developed flow in a rectangular duct may be studied by the use of the integral method. In this section, the uniform wall temperature and the uniform wall heat flux cases are discussed. The physical model is based on the following assumptions:

1.   All fluid properties are constant.
2.   The flow is laminar.
3.   The viscous dissipation and the work of compression are both negligible.

4.   Both walls of the duct either have the same uniform temperature T_w, or the same uniform heat flux q_w.
5.   The thermal boundary layer thickness is zero at the entrance where $x = 0$.
6.   The effects of heat transfer are found only within the thermal boundary layer. The fluid outside the thermal boundary layer will be unaffected by the heat transfer, and have a uniform temperature T_{in} at the entrance where $x = 0$.

The velocity profile is assumed fully developed at $x = 0$. The heating (or cooling) section starts at $x = 0$; the thermal boundary layer grows in thickness as x increases until it reaches the center line where it meets the boundary layer from the other wall of the duct.

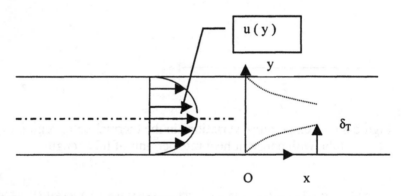

Figure 8.7   Development of the thermal boundary layer along the walls of a rectangular duct.

8.4.1 Constant Wall Heat Flux

If we neglect the viscous dissipation term, the energy integral equation, Eq. (8.19) becomes

$$q_w = \frac{d}{dx}\{\int_0^{\delta_T} \rho C_p u(T - T_i)dy\}. \tag{8.88}$$

The fully developed velocity profile for flow between two flat plates can be used,

$$\frac{u}{U_c} = 2\left(\frac{y}{d}\right) - \left(\frac{y}{d}\right)^2.$$ (8.89)

For the temperature distribution within the boundary layer, the following polynomial is selected:

$$T - T_{in} = \frac{q_w \delta_T}{3k}\{2 - 3\frac{y}{\delta_T} - \left(\frac{y}{\delta_T}\right)^3\} \quad \text{in} \quad 0 \le y \le \delta_T \quad (8.90)$$

Substituting Eqs. (8.89) and (8.90) into the energy integral equation (8.88),

$$q_w = \frac{d}{dx}\{\frac{\rho C_p U_c q_w}{3k}\delta_T \int_0^1 (2\eta\xi - \eta^2\xi^2)(2 - 3\eta + \eta^3)d\eta\} \quad (8.91)$$

where $\eta = \dfrac{y}{\delta_T}$ and $\xi = \dfrac{\delta_T}{d}$. (8.92a,b)

Integrating Eq. (8.91),

$$\frac{d\xi^3}{dx} = \frac{15}{2}\frac{k}{\rho C_p U_c d^2}x + c_1. \quad (8.93)$$

The constant of integration $c_1 = 0$ because $\xi = 0$ at $x = 0$. So,

$$\xi = \left(\frac{80x/d_H}{Pe}\right)^{\frac{1}{3}} \quad (8.94)$$

where $d_H = 4d$, $Pe = \dfrac{U_m d_H}{\alpha}$ and $U_m = \dfrac{2}{3}U_c$.

From Eq. (8.90), $T_w - T_{in} = \dfrac{2q_w}{3k}\delta_T.$ (8.95)

Since $\delta_T = \xi d$, Eq. (8.95) may also be expressed as

$$T_w - T_{in} = \frac{2q_w d}{3k}\left(\frac{80x/d_H}{Pe}\right)^{\frac{1}{3}}.$$ (8.96)

The local heat transfer coefficient based on $T_w - T_{in}$ is expressed as

$$h_x = \frac{q_w}{T_w - T_{in}}.$$ (8.97)

Substituting Eq. (8.96) into Eq. (8.97),

$$h_x = \frac{64}{d_H}\left(\frac{Pe}{80x/d_H}\right)^{\frac{1}{3}}.$$ (8.98)

The Nusselt number based on $T_w - T_{in}$ is expressed as

$$Nu_x = \frac{hd_H}{k} = 6\left(\frac{Pe}{80x/d_H}\right)^{\frac{1}{3}}.$$ (8.99)

The bulk (mean) temperature is defined as

$$T_m = \frac{\int_A TudA}{\int_A udA}$$ (8.100)

Hence,

$$T_m - T_{in} = \frac{4}{d_H} \int_0^{\delta_T} \frac{u}{U_m}(T - T_{in})dy = \frac{4q_w}{3kd_H}\delta_T \int_0^1 (2\eta\xi - \eta^2\xi^2)(\eta - 3\eta + \eta^3)d\eta.$$

(8.101)

Therefore,

$$T_m - T_{in} = \frac{2q_w d_H}{k}\left\{\frac{\xi^3}{10} - \frac{\xi^4}{48}\right\}.$$

(8.102)

Combining Eqs. (8.96) and (8.102),

$$T_w - T_m = \frac{2q_w d_H}{k}\left\{\frac{\xi}{3} - \frac{\xi^3}{10} + \frac{\xi^4}{48}\right\}.$$

(8.103)

Thus, the local heat transfer coefficient based on $T_w - T_{in}$ can be expressed as

$$h_x = \frac{q_w}{T_w - T_m} = \frac{k}{d_H}\frac{2}{\left\{\dfrac{\xi}{3} - \dfrac{\xi^3}{10} + \dfrac{\xi^4}{48}\right\}}.$$

(8.104)

The Nusselt number based on $T_w - T_m$ is then

$$Nu_x = \frac{hd_H}{k} = \frac{2}{\left\{\dfrac{\xi}{3} - \dfrac{\xi^3}{10} + \dfrac{\xi^4}{48}\right\}}$$

(8.105)

8.4.2 Constant Wall Temperature

Consider laminar fluid flow between parallel plates with a uniform wall temperature. A fully developed parabolic velocity profile, Eq. (8.89), is assumed as in the previous case. For the temperature profile, the following polynomial is assumed:

$$\frac{T-T_{in}}{T_w-T_{in}} = 1 - \frac{1}{2}\frac{y}{\delta_T}\left[3-\left(\frac{y}{\delta_T}\right)^2\right] \qquad \text{in} \qquad 0 \le y \le \delta_T.$$

$$(8.106)$$

Substituting the velocity and temperature distributions into the energy integral equation (8.19), and integrating, a differential equation in $\xi = \delta_T / d$ is obtained. The solution of this differential equation is

$$\xi = \left(\frac{120x/d_H}{Pe}\right)^{\frac{1}{3}}.$$

$$(8.107)$$

The heat transfer coefficient based on $T_w - T_{in}$ is

$$h_x = 6\left(\frac{Pe}{120x/d_H}\right)^{\frac{1}{3}}.$$

$$(8.108)$$

The Nusselt number based on $T_w - T_m$ is

$$Nu_x = \frac{hd_H}{k} = \frac{6}{\xi\left[1 - \frac{3}{2}\left(\frac{\xi^2}{5}+\frac{\xi^3}{24}\right)\right]}.$$

PROBLEMS

8.1.   Consider the fully developed laminar flow between two parallel plates at a distance 2a apart. Find the expression for the velocity profile and the friction factor.

8.2.   Obtain the steady-state, fully developed velocity distribution for laminar flow between two parallel plates, in the absence of body forces.

8.3.   Obtain the steady-state, fully developed velocity distribution for laminar flow between two parallel plates, in the presence of a body force due to gravity, such that the body force is equal to $-\vec{j}\,g$, where g is the gravitational acceleration.

8.4. Systematically find the expression for the temperature distribution and the Nusselt number for laminar flow between two large parallel plates in the region of fully developed velocity and temperature profiles for a uniformly applied wall heat flux.

8.5. Consider a fully developed steady-state laminar flow of a constant-property fluid through a circular duct with a constant heat flux condition imposed at the duct wall. Neglect axial conduction and assume that the velocity profile may be approximated by a uniform velocity across the entire flow area (i.e., slug flow). Obtain an expression for the Nusselt number.

8.6. Consider a fully-developed steady-state laminar flow of a constant-property fluid through a circular pipe with a constant heat flux condition imposed at the duct wall. Neglect axial conduction, but include the effect of viscous dissipation. Obtain an expression for the Nusselt number.

8.7. Consider the fully-developed flow of a viscous fluid in a circular duct of radius a_o. Without neglecting viscous dissipation, derive an expression for the Nusselt number if the boundary condition at $r = a_o$, is $T = T_w < T_m$, where T_m is the mean temperature of the fluid.

8.8.

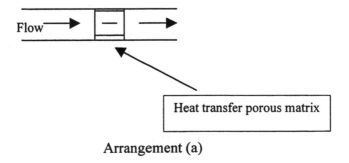

Arrangement (a)

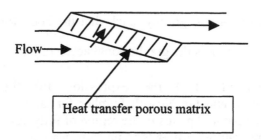

Arrangement (b)

Arrangements (a) and (b) are two possibilities such that heat is exchanged between the solid porous matrix and the fluid. Which is a more efficient heat transfer arrangement? Discuss.

8.9. In Ulrichson and Schmit's work on laminar flow heat transfer in the entrance region of circular tubes the following results were obtained.

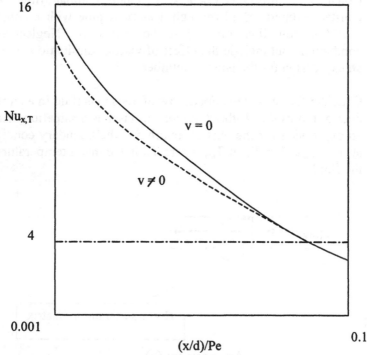

Figure for Prob. 8.9.

$Nu_{x,T}$ = local Nusselt number

$\dfrac{x/d}{Pe}$ = dimensionless axial coordinate

v = radial velocity component

The two different curves represent the following:

(i)   v = 0, radial velocity component neglected

(ii)   v $\neq$ 0, radial velocity component not neglected.

In both cases the axial velocity component, u, has been considered. Give a physical explanation for the finding that curve (i) is consistently higher than curve (ii).

8.10.   A constant property fluid flows between two horizontal, semi-infinite, parallel plates, kept at a distance 2m apart. The upper plate is at a constant temperature T_1 and the lower plate is at a constant temperature T_2. Consider the fully developed velocity and temperature profiles region for laminar flow. Include viscous dissipation. Find the heat flux to each of the plates.

8.11.   Consider fully developed flow in a pipe. A thermal boundary condition is applied, starting at a distance $x = x_o$. For the four different cases listed below, sketch the temperature profiles in the pipe as the thermal boundary layer develops, and the temperature profiles after the thermal boundary layer has fully developed:

(i)   Constant heat flux input at the walls; at the entrance, $T_{wall} > T_{fluid}$;

(ii)   Constant heat flux outflow at the walls; at the entrance, $T_{wall} < T_{fluid}$;

(iii)   Constant wall temperature; at the entrance $T_{wall} > T_{fluid}$;

(iv)   Constant wall temperature; at the entrance $T_{wall} < T_{fluid}$.

8.12.   A constant-property fluid flows in a laminar manner in the x direction between two large parallel plates. The same constant heat flux q_w is maintained from the plates to the fluid for all $x \geq 0$. The fluid temperature is T_{in} at x = 0. Find an expression of the local Nusselt number by the integral method. What is this expression if Pr = 1?

Internal Flows

Flow between parallel plates is modeled by Couette flow
Flow in tubes, pipes and blood vessels is modeled by Poiseuille flow
The moving plate is the driving factor in Couette flow
The pressure gradient is the driving factor in Poiseuille flow.

We consider viscous dissipation in Couette flow
We consider the pressure gradient in Poiseuille flow
Viscous dissipation acts as heat source in Couette flow
For moderate flow, no heat source within Poiseuille flow.

K.V. Wong

9

Natural Convection

Free or natural convection occurs when fluid motion is generated predominantly by body forces caused by density variations, under the earth's gravitational field. In the absence of the gravitational field, body forces may be caused by surface tension. The subject material here is focussed on heat transfer with motion produced by buoyancy forces.

9.1. Boundary Layer Concept for Free Convection

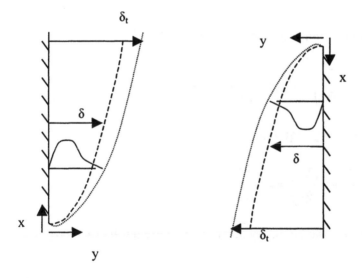

(a) Vertical plate is hot compared to the environmental temperature.

(b) Vertical plate is cold compared to the environmental temperature.

Figure 9.1   Boundary layer concept for free convection.

In Fig. 9.1, the vertical plate is at a much different temperature from that of the environment. In Fig. 9.1(a), the plate is hotter than the environment, hence the air in contact with the plate gets hotter, less

dense and rises. In Fig. 9.1(b), the plate is colder than the environment, hence the air in contact with it gets colder, more denser and falls.
The momentum boundary layer thickness is represented by δ, and the thermal boundary layer thickness is represented by δ_t.

Consider laminar, steady, two-dimensional free convection with no viscous dissipation of energy of an incompressible fluid. The concept used here is that the fluid has constant properties, but the body force is produced by a difference in density caused by the temperature distribution. For the continuity equation, we have

$$\frac{\partial u}{\partial x} + \frac{\partial v}{\partial y} = 0 \qquad (9.1)$$

for the momentum equation,

$$\rho\left(u\frac{\partial u}{\partial x} + v\frac{\partial u}{\partial y} \right) = -\rho g - \frac{\partial p}{\partial x} + \mu\frac{\partial^2 u}{\partial y^2} \qquad (9.2)$$

and for the energy equation,

$$\rho C_P\left(u\frac{\partial T}{\partial x} + v\frac{\partial T}{\partial y} \right) = k\frac{\partial^2 T}{\partial y^2}. \qquad (9.3)$$

If we examine the boundary layer edge where u = 0 at y → ∞,

$$\frac{\partial p}{\partial x} = -\rho_\infty g \qquad (9.4)$$

where ρ_∞ is the fluid density outside the boundary layer. The pressure field is given by hydrostatics. Since $\frac{\partial p}{\partial y} \approx 0$, then

$$-\rho g - \frac{\partial p}{\partial x} = (\rho_\infty - \rho)g. \qquad (9.5)$$

To express the change in fluid density as a result of the fluid temperature, we can define the volumetric coefficient of thermal expansion, β, as

$$\beta = -\frac{1}{\rho}\frac{\partial \rho}{\partial T}\bigg|_{pressure} \tag{9.6}$$

or $\Delta \rho = -\beta \rho \Delta T$ (9.7)

or $\rho_\infty - \rho = -\beta \rho (T_\infty - T)$. (9.8)

Hence, $-\rho g - \dfrac{\partial p}{\partial x} = -\beta \rho (T_\infty - T)g$. (9.9)

The boundary layer momentum equation becomes

$$u\frac{\partial u}{\partial x} + v\frac{\partial u}{\partial y} = g\beta(T - T_\infty) + \upsilon\frac{\partial^2 u}{\partial y^2}. \tag{9.10}$$

We define the following dimensionless quantities:

$$X = \frac{x}{L} \qquad\qquad Y = \frac{y}{L} \tag{9.11}$$

$$U = \frac{u}{u_o} \qquad\qquad V = \frac{v}{u_o} \tag{9.12}$$

$$\theta = \frac{T - T_\infty}{T_w - T_\infty} \tag{9.13}$$

The dimensionless governing equations are then as follows:

$$\frac{\partial U}{\partial X} + \frac{\partial V}{\partial Y} = 0 \tag{9.14}$$

$$U\frac{\partial U}{\partial X}+V\frac{\partial U}{\partial Y}=\frac{Gr}{Re^2}\theta+\frac{1}{Re}\frac{\partial^2 U}{\partial Y^2} \tag{9.15}$$

$$U\frac{\partial \theta}{\partial X}+V\frac{\partial \theta}{\partial Y}=\frac{1}{Re\,Pr}\frac{\partial^2 \theta}{\partial Y^2} \tag{9.16}$$

where Gr = Grashof number = $\dfrac{g\beta(T_w-T_\infty)L^3}{v^2}$. $\tag{9.17}$

The ratio $\dfrac{Gr}{Re^2}=\dfrac{bouyancy\ forces}{inertial\ forces}$, since in free convection, the buoyancy forces are the forces causing action. In practical engineering, the Nusselt number may be correlated to the Grashof number and the Prandtl number. Often, for simplicity, the product of Grashof and Prandtl number is used, that is, the Raleigh number, Ra,

$$Ra = GrPr = g\beta L^3(T_w-T_\infty)/(\alpha v). \tag{9.18}$$

In the foregoing discussion, Eqs. (9.14)-(9.16) are for free convection from a hot vertical plate with x,y coordinates selected as depicted in Fig. 9.1(a). The same equations are suitable for a cold vertical plate (i.e., $T_\infty > T_w$) if the coordinates are selected as shown in Fig. 9.1(b). In Fig. 9.1(a), g acts in the negative x direction, so the product $g\beta(T-T_\infty)$ is a negative quantity since $T > T_\infty$ for a hot plate. In Fig. 9.1(b), $g\beta(T-T_\infty)$ is also a negative quantity since $T < T_\infty$ for a cold plate.

9.2 Similarity Solution: Boundary Layer with Uniform Temperature

The equations (9.14)-(9.16) have been solved using a similarity solution by Ostrach (1953). The solution method hinges on the introduction of a similarity parameter of the form

$$\eta \equiv \frac{y}{x} \left(\frac{Gr_x}{4} \right)^{\frac{1}{4}} . \qquad (9.19)$$

The velocity components are represented in terms of a stream function defined as

$$\psi(x, y) \equiv f(\eta) \left[4\nu \left(\frac{Gr_x}{4} \right)^{\frac{1}{4}} \right] . \qquad (9.20)$$

With the definition of the stream function in Eq. (9.20), the x-velocity component may be represented by

$$u = \frac{\partial \psi}{\partial y} = \frac{\partial \psi}{\partial \eta} \frac{\partial \eta}{\partial y} = 4\nu \left(\frac{Gr_x}{4} \right)^{\frac{1}{4}} f'(\eta) \frac{1}{x} \left(\frac{Gr_x}{4} \right)^{\frac{1}{4}} = \frac{2\nu}{x} Gr_x^{\frac{1}{2}} f'(\eta). \qquad (9.21)$$

The three original partial differential equations may then be reduced to two ordincary differential equations of the form

$$f''' + 3ff'' - 2(f')^2 + \theta = 0 \qquad (9.22)$$

$$\theta'' + 3Prf\theta' = 0 \qquad (9.23)$$

where f and θ are functions of only η and the double and triple primes, respectively, indicate the second and third derivatives with respect to η. The function f takes on the role of the dependent variable for the velocity boundary layer. The continuity equation is automatically satisfied by introducing the stream function.

The transformed boundary conditions required to solve the momentum and energy equations, are as follows:

$$\eta = 0, \quad f = f' = 0, \quad \theta = 1 \tag{9.24}$$

$$\eta \rightarrow \infty, \quad f \rightarrow 0, \quad \theta \rightarrow 0. \tag{9.25}$$

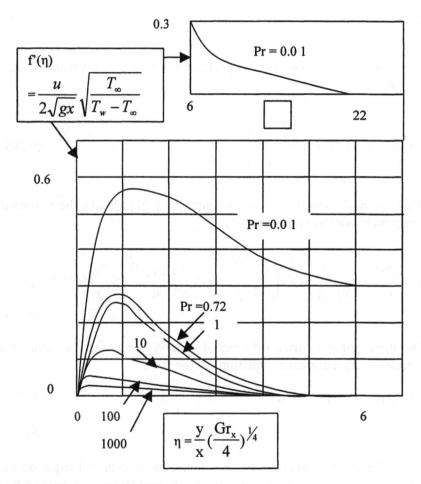

Figure 9.2   Dimensionless velocity distribution for laminar free convection on a vertical flat plate. Ostrach, 1953 [3].

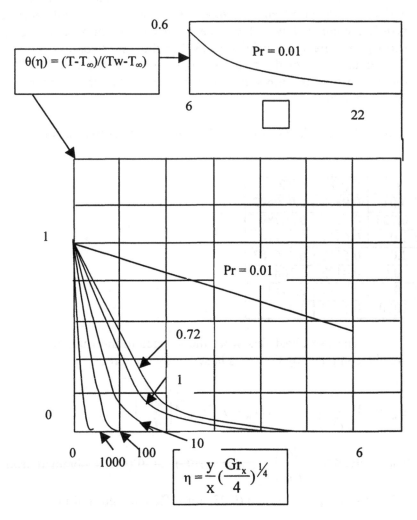

Figure 9.3 Dimensionless temperature distribution for laminar free convection on a vertical flat plate. (Ostrach, 1953 [3].)

Pohlhausen, 1911 [1] solved these equations first, whereas Schmidt and Beckmann, 1930 [2] solved them for Pr = 0.733 in 1930. Ostrach, 1953 [3], solved the same equations for the range 0.01 to 1000. For free convection laminar boundary layer on a heated vertical plate in that range of Pr, the velocity and the temperature distributions are shown in Figs. 9.2 and 9.3, respectively.

The values of θ'(0) and f''(0) are obtained from the solutions, and these are provided in Table 9.1. The velocity and temperature profiles, compared with the experiments of Schmidt and Beckmann, show good agreement at Pr = 0.733. For the dimensionless velocity, the maximum values of the distributions increase with decrease in Pr. For the dimensionless temperature, at any η the value of θ increases with a decrease in Pr.

Table 9.1   Calculated values of f''(0) and θ'(0) at different values of Pr

Pr	f''(0)	Θ'(0)
0.01	0.9862	0.080592
0.733	0.6741	0.50789
1.0	0.6421	0.56714
2.0	0.5713	0.716483
10.0	0.4192	1.168
100	0.2517	2.1914
1000	0.1450	3.97

The local heat flux from the surface to the fluid at any x value may be computed by Fourier's heat conduction law,

$$q_w = -k\left(\frac{\partial T}{\partial y}\right)_{y=0} = -k(T_w - T_\infty)Cx^{-\frac{1}{4}}\left(\frac{d\theta}{d\eta}\right)_{\eta=0} \tag{9.26}$$

The derivative $\left(\dfrac{d\theta}{d\eta}\right)_{\eta=0}$, also written as θ'(0), is obtained from the solutions of Eqs. (9.22) and (9.23) for different values of Pr.

For heat transfer considerations, the local heat transfer coefficient and the local Nusselt number are written in the usual way as

$$h_x = \frac{q_w}{T_w - T_\infty}, \quad Nu_x = \frac{h_x x}{k}. \tag{9.27}$$

From Eq. (9.26), we obtain

$$Nu_x = \frac{\theta'(0)Gr_x^{\frac{1}{4}}}{\sqrt{2}} \tag{9.28}$$

where Gr_x is the local Grashof number. This equation is suitable for both $T_w > T_\infty$ and $T_w < T_\infty$. For an ideal gas, the local Grashof number may be written as

$$Gr_x = \frac{g(T_w - T_\infty)x^3}{T_\infty \upsilon^2}. \tag{9.29}$$

By convention, only positive dimensionless numbers are used. This means that in Eq. (9.29), it is the modulus of $(T_w - T_\infty)$ that is used. A large value for Gr implies that the effects in the momentum equation are not very large.

For a vertical surface, Ostrach's computations are approximated by

$$\frac{Nu_x}{\left(\dfrac{Gr_L}{4}\right)^{\frac{1}{4}}} = \frac{0.676\,Pr^{\frac{1}{2}}}{(0.861 + Pr)^{\frac{1}{4}}}. \tag{9.30}$$

As $h \sim x^{0.25}$, the average heat transfer coefficient from 0 to L is given by $h = 4h_L/3$. The average Nusselt number is then

$$\frac{\overline{Nu}}{\left(\dfrac{Gr_L}{4}\right)^{\frac{1}{4}}} = \frac{0.902\,Pr^{\frac{1}{2}}}{(0.861 + Pr)^{\frac{1}{4}}}. \tag{9.31}$$

A Nusselt number relation that is used in practice is that given in McAdams, 1954 [4]. It is a semiempirical equation relating the average (over the length L) Nusselt number to Pr and Gr,

$$Nu = 0.548(Gr_L \text{ Pr})^{\frac{1}{4}}, \ 10 \le Gr_L \le 10^9. \tag{9.32}$$

For air, the above equation can be used. For oils, the constant should be replaced by 0.555 [3]; for mercury, the constant should be about 0.33 [5]. The results of Eq. (9.31) and those of Eq. (9.32) agree only for a limited range of Pr.

The asymptotic forms of Eqs. (10.22) and (10.23) were solved for very small and infinite values of Pr by Le Fevre, 1956 [6]. He developed the following correlation for the mean Nu which agrees well with the exact results of [3]:

$$Nu = \left(\frac{Gr_L \text{ Pr}^2}{2.43478 + 4.884 \text{ Pr}^{0.5} + 4.95283 \text{ Pr}} \right)^{\frac{1}{4}}. \tag{9.33}$$

The above discussion is for laminar flows. In practice, the transition from laminar to turbulent flow in free convection typically occurs when the Rayleigh number Ra = GrPr = 10^9.

9.3   Similarity Solution: Boundary Layer with Uniform Heat Flux

Sparrow and Gregg, 1958 [7], obtained a similarity solution for a vertical plate with uniform heat flux boundary condition. The range of Pr investigated was from 0.1 to 100.

Equations (9.10) and (9.3) can be transformed to ordinary differential equations by the following similarity parameter:

$$\eta = D_1 y x^{-\frac{1}{5}} \quad \text{where} \quad D_1 = \left(\frac{g\beta q_w}{5k\upsilon^2} \right)^{\frac{1}{5}}, \ q_w = \text{heat flux at the wall.} \tag{9.34}$$

The velocity components are represented in terms of a stream function defined as

$$\psi(x,y) = D_2 x^{\frac{4}{5}} f(\eta) \quad \text{where} \quad D_2 = \left(\frac{5^4 \, g\beta q_w \upsilon^3}{k}\right)^{\frac{1}{5}}. \tag{9.35}$$

The temperature is represented by

$$\theta(\eta) = \frac{D_1(T_\infty - T)}{x^{\frac{1}{5}} q_w / k}. \tag{9.36}$$

The velocity components can be derived from

$$u = \frac{\partial \psi}{\partial y} \quad \text{and} \quad v = -\frac{\partial \psi}{\partial x} \tag{9.37}$$

and Eq. (9.35) to give

$$u = D_1 D_2 x^{\frac{3}{5}} f'(\eta), \quad v = \frac{D_2}{5x^{\frac{1}{5}}} [\eta f'(\eta) - 4 f(\eta)]. \tag{9.38}$$

The three original partial differential equations may then be reduced to two ordinary differential equations of the form

$$f''' - 3(f')^2 + 4ff'' - \theta = 0 \tag{9.39}$$

$$\theta'' + \Pr(4f\theta' - \theta f') = 0. \tag{9.40}$$

The transformed boundary conditions required to sove the momentum and energy equations, are as follows :

$$\eta = 0, \quad f = f' = 0, \quad \theta' = 1 \tag{9.41}$$
$$\eta \to \infty, \quad f' \to 0, \quad \theta \to 0. \tag{9.42}$$

Numerical solutions to the above problem have been obtained.

Evaluating Eqs. (9.34)-(9.36) at the surface, i.e., $\eta = 0$, the surface temperature is represented by

$$T_w - T_\infty = -5^{\frac{1}{5}}\theta(0)\frac{q_w x}{k}\left(\frac{\beta g q_w x^4}{v^2 k}\right)^{-\frac{1}{5}} = -\theta(0)\left(\frac{k^4 g\beta}{5v^2 q_w^4 x}\right)^{-\frac{1}{5}}. \quad (9.43)$$

Written in terms of the modified Grashof number Gr*, Eq. (9.43) becomes

$$\frac{T_w - T_\infty}{\dfrac{q_w x}{k}}Gr_x^{*\,\frac{1}{5}} = -5^{\frac{1}{5}}\theta(0) \qquad \text{or} \qquad \frac{Nu_x}{Gr_x^{*\,\frac{1}{5}}} = -\frac{1}{5^{\frac{1}{5}}\theta(0)} \qquad (9.44)$$

where

$$Gr_x^* = \frac{\beta g q_w x^4}{v^2 k}, \quad Nu_x = \frac{hx}{k} = \frac{q_w x}{k(T_w - T_\infty)}. \quad (9.45)$$

The solution of Eqs. (9.39) and (9.40) subjected to the boundary conditions, Eqs. (9.41) and (9.42) give rise to values of $\theta(0)$ and f'(0), which are listed in Table 9.2.

Table 9.2   Values of f''0) and $\theta(0)$, Sparrow and Gregg [7].

Pr	f'(0)	$\Theta(0)$
0.1	1.6434	-2.7507
1	0.72196	-1.3574
10	0.30639	-0.76746
100	0.12620	-0.46566

There is no obvious characteristic temperature difference in the problem. In the literature, the average Nusselt numbers are often defined using the average plate temperature minus the environmental temperature, that is,

$$\overline{T_w - T_\infty} = \frac{1}{L}\int_0^L (T_w - T_\infty)dx \qquad (9.46)$$

The left-hand side of Eq. (9.46) may be obtained from the expression in Eq. (9.43). Hence, the average Nu based on this temperature difference is

$$Nu = \frac{hL}{k} = -\frac{6}{5^{\frac{6}{5}}\theta(0)} Gr_L^{*\frac{1}{5}}. \tag{9.47}$$

Since

$$Gr_L^* = \left(\frac{g\beta q_w L^4}{k\upsilon^2}\right) = \left\{\frac{g\beta \overline{(T_w - T_\infty)}L^3}{\upsilon^2}\right\}\frac{hL}{k}, \tag{9.48}$$

then

$$\frac{Nu}{Gr_L^{\frac{1}{4}}} = \left\{-\frac{6}{5^{\frac{6}{5}}\theta(0)}\right\}^{\frac{5}{4}}. \tag{9.49}$$

The values of $Nu/Gr_L^{1/4}$ for a flat plate with uniform surface temperature have been computed from the results of Ostrach [3] and are tabled with the values from Eq. (9.49) in Table 9.3. The average Nu have been defined using the temperature difference midway along the plate.

Table 9.3   Average Nu for uniform wall heat flux and constant wall temperature, Ostrach [3].

Pr	$Nu/Gr_L^{\frac{1}{4}}$ [3]	$Nu/Gr_L^{\frac{1}{4}}$ (Eq.9.49)
0.1	0.219	0.237
1.0	0.535	0.573
10	1.10	1.17
100	2.07	2.18

9.4   Integral Method of Solution

The momentum integral equation for natural convection is

$$\frac{d}{dx}\left(\int_0^\delta \rho u^2\, dy\right) = -\tau_w + \int_0^\delta \rho g \beta (T - T_\infty)\, dy. \qquad (9.50)$$

Rate of change Shear   Bouyancy force
of momentum   stress   per unit element

Replacing the shear stress at the wall by the velocity gradient,

$$\frac{d}{dx}\left(\int_0^\delta \rho u^2\, dy\right) = -\mu \frac{\partial u}{\partial y}\bigg|_{y=0} + \int_0^\delta \rho g \beta (T - T_\infty)\, dy. \qquad (9.51)$$

The thermal boundary conditions are

$T = T_w$ at $y = 0$ (9.52a)

$T = T_\infty$ at $y = \delta$ (9.52b)

$\dfrac{\partial T}{\partial y} = 0$ at $y = \delta$ (9.52c)

It is assumed that the temperature, T, takes the form of a second-degree polynomial. In other words,

$T = a + by + cy^2.$ (9.53)

Then, the dimensionless temperature, θ, is given by

$$\theta = \frac{T - T_\infty}{T_w - T_\infty} = \left(1 - \frac{y}{\delta}\right)^2. \qquad (9.54)$$

The velocity boundary conditions are

$u = 0$ at $y = 0$ (9.55a)
$u = 0$ at $y = \delta$ (9.55b)

$$\frac{\partial u}{\partial y} = 0 \text{ at } y = \delta. \tag{9.55c}$$

From the momentum equation,

$$\frac{\partial^2 u}{\partial y^2} = -g\beta \frac{T_w - T_\infty}{v} \quad \text{at} \quad y = 0. \tag{9.56}$$

If u_x is an undefined fictitious velocity, we can express the dimensionless velocity as

$$U = \frac{u}{u_x} = a_1 + b_1 y + c_1 y^2 + d_1 y^3. \tag{9.57}$$

Then, $\frac{u}{u_x} = \frac{y}{\delta}\left(1 - \frac{y}{\delta}\right)^2$ where $u_x \sim \delta^2$. \hfill (9.58)

Hence, $\dfrac{1}{105}\dfrac{d}{dx}\left(u_x^2 \delta\right) = \dfrac{1}{3}g\beta(T_w - T_\infty)\delta - v\dfrac{u_x}{\delta}.$ \hfill (9.59)

The energy integral equation is

$$\frac{d}{dx}[\int_0^\delta u(T - T_\infty)dy] = -\alpha\frac{dT}{dy}\Big|_{y=0} \tag{9.60}$$

from which, we obtain

$$\frac{1}{30}(T_w - T_\infty)\frac{d}{dx}(u_x \delta) = 2\alpha\frac{(T_w - T_\infty)}{\delta}. \tag{9.61}$$

From Eq. (9.59),

$$\frac{d}{dx}(\delta^4 . \delta) \sim \frac{1}{3}g\beta(T_w - T_\infty)\delta - v.\delta. \tag{9.62}$$

Since the right-hand side of Eq. (9.62) is independent of x, the left-hand side has to be independent of x. For this to happen,

$$\delta^4 \sim x \quad \text{or} \quad \delta \sim x^{0.25} \tag{9.63}$$

It is then assumed that $u_x = C_1 x^{\frac{1}{2}}, \delta = C_2 x^{\frac{1}{4}}$. $\tag{9.64}$

The constants C_1 and C_2 can be found from Eqs. (9.59) and (9.61). Solving,

$$\frac{\delta}{x} = 3.93 \, Pr^{-\frac{1}{2}} (0.952 + Pr)^{\frac{1}{4}} \, Gr_x^{-\frac{1}{4}} \tag{9.65}$$

where $Gr_x = \dfrac{g\beta(T_w - T_\infty)x^3}{\upsilon^2}$.

     In considering the energy heat transfer, the heat transfer coefficient may be evaluated from

$$q_w = -kA\frac{dT}{dy}\bigg|_w = hA(T_w - T_\infty) \tag{9.66}$$

It can be shown that $h = 2k/\delta$, or that $Nu_x = hx/k = 2x/\delta$. Hence, the Nusselt number can be expressed as

$$Nu_x = 0.508 \, Pr^{\frac{1}{2}} (0.952 + Pr)^{-\frac{1}{4}} \, Gr_x^{\frac{1}{4}} \tag{9.67}$$

For the vertical plate, the average heat transfer coefficient can be found from the relation

$$\bar{h} = \frac{1}{L} \int_0^L h_x \, dx = \frac{4}{3} h_{x=L}. \tag{9.68}$$

Equation (9.67) agrees well with Eq. (9.30). Equation (9.67) can also be expressed as

$$Nu_x = 0.508 Ra_x^{\frac{1}{4}} \left(\frac{Pr}{0.952 + Pr} \right)^{\frac{1}{4}} \qquad (9.69)$$

where Ra_x is the local Rayleigh number,

$$Ra_x = \frac{g\beta(T_x - T_\infty)x^3}{\alpha \upsilon}. \qquad (9.70)$$

Hence, the average Nusselt number for x = 0 to x = L along the flat plate

is $\overline{Nu_L} = 0.68 Ra_L^{\frac{1}{4}} \left(\dfrac{Pr}{0.952 + Pr} \right)^{\frac{1}{4}}$ \qquad (9.71)

Comparison of the results of the integral method with experimental results and exact solutions show that the prediction of the heat transfer coefficient with the integral method is satisfactory.

If Pr is approximately equal to one, Eq. (9.71) gives $\overline{Nu_L} \approx 0.57 Ra_L^{\frac{1}{4}}$.

If Pr $\rightarrow \infty$, Eq. (9.71) gives $\overline{Nu_L} \approx 0.68 Ra_L^{\frac{1}{4}}$. If Pr $\rightarrow 0$, $\overline{Nu_L} \approx 0.688 Ra_L^{\frac{1}{4}} Pr^{\frac{1}{4}}$.

It can be seen that the expression for the average Nusselt number for Pr ~ 1 is closer in form to the case where Pr $\rightarrow \infty$, than the case where Pr $\rightarrow 0$. The reason for this is that in natural convection, the driving force is caused by the temperature gradients, and thus defined by the thermal boundary layer. When Pr ~ 1 and when Pr $\rightarrow \infty$, the thermal boundary layer is thicker than the velocity boundary layer. Hence, the behavior of the Nusselt number would be similar in form for both cases. When Pr $\rightarrow 0$, the behavior of the kinematic viscosity relative to the thermal diffusivity is going to be different from that of the other two cases. In addition, the right-hand side of the expression for Pr $\rightarrow 0$ is independent of υ, as one would expect for this case where the effects of the kinematic viscosity are very small or negligible.

Example 9.1

Problem: . Derive an expression for the maximum velocity in the free convection boundary layer on a vertical flat plate. At what position in the boundary layer does this maximum velocity occur?

Solution
The velocity profile given by Eq. (9.58) is assumed. The profile satisfies the velocity boundary conditions. Hence,

$$\frac{u}{u_x} = \frac{y}{\delta}\left(1 - \frac{y}{\delta}\right)^2 .$$

Taking the derivative with respect to y,

$$\frac{du}{dy} = u_x[\frac{y}{\delta}(-\frac{2}{\delta})(1-\frac{y}{\delta}) + \frac{1}{\delta}(1-\frac{y}{\delta})^2]$$

$$-\frac{2}{\delta^2}(y-\frac{y^2}{\delta}) + \frac{1}{\delta}(1-\frac{2y}{\delta}+\frac{y^2}{\delta^2}) = 0 .$$

Hence, $\qquad \dfrac{3y^2}{\delta^3} - \dfrac{4y}{\delta^2} + \dfrac{1}{\delta} = 0 .$

From which we get $y = \delta$ or $\dfrac{\delta}{3}$.

Since u = 0 at y = δ, the maximum velocity occurs at y = δ/3. This maximum velocity is

$$u_{max} = \frac{4}{27}u_x ,$$ which may be expressed as $\dfrac{4}{27}C_1 x^m$.

PROBLEMS

9.1.   Most of the correlations for the average Nusselt number used in free convection are expressed in the form

$$Nu = D(Gr.Pr)^m .$$

By choosing the appropriate average property values in these correlations, demonstrate that the average heat transfer coefficient may be expressed as

$$h = D' \left(\frac{\Delta T}{L} \right)^m, \text{ where D and } D' \text{ are constants.}$$

9.2.   Consider Eq. (9.31), which is an expression for the average Nusselt number in terms of Gr and Pr. Obtain expressions for this average Nu for (a) Pr >>1 and (b) Pr <<1.

9.3.   Consider Eq. (9.71), which is an expression for the average Nusselt number in terms of Ra and Pr. Obtain expressions for this average Nu for (a) Pr → ∞ and (b) Pr→0.

9.4.   If a flat plate is inclined with an angle β from the body force direction, show that the Nusselt number for free convection on this inclined plate is a function of Pr and Grcosβ.

9.5.   Consider Eq. (9.31), which is an expression for the average Nusselt number in terms of Gr and Pr. Obtain an expression for this average Nu for Pr ~ 1.

9.6.   Consider Eq. (9.71), which is an expression for the average Nusselt number in terms of Ra and Pr. Obtain an expression for this average Nu for Pr ~ 1.

9.7.   Show that the solution to Prob. 9.5 is approximately the same as that to Prob. 9.6.

9.8.   Consider Eq. (9.31), which is an expression for the average Nusselt number in terms of Gr and Pr. When Pr ~ 1, is the expression more similar to the case where Pr >>1 or to that for the case where Pr << 1? Explain.

9.9.   An empirical equation proposed by Heilman for the coefficient of heat transfer in free convection from a long horizontal cylinder to air is

$$h = \frac{1.016(T_s - T_\infty)^{0.266}}{D^{0.2}T_f^{0.181}}.$$

The corresponding equation in dimensionless form is

$$\frac{hD}{k_f} = cGr_f^m \, Pr_f^n.$$

Determine the values of the indices m and n in the dimensionless form which corresponds to Heilman's equation.

REFERENCES

1. E Pohlhausen. Der Warmeaustausch Zwischen Festen Korpen und Flussigkeiten mit kleiner Reiburg and kleiner warmeleitung, Z Angew Math Mech, Vol 1: 115, 1911.

2. E Schmidt and W Beckmann. Das Temperatur-und Geschwindigkeitsfeld Von einer Warme Abgebenden Senkrechten Platte bei naturlicher Konvektion. Forsch-Ing-Wes. Vol 1: 391, 1930.

3. S Ostrach. An Analysis of Laminar Free-Convection Flow and Heat Transfer About a Flat Plate Parallel to the Direction of the Generating Body Force. NACA Report 1111, 1953.

4. W H McAdams. Heat Transmission, 3rd ed. NewYork: McGraw-Hill, 1954.

5. O A Saunders. Natural Convection in Liquids. Proc Roy Soc London A, 172: 55-71, 1939.

6. E J Le Fevre. Laminar Free Convection from a Vertical Plane Surface. Proc 9th Int Con Appl Mech, Brussels, Belgium, Vol 4:168, 1956.

7. E M Sparrow and J L Gregg. Laminar Free Convection from a Vertical Flat Plate with Uniform Surface Het Flux. Trans ASME, Vol 78: 435-440, 1956.

Natural Convection

When a fluid moves under natural convection
Because of body forces caused by density variations
This phenomenon results from the earth's field of gravitation
Body forces may also be caused by surface tension.

Velocity and temperature found by similarity method
Also solved by use of the integral method
Then heat transfer coefficient found in Nusselt number
Correlation between Nusselt number and Grashof number.

K.V. Wong

10

Numerical Analysis in Convection

10.1 Introduction

In problems of heat convection, the most complex equations to solve are the fluid flow equations. Often times, the governing equations for the fluid flow are the Navier-Stokes equations. It is useful, therefore, to study a model equation that has similar characteristics to the Navier-Stokes equations. This model equation has to be time-dependent and include both convection and diffusion terms. The viscous Burgers equation is an appropriate model equation. In the first few sections of this chapter, several important numerical schemes for the Burgers equation will be discussed. A simple physical heat convection problem is solved as a demonstration.

After the Burgers equation, the numerical analysis of the incompressible boundary layer equations for convection heat transfer are discussed. A few important numerical schemes are discussed. The classic solution for flow in a laminar boundary layer is then presented in the example.

In the next section, incompressible flow with constant properties and no body forces is discussed. Under such conditions, the governing momentum equations are decoupled from the governing energy equation. Once the flow field is known, different temperature distributions may be computed with different types of thermal boundary conditions.

In the last section, convection in a two-dimensional porous medium is presented as a physical problem. Porous media is important in environmental heat transfer studies, transpiration cooling, and fuel cells, as some examples. Using the slug flow assumption, the energy equation is solved using an alternating implicit method to show its effectiveness.

There is no attempt to be exhaustive in the discussion of numerical analysis for convective heat transfer in this chapter. The aim is to help the reader appreciate the significance of this method of analysis for the study of heat convection in this modern age of computers.

10.2 Burgers Equation

In the general form, the viscous Burgers equation may be written as

$$\frac{\partial u}{\partial t} + (\alpha + \beta u)\frac{\partial u}{\partial x} = \upsilon\frac{\partial^2 u}{\partial x^2} \tag{10.1}$$

where α and β are prescribed parameters; α and v are assumed constant. When $\alpha = c =$ speed of sound and $\beta = 0$, the linear Burgers equation is obtained:

$$\frac{\partial u}{\partial t} + c\frac{\partial u}{\partial x} = \upsilon\frac{\partial^2 u}{\partial x^2} \tag{10.2}$$

When $\alpha = 0$ and $\beta = 1$, the standard nonlinear Burgers equation results:

$$\frac{\partial u}{\partial t} + u\frac{\partial u}{\partial x} = \upsilon\frac{\partial^2 u}{\partial x^2}. \tag{10.3}$$

Putting $E = \frac{1}{2}u^2$,

$$\frac{\partial u}{\partial t} + \frac{\partial E}{\partial x} = \upsilon\frac{\partial^2 u}{\partial x^2}. \tag{10.4}$$

If $F = \frac{1}{2}v^2$, $G = \frac{1}{2}w^2$, a multidimensional form of this equation may be expressed as

$$\frac{\partial u}{\partial t} + \frac{\partial E}{\partial x} + \frac{\partial F}{\partial y} + \frac{\partial G}{\partial z} = \upsilon\left(\frac{\partial^2 u}{\partial x^2} + \frac{\partial^2 u}{\partial y^2} + \frac{\partial^2 u}{\partial z^2}\right). \tag{10.5}$$

10.3 Convection Equations

The convection equations include conservation of mass, momentum and energy. The dimensionless form of the equations are listed below.

Continuity

$$\frac{\partial \rho}{\partial t} + \frac{\partial}{\partial x}(\rho u) + \frac{\partial}{\partial y}(\rho v) + \frac{\partial}{\partial z}(\rho w) = 0 \tag{10.6}$$

X-Momentum Equation

$$\frac{\partial}{\partial t}(\rho u) + \frac{\partial}{\partial x}(\rho u^2 + p) + \frac{\partial}{\partial y}(\rho u v) + \frac{\partial}{\partial z}(\rho u w) =$$

$$\frac{\partial}{\partial x}(\tau_{xx}) + \frac{\partial}{\partial y}(\tau_{xy}) + \frac{\partial}{\partial z}(\tau_{xz}) \tag{10.7}$$

Y-Momentum Equation

$$\frac{\partial}{\partial t}(\rho v) + \frac{\partial}{\partial x}(\rho u v) + \frac{\partial}{\partial y}(\rho v^2 + p) + \frac{\partial}{\partial z}(\rho v w) =$$

$$\frac{\partial}{\partial x}(\tau_{xy}) + \frac{\partial}{\partial y}(\tau_{yy}) + \frac{\partial}{\partial z}(\tau_{yz}) \tag{10.8}$$

Z-Momentum Equation

$$\frac{\partial}{\partial t}(\rho w) + \frac{\partial}{\partial x}(\rho u w) + \frac{\partial}{\partial y}(\rho v w) + \frac{\partial}{\partial z}(\rho w^2 + p) =$$

$$\frac{\partial}{\partial x}(\tau_{xz}) + \frac{\partial}{\partial y}(\tau_{yz}) + \frac{\partial}{\partial z}(\tau_{zz}) \tag{10.9}$$

Energy Equation

$$\frac{\partial}{\partial t}(\rho e_t)+\frac{\partial}{\partial x}(\rho u e_t + pu)+\frac{\partial}{\partial y}(\rho v e_t + pv)+\frac{\partial}{\partial z}(\rho w e_t + pw)=$$

$$\frac{\partial}{\partial x}(u\tau_{xx}+v\tau_{xy}+w\tau_{xz}-q_x)+\frac{\partial}{\partial y}(u\tau_{yx}+v\tau_{yy}+w\tau_{yz}-q_y)+$$

$$\frac{\partial}{\partial z}(u_{zx}+v\tau_{zy}+w\tau_{zz}-q_z) \tag{10.10}$$

These convection equations may be written in vector form as

$$\frac{\partial Q}{\partial t}+\frac{\partial E}{\partial x}+\frac{\partial F}{\partial y}+\frac{\partial G}{\partial z}=\frac{\partial E_\upsilon}{\partial x}+\frac{\partial F_\upsilon}{\partial y}+\frac{\partial G_\upsilon}{\partial z} \tag{10.11}$$

where

$$Q=\begin{bmatrix}\rho\\\rho u\\\rho v\\\rho w\\\rho e_t\end{bmatrix}$$

$$E=\begin{bmatrix}\rho u\\\rho u^2+p\\\rho uv\\\rho uW\\(\rho e_t+p)u\end{bmatrix}\qquad E_\upsilon=\begin{bmatrix}0\\\tau_{xx}\\\tau_{xy}\\\tau_{xz}\\u\tau_{xx}+v\tau_{xy}+w\tau_{xz}-q_x\end{bmatrix}$$

$$F = \begin{bmatrix} \rho v \\ \rho v u \\ \rho v^2 + p \\ \rho v w \\ (\rho e_t + p)v \end{bmatrix} \qquad F_v = \begin{bmatrix} 0 \\ \tau_{yx} \\ \tau_{yy} \\ \tau_{yz} \\ u\tau_{yx} + v\tau_{yy} + w\tau_{yz} - q_y \end{bmatrix}$$

$$G = \begin{bmatrix} \rho w \\ \rho w u \\ \rho w v \\ \rho w^2 + p \\ (\rho e_t + p)w \end{bmatrix} \qquad G_v = \begin{bmatrix} 0 \\ \tau_{zx} \\ \tau_{zy} \\ \tau_{zz} \\ u\tau_{zx} + v\tau_{zy} + w\tau_{zz} - q_z \end{bmatrix}.$$

The convection equations represented by Eq. (10.11) are well modeled by Eq. (10.5). If the right-hand side of Eq. (10.11) is set to zero, the Euler equation is obtained,

$$\frac{\partial Q}{\partial t} + \frac{\partial E}{\partial x} + \frac{\partial F}{\partial y} + \frac{\partial G}{\partial z} = 0 \qquad\qquad (10.12)$$

This vector equation is a hyperbolic-type equation. In general, the Navier-Stokes equations are a mixed hyperbolic (in inviscid region), parabolic (in viscous region) equation. Equations (10.11) and (10.12) are solved by marching in time. By assuming no time dependence, the Navier-Stokes equations are a mixed hyperbolic (in inviscid region), elliptic (in viscous region) equation. In the time-independent equations, solving involves integration with respect to space coordinates.

10.4 Numerical Algorithms

In this section, a few selected numerical schemes for the solution of the model scalar equation (10.2) is investigated.

10.4.1   Forward Time Central Space (FTCS) Explicit Scheme

In this explicit scheme, the first-order forward difference approximation is used for the time derivative. The second-order central difference approximation is used for the spatial derivatives. Hence, the finite difference equation (FDE) of the partial differential equation (PDE) Eq. (10.2) is

$$\frac{u_i^{n+1} - u_i^n}{\Delta t} + c\frac{u_{i+1}^n - u_{i-1}^n}{2\Delta x} = \upsilon\frac{u_{i+1}^n - 2u_i^n + u_{i-1}^n}{(\Delta x)^2}. \tag{10.13}$$

This FDE has a truncation error of the order of $[(\Delta t),(\Delta x)^2]$.

10.4.2   Forward Time Backward Central Space (FTBCS) Explicit Scheme

In this explicit scheme, the first-order forward difference approximation is used for the time derivative. The second-order central difference approximation is used for the spatial derivatives. When a first-order backward difference approximation (c > 0) for the convective term is used, then the FDE of the PDE Eq. (10.2) is

$$\frac{u_i^{n+1} - u_i^n}{\Delta t} + c\frac{u_i^n - u_{i-1}^n}{\Delta x} = \upsilon\frac{u_{i+1}^n - 2u_i^n + u_{i-1}^n}{(\Delta x)^2}. \tag{10.14}$$

This first-order approximation of the convective term may introduce too much dissipation error so that it is of the same order of magnitude as the viscosity. Then an accurate solution is not obtained. An alternative is to use a third-order scheme resulting in the following FDE for Eq. (10.2):

$$\frac{u_i^{n+1} - u_i^n}{\Delta t} + c\left(\frac{u_{i+1}^n - u_{i-1}^n}{2\Delta x} - \frac{u_{i+1}^n - 3u_i^n + 3u_{i-1}^n - u_{i-2}^n}{6\Delta x}\right) = \upsilon\frac{u_{i+1}^n - 2u_i^n + u_{i-1}^n}{(\Delta x)^2}.$$
$$\tag{10.15}$$

10.4.3   DuFort-Frankel Explicit

All derivatives are represented by second-order central difference approximations. The FDE is

$$\frac{u_i^{n+1} - u_i^{n-1}}{2\Delta t} + c\frac{u_{i+1}^n - u_{i-1}^n}{2\Delta x} = \upsilon\frac{u_{i+1}^n - \left(u_i^{n-1} + u_i^{n+1}\right) + u_{i-1}^n}{(\Delta x)^2}. \qquad (10.16)$$

This FDE has a truncation error of the order of $[(\Delta t)^2, (\Delta x)^2, \left(\frac{\Delta t}{\Delta x}\right)^2]$.

The equation may be rearranged as

$$u_i^{n+1} = \left(\frac{1-2d}{1+2d}\right)u_i^{n-1} + \left(\frac{a+2d}{1+2d}\right)u_{i-1}^n - \left(\frac{a-2d}{1+2d}\right)u_{i+1}^n. \qquad (10.17)$$

The stability criterion of the scheme is that a $\leq$ 1.

10.4.4 MacCormack Explicit

This predictor-corrector scheme or double-step scheme is done in two steps:

$$u_i' = u_i^n - c\frac{\Delta t}{\Delta x}\left(u_{i+1}^n - u_i^n\right) + \upsilon\frac{\Delta t}{(\Delta x)^2}\left(u_{i+1}^n - 2u_i^n + u_{i-1}^n\right) \qquad (10.18)$$

$$u_i^{n+1} = \frac{1}{2}\left[u_i^n + u_i' - c\frac{\Delta t}{\Delta x}\left(u_i' - u_{i-1}'\right) + \upsilon\frac{\Delta t}{(\Delta x)^2}\left(u_{i+1}' - 2u_i' + u_{i-1}'\right)\right].$$
$$(10.19)$$

Equations (10.18) and (10.19) may be written in incremental form as follows:

$$\Delta u_i^n = -c\frac{\Delta t}{\Delta x}\left(u_{i+1}^n - u_i^n\right) + \upsilon\frac{\Delta t}{(\Delta x)^2}\left(u_{i+1}^n - 2u_i^n + u_{i-1}^n\right)$$

$$u_i' = u_i^n + \Delta u_i^n \qquad (10.20)$$

$$\Delta u_i' = \left[-c\frac{\Delta t}{\Delta x}\left(u_i' - u_{i-1}'\right) + \upsilon\frac{\Delta t}{(\Delta x)^2}\left(u_{i+1}' - 2u_i' + u_{i-1}'\right) \right]$$

$$u_i^{n+1} = \frac{1}{2}\left(u_i^n + u_i' + \Delta u_i'\right) \qquad\qquad (10.21)$$

The MacCormack explicit scheme is second-order accurate with the stability criterion of

$$\Delta t \le \frac{1}{\dfrac{c}{\Delta x} + \dfrac{\upsilon}{(\Delta x)^2}}.$$

10.4.5 MacCormack Implicit

As in the explicit method, the formulation of this implicit method is in two steps, as follows:

$$\left(1 + \lambda\frac{\Delta t}{\Delta x}\right)\delta u_i' = \Delta u_i' + \lambda\frac{\Delta t}{\Delta x}\delta u_{i+1}'$$

$$u_i' = u_i^n + \delta u_i' \qquad\qquad (10.22)$$

where Δu_i^n is computed from Eq. (10.20)

$$\left(1 + \lambda\frac{\Delta t}{\Delta x}\right)\delta u_i^{n+1} = \Delta u_i + \lambda\frac{\Delta t}{\Delta x}\delta u_{i-1}^{n+1} \qquad\qquad (10.23)$$

$$u_i^{n+1} = \frac{1}{2}\left(u_i^n + u_i' + \delta u_i^{n+1}\right) \qquad\qquad (10.24)$$

Equation (10.21) provides $\Delta u_i'$. In Eqs. (10.23) and (10.24), the parameter λ is chosen such that

$$\lambda \geq \max\left[\frac{1}{2}\left(|c| + 2\frac{\upsilon}{\Delta x} - \frac{\Delta x}{\Delta t}, 0.0\right)\right].$$

Bidiagonal systems result from Eqs. (10.22) and (10.23). These can be solved efficiently using various routines. The algorithm is unconditionally stable and second-order accurate if the diffusion number

$\upsilon\dfrac{\Delta t}{(\Delta x)^2}$ is bounded for the limiting process as Δx, Δt goes to zero.

10.4.6 Backward Time Central Space (BTCS) Implicit Scheme

This implicit method uses a first-order backward difference approximation for the time derivative and a second-order central difference approximation for the spatial derivatives. The FDE is

$$\frac{u_i^{n+1} - u_i^n}{\Delta t} + c\frac{u_{i+1}^{n+1} - u_{i-1}^{n+1}}{2\Delta x} = \upsilon\frac{u_{i+1}^{n+1} - 2u_i^{n+1} + u_{i-1}^{n+1}}{(\Delta x)^2}.$$

The above can be written as

$$-(0.5a + d)u_{i-1}^{n+1} + (1 + 2d)u_i^{n+1} + (0.5a - d)u_{i+1}^{n+1} = u_i^n \qquad (10.24a)$$

Tridiagonal systems result from Eq. (10.24a). These can be solved efficiently using the Thomas algorithm, as discussed in the last section of this chapter.

Consider the linear Burgers equation:

$$\frac{\partial T}{\partial t} + u\frac{\partial T}{\partial x} = \alpha\frac{\partial^2 T}{\partial x^2} \qquad (10.25)$$

The equation is linear if u is a known function. This may be considered to describe a time-dependent one-dimensional heat convection equation for a problem with a known flow field. For a fluid with constant properties in the temperature range considered, the momentum equation is decoupled from the energy equation. In the following example, the linear Burgers equation will be applied to a simple physical problem where the approximate solution is rather intuitive. The example serves to illustrate one of the numerical schemes presented above.

Example 10.1

Problem: Use the linear Burgers equation for heat convection in a channel where the water is flowing with uniform velocity of 0.1 m/s across the cross-section of the channel (boundary layers are neglected). The water is initially at 25°C throughout. At time t = 0 sec, waste heat is continuously rejected at x = 0 m, and the channel is long such that dT/dx = 0 for x ≥ 1 m. The amount of heat rejected is 6.23 W/m² for t > 0. Using the FTCS explicit scheme, calculate the first 9 time steps, to show the transient temperature distributions.

Solution

The linear Burgers equation is $\dfrac{\partial T}{\partial t} + u \dfrac{\partial T}{\partial x} = \alpha \dfrac{\partial^2 T}{\partial x^2}$.   (i)

Using the FTCS explicit scheme, the FDE of (i) is

$$\frac{T_i^{n+1} - T_i^n}{\Delta t} + U_\infty \frac{T_{i+1}^n - T_{i-1}^n}{2\Delta x} = \alpha \frac{T_{i+1}^n - 2T_i^n + T_{i-1}^n}{(\Delta x)^2}$$ (ii)

Hence, $T_i^{n+1} = T_i^n - \dfrac{U_\infty \Delta t}{2\Delta x}\left(T_{i+1}^n - T_{i-1}^n\right) + \dfrac{\alpha \Delta t}{(\Delta x)^2}\left(T_{i+1}^n - 2T_i^n + T_{i-1}^n\right)$

(iii)

For water (at 25°C), $\alpha \approx 1.5 \times 10\text{-}4 \ \text{m/s}^2$, $k \approx 0.623 \ \text{W/m.°C}$

Choose $\Delta t = 0.1$ sec, $\Delta x = 0.01$ m.

Hence, $\dfrac{U_\infty \Delta t}{2\Delta x} = 0.05$, $\dfrac{\alpha \Delta t}{(\Delta x)^2} = 0.000015$

Eq. (ii) becomes $T_i^{n+1} = T_i^n - 0.5\left(T_{i+1}^n - T_{i-1}^n\right) + 0.000015\left(T_{i+1}^n - 2T_i^n + T_{i-1}^n\right)$

(iv)

At x = 0, the boundary condition is such that $k\dfrac{T_1^n - T_2^n}{\Delta x} = 6.23\dfrac{W}{m^2}$.

Hence, $T_1^n = T_2^n + 0.1°C$

We assume that $T_1^{n+1} = T_2^n + 0.1°C$.

Using an Excel spreadsheet, with a hundred and one points in a row, the following results were obtained for the first nine time-steps.

	X=0.0	0.01	0.02	0.03	0.04	0.05	0.06	0.07
t= 0	25	25	25	25	25	25	25	25
0.01	25.1	25	25	25	25	25	25	25
0.02	25.15	25.05	25	25	25	25	25	25
0.03	25.225	25.125	25.025	25	25	25	25	25
0.04	25.325	25.225	25.0875	25.0125	25	25	25	25
0.05	25.44375	25.34375	25.19376	25.05625	25.00625	25	25	25
0.06	25.56875	25.46875	25.33751	25.15001	25.03438	25.00313	25	25
0.07	25.68437	25.58437	25.49688	25.30157	25.10782	25.02032	25.00156	25
0.08	25.77812	25.67812	25.63827	25.4961	25.24845	25.07345	25.01172	25.00078
0.09	25.84804	25.74804	25.72928	25.69101	25.45978	25.19181	25.04805	25.00664

10.5 Boundary Layer Equations

The boundary layer equations were derived in a previous chapter, or may be deduced from the general convection equations in the early part of this chapter. For two-dimensional, steady flow over a flat plate of an incompressible, constant-property fluid, the continuity, x-momentum and the energy equations are as follows:

$$\frac{\partial u}{\partial x} + \frac{\partial v}{\partial y} = 0 \tag{10.26}$$

$$\rho\left(u\frac{\partial u}{\partial x} + v\frac{\partial u}{\partial y}\right) = -\frac{dp}{dx} + \mu\frac{\partial^2 u}{\partial y^2} \tag{10.27}$$

$$\rho C_P\left(u\frac{\partial T}{\partial x} + v\frac{\partial T}{\partial y}\right) = k\frac{\partial^2 T}{\partial y^2} + \mu\left(\frac{\partial u}{\partial y}\right)^2. \tag{10.28}$$

The pressure gradient term in the x-momentum equation has to be known for the solution of the equation. This may be obtained from Bernoulli's equation, and seen to be equal to $-u_\infty \dfrac{\partial u_\infty}{\partial x}$, where u_∞ is the velocity far away from the plate, and it may be a function of x. Equation (10.27) becomes

$$\rho\left(u\frac{\partial u}{\partial x} + v\frac{\partial u}{\partial y}\right) = -u_\infty \frac{du_\infty}{dx} + \mu\frac{\partial^2 u}{\partial y^2} \tag{10.29}$$

The time-independent equations above are elliptic in nature. In a rectangular grid system, the finite difference forms of Eqs. (10.29) and (10.26) can be written as follows:

$$u_{ij}\frac{\left(u_{i+1,j} - u_{ij}\right)}{\Delta x} + v_{ij}\frac{\left(u_{i,j+1} - u_{i,j-1}\right)}{2\Delta y}$$

$$= u_{i,\infty}\frac{\left(u_{i+1,\infty} - u_{i,\infty}\right)}{\Delta x} + \frac{v}{(\Delta y)^2}\left(u_{i,j+1} - 2u_{ij} + u_{i,j-1}\right) \tag{10.30}$$

$$\frac{v_{i+1,j} - v_{i+1,j-1}}{\Delta y} + \frac{u_{i+1,j} + u_{i+1,j-1} - u_{ij} - u_{i,j-1}}{2\Delta x} = 0. \tag{10.31}$$

The truncation error for Eq. (10.30) is of the order (Δx) plus order $(\Delta y)^2$. It is the same for Eq. (10.31). The explicit method shown by these equations are no longer used extensively because of the restrictive stability constraint. It is shown here for simplicity, and for discussion purposes.

Consider flow over a flat plate. The computation is started by assuming that $u_{ij} = u_\infty$, at the leading edge and $v_{ij} = 0$. The value of v_{ij} is needed in the explicit algorithm to move on to the i+1 level. It is not required to specify the initial values of v_{ij} in the formal mathematical formulation of the partial differential equation. A suitable initial distribution for v_{ij} can be obtained by using the continuity equation to eliminate $\partial u/\partial x$ from the x-momentum equation. For a laminar, incompressible flow, this means that

$$-u\frac{\partial v}{\partial y}+v\frac{\partial u}{\partial y}=u_{\infty}\frac{du_{\infty}}{dx}+v\frac{\partial^2 u}{\partial y^2}. \tag{10.32}$$

So, $$-u\frac{\partial v}{\partial y}+v\frac{\partial u}{\partial y}=-u^2\frac{\partial}{\partial y}\left(\frac{v}{u}\right). \tag{10.33}$$

Hence $$\frac{\partial}{\partial y}\left(\frac{v}{u}\right)=-\frac{1}{u^2}\left(u_{\infty}\frac{du_{\infty}}{dx}+v\frac{\partial^2 u}{\partial y^2}\right). \tag{10.34}$$

Using the boundary condition that $v = 0$ at $y = 0$,

$$v(y)=-u\int_0^y\frac{1}{u^2}\left(u_{\infty}\frac{du_{\infty}}{dx}+v\frac{\partial^2 u}{\partial y^2}\right)dy \tag{10.35}$$

For the flat plate problem, we can assume that at $x = 0$, $u_{ij} = u_{incoming}$ except at the plate, where it is equal to zero. We can also use a numerical calculation of Eq. (10.35) to obtain an estimate of a compatible initial distribution of v_{ij}. In practice, letting v_{ij} be zero everywhere initially also works.

       With the initial values for u_{ij}, Eq. (10.30) may be solved for $u_{i+1,j}$ explicitly, usually by starting from the flat plate and working outward until $u_{i,j+1}/u_{i+1,\infty} = 1- \varepsilon = 0.995$ or some other predetermined value of ε. Because of the asymptotic nature of the boundary layer condition, the location of the outer boundary is found as the solution proceeds. The values of $v_{i+1,j}$ can be computed from Eq. (10.31), starting at the point next to the lower boundary and computing upwards in the positive y direction. The stability criteria for this method are

$$\frac{2v\Delta x}{u_{ij}(\Delta y)^2}\le 1 \text{ and } \frac{\left(v_{ij}\right)^2\Delta x}{u_{ij}v}\le 2. \tag{10.36}$$

The second term in the momentum equation, Eq. (10.30), is principally responsible for the difference between the stability constraints of Eq.

(10.30) and the energy equation. The discussion following also gives an alternative treatment for this term.

Alternative Explicit Method

To have a better control on the stability of the explicit method by monitoring a single criterion, the second term in the x-momentum boundary layer equation can be depicted as

$$v_{ij} \frac{u_{ij} - u_{i,j-1}}{\Delta y} \quad \text{when} \quad v_{ij} > 0$$

and $\quad v_{ij} \dfrac{u_{i,j+1} - u_{ij}}{\Delta y} \quad \text{when} \quad v_{ij} < 0.$ \hfill (10.37)

The stability criterion is then

$$\Delta x \le \frac{1}{\dfrac{2v}{u_{ij}(\Delta y)^2} + \dfrac{|v_{ij}|}{u_{ij}\Delta y}}. \tag{10.38}$$

The truncation error becomes of order $[(\Delta x),(\Delta y)]$.

Example 10.2

Problem: Consider laminar flow of a fluid over a flat plate. Use the explicit method of finite differencing to compute the x-component velocity profile within the boundary layer.

Solution
Equations (10.30) and (10.31) are used for the finite difference scheme. A rectangular grid is suggested, with $\Delta x = 0.01$ and $\Delta y = 0.001$. The solution for the boundary layer velocity distribution is given by Howarth, 1938 [1].

$\eta = y\sqrt{\dfrac{U_\infty}{\upsilon x}}$	$\dfrac{u}{U_\infty}$	$\eta = y\sqrt{\dfrac{U_\infty}{\upsilon x}}$	$\dfrac{u}{U_\infty}$
0	0		
0.2	0.06641	4.2	0.96696
0.4	0.13277	4.4	0.97587
0.6	0.19894	4.6	0.98269
0.8	0.26471	4.8	0.98779
1.0	0.32979	5.0	0.99155
1.2	0.39378	5.2	0.99425
1.4	0.45627	5.4	0.99616
1.6	0.51676	5.6	0.99748
1.8	0.57477	5.8	0.99838
2.0	0.62977	6.0	0.99898
2.2	0.68132	6.2	0.99937
2.4	0.72899	6.4	0.99961
2.6	0.77246	6.6	0.99977
2.8	0.81152	6.8	0.99987
3.0	0.84605	7.0	0.99992
3.2	0.87609	7.2	0.99996
3.4	0.90177	7.4	0.99998
3.6	0.92333	7.6	0.99999
3.8	0.94112	7.8	1.00000
4.0	0.95552	8.0	1.00000

In the appendix, is listed a program in FORTRAN that is a starting point in obtaining the solution shown in Example 10.2 above. The program computes both the x- and y-components of velocity.

10.5.1 Fully Implicit and Crank-Nicholson Methods

For a mesh with a constant rectangular grid, the incompressible laminar boundary layer equations include the momentum equation as

$$\frac{\rho\{\beta u_{i+1,j}+(1-\beta)u_{ij}\}(u_{i+1,j}-u_{ij})}{\Delta x}+\frac{\rho\beta v_{i+1,j}(u_{i+1,j+1}-u_{i+1,j-1})+\rho(1-\beta)v_{ij}(u_{i,j+1}-u_{i,j-1})}{2\Delta y}$$

$$=\frac{\{\rho\beta u_{i+1,\infty}+\rho(1-\beta)u_{i,\infty}\}(u_{i+1,\infty}-u_{i,\infty})}{\Delta x}$$

$$+\frac{1}{(\Delta y)^2}[\beta\mu\{(u_{i+1,j+1}-u_{i+1,j})-(u_{i+1,j}-u_{i+1,j-1})\}$$

$$+(1-\beta)\mu\{(u_{i,j+1}-u_{ij})-(u_{ij}-u_{i,j-1})\}] \qquad (10.39)$$

In Eq. (10.39), β is a weighting factor. When $\beta = 0$, the method is explicit. The truncation error is of the order (Δx) plus order $(\Delta y)^2$. The von Neumann stability constraint, Eq. (10.36), limits severely the step size.

When $\beta = 0.5$, the method is the Crank-Nicholson implicit method. The expansion point should be taken at $(i+1/2,j)$. The truncation error is of the order $(\Delta x)^2$ plus order $(\Delta y)^2$. No stability criterion comes out of the von Neumann analysis, but difficulties can come about if diagonal dominance is not kept for the tridiagonal algorithm.

When $\beta = 1$, the method is the fully implicit method. The expansion point should be taken at $(i+1,j)$. The truncation error is of the order (Δx) plus order $(\Delta y)^2$. No stability criterion comes out of the von Neumann analysis, but difficulties can come about if diagonal dominance is not kept for the tridiagonal algorithm.

It is observed that the above finite difference scheme is implicit if $\beta \geq$ ½. The finite difference equation (10.41) may be used as the continuity equation for both the fully implicit and the explicit methods.

Finite differencing of the energy equation uses the same general procedure employed for the momentum equation. The energy equation with a non-negligible viscous dissipation term, may be written as

$$\rho C_P \{\beta u_{i+1,j} + (1-\beta)u_{ij}\} \frac{T_{i+1,j} - T_{ij}}{\Delta x}$$

$$+ \frac{\rho C_P \beta v_{i+1,j}(T_{i+1,j+1} - T_{i+1,j-1}) + \rho C_P (1-\beta)v_{ij}(T_{i,j+1} - T_{i,j-1})}{2\Delta y}$$

$$= \frac{1}{(\Delta y)^2}[k\beta\{(T_{i+1,j+1} - T_{i+1,j-1}) - (T_{i+1,j} - T_{i+1,j-1})\}$$

$$+ k(1-\beta)\{(T_{i+1,j+1} - T_{i+1,j}) - (T_{ij} - T_{i,j-1})\}]$$

$$+ \mu\beta\left(\frac{u_{i+1,j+1} - u_{i+1,j-1}}{2\Delta y}\right)^2 + \mu(1-\beta)\left(\frac{u_{i,j+1} - u_{i,j-1}}{2\Delta y}\right)^2 \quad (10.40)$$

The truncation error for Eq. (10.40) is the same as those stated for the momentum equation for $\beta = 0$, ½, 1. The fully implicit scheme can be increased to a formal second-order accuracy by representing the streamwise derivatives with three-level (i-1,i,i+1) second-order differences. For any implicit method, the finite difference momentum and energy equations are algebraically nonlinear in the unknowns because of the quantities unknown at the i+1 level in the coefficients. Linearizing procedures can and have been used, but are beyond the scope of this book.

10.6 Convection with Incompressible Flow

In incompressible flow with constant properties and no body forces, the dynamics are independent of the thermodynamics. Once the kinematic flow field is described by the stream function ψ, any number of temperature distributions may be solved with different thermal boundary conditions.

Consider the situation where the velocity field $\vec{V}$ is known. For a fluid with constant properties, the energy equation is given by

$$\rho C_P \left(\bar{u} \frac{\partial \bar{T}}{\partial x} + \bar{v} \frac{\partial \bar{T}}{\partial y} + \bar{w} \frac{\partial \bar{T}}{\partial z} + \frac{\partial \bar{T}}{\partial t} \right) = k \bar{\nabla}^2 \bar{T} + \mu \bar{\phi} \quad (10.41)$$

where all the variables are dimensional. In another mathematical form, Eq. (10.41) may be written as

$$\rho C_P \frac{D\bar{T}}{D\bar{t}} = k \bar{\nabla}^2 \bar{T} + \mu \bar{\phi}. \tag{10.42}$$

We define dimensionless quantities as

$$x = \frac{\bar{x}}{L}, \quad y = \frac{\bar{y}}{L}, \quad z = \frac{\bar{z}}{L}, \quad u = \frac{\bar{u}}{u_\infty}, \quad v = \frac{\bar{v}}{u_\infty}, \quad w = \frac{\bar{w}}{u_\infty} \tag{10.43}$$

$$T = \frac{\bar{T} - T_\infty}{T_1 - T_\infty}, \quad t = \frac{\bar{t}}{L/u_\infty}. \tag{10.44}$$

Hence, the dimensionless form of Eq. (10.42) is

$$\frac{DT}{Dt} = \frac{k}{\rho C_p u_\infty L} \nabla^2 T + \frac{\mu u_\infty}{\rho C_p L (T_1 - T_\infty)} \bar{\phi}. \tag{10.45}$$

From continuity, $\vec{\nabla} \cdot \vec{v} = 0$. $\tag{10.46}$

$$\nabla \cdot \left(\vec{v} T \right) = \vec{v} \cdot \nabla T + T \vec{\nabla} \cdot \vec{v} = \vec{v} \cdot \nabla T \tag{10.47}$$

Hence,

$$\frac{\partial T}{\partial t} = -\vec{v}.(\nabla T) + \frac{1}{Pe}\nabla^2 T + \frac{E}{Re}\bar{\phi} = -\nabla.(\vec{v}T) + \frac{1}{Pe}\nabla^2 T + \frac{E}{Re}\bar{\phi}.$$

$$(10.48)$$

Neglecting viscous dissipation for moderate velocities,

$$\frac{\partial T}{\partial t} = -\nabla.(\vec{v}T) + \frac{1}{Pe}\nabla^2 T.$$

$$(10.49)$$

10.7 Two-Dimensional Convection with Incompressible Flow

In two dimensions, Eq. (10.45) reduces to

$$\frac{\partial T}{\partial t} = -u\frac{\partial T}{\partial x} - v\frac{\partial T}{\partial y} + \frac{1}{Pe}\frac{\partial^2 T}{\partial x^2} + \frac{1}{Pe}\frac{\partial^2 T}{\partial y^2}.$$

$$(10.50)$$

The alternating direction implicit methods, or ADI methods, is a method of variable direction. This method employs a splitting of the time step to obtain a multi-dimensional implicit method which requires only the inversion of a tridiagonal matrix.

The advancement over the time step Δt is accomplished in 2 steps.

Step (1)
$$\frac{T^{n+\frac{1}{2}}-T^n}{\frac{\Delta t}{2}} = -u\frac{\delta T^{n+\frac{1}{2}}}{\delta x} - v\frac{\delta T^n}{\delta y} + \frac{1}{Pe}\frac{\delta^2 T^{n+\frac{1}{2}}}{\delta x^2} + \frac{1}{Pe}\frac{\delta^2 T^n}{\delta y^2}$$

$$(10.51)$$

Step (2)

$$\frac{T^{n+1}-T^{n+\frac{1}{2}}}{\frac{\Delta t}{2}} = -u\frac{\delta T^{n+\frac{1}{2}}}{\delta x} - v\frac{\delta T^{n+1}}{\delta y} + \frac{1}{Pe}\frac{\delta^2 T^{n+\frac{1}{2}}}{\delta x^2} + \frac{1}{Pe}\frac{\delta^2 T^{n+1}}{\delta y^2}$$

$$(10.52)$$

Other x-y permutations different from the above have also been used successfully.

The main advantage of the ADI method is that the stability of this two-dimensional method is unconditional, as is the fully implicit method.   In addition, each of Eqs. (10.51) and (10.52) are only tridiagonal. Equation (10.51) contains implicit unknowns $T_{ij}^{n+\frac{1}{2}}, T_{i\pm1,j}^{n+\frac{1}{2}}$. Equation (10.52) contains implicit unknowns $T_{ij}^{n+1}, T_{ij\pm1}^{n+1}$. This requires a solution of tridiagonal system, which occurs only for usual implicit methods in one dimension, not usually in two dimensions. The linear forms of the equations (10.51) and (10.52) have a truncation error of the order of $[(\Delta t)^2, (\Delta x)^2, (\Delta y)^2]$.

10.8   Convection in a Two-Dimensional Porous Medium

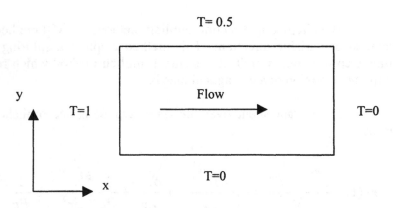

Figure 10.1   Flow through a rectangular porous medium.

Consider a rectangular porous medium shown in Fig. 10.1. The temperatures are prescribed in all four boundaries as shown. The porous medium is a solid through which fluid can flow. The principal flow is from left to right, parallel to the longer side of the rectangle. Dimensionless velocity components can be defined such that $u = u'/u_\infty$ and $v = v'/u_\infty$ where u_∞ is the average velocity in the principal flow direction.   If the dimensional temperature is denoted by T', a dimensionless temperature T can be defined such that

$$T = \frac{T'-T_i}{T_o - T_i} \tag{10.53}$$

where T_i is the initial temperature of the porous medium, and T_o is the temperature of the face at $x = 0$ for all time $t' > 0$. Defining dimensionless space coordinates as $x = x'/L$ and $y = y'/L$, where x' and y' are the dimensional coordinates and L is a reference length, and the dimensionless time $t = t'/t_{ref}$ where t_{ref} is a reference time quantity, the governing energy equation for convection in the two-dimensional porous medium shown in the figure is

$$\frac{\partial T}{\partial t} = -(u\frac{\partial T}{\partial x} + v\frac{\partial T}{\partial y}) + \frac{1}{Pe}\frac{\partial^2 T}{\partial x^2} + \frac{1}{Pe}\frac{\partial^2 T}{\partial y^2}. \tag{10.54}$$

Equation (10.54) is a parabolic-type equation. The Peclet number is defined as

$$Pe = \frac{\rho_e C_e u_\infty L}{k_e} \tag{10.55}$$

where ρ_e is the effective density of the porous medium, C_e is effective specific heat at constant pressure of the fluid and k_e is the effective thermal conductivity of the fluid.

If a rectangular grid is chosen with Δx and Δy as the dimensions of the individual rectangles, then the finite difference of Eq. (10.54) over the time step Δt, using the ADI method, is given in 2 steps by

$$\left(T_{ij}^{n+\frac{1}{2}} - T_{ij}^n\right)\frac{2}{\Delta t} = -u_{ij}^{n+\frac{1}{2}}\frac{T_{i+1,j}^{n+\frac{1}{2}} - T_{i-1,j}^{n+\frac{1}{2}}}{2\Delta x} - v_{ij}^n\frac{T_{i,j+1}^n - T_{i,j-1}^n}{2\Delta y}$$

$$+ \frac{1}{Pe}\frac{T_{i+1,j}^{n+\frac{1}{2}} - 2T_{ij}^{n+\frac{1}{2}} + T_{i-1,j}^{n+\frac{1}{2}}}{\Delta x^2} + \frac{1}{Pe}\frac{T_{i+1,j}^n - 2T_{ij}^n + T_{i,j+1}^n}{\Delta y^2} \tag{10.56}$$

$$\left(T_{ij}^{n+1} - T_{ij}^{n+\frac{1}{2}}\right)\frac{2}{\Delta t} = -u_{ij}^{n+\frac{1}{2}}\frac{T_{i+1,j}^{n+\frac{1}{2}} - T_{i-1,j}^{n+\frac{1}{2}}}{2\Delta x} - v_{ij}^{n+1}\frac{T_{i,j+1}^{n+1} - T_{i,j-1}^{n+1}}{2\Delta y}$$

$$+ \frac{1}{Pe}\frac{T_{i+1,j}^{n+\frac{1}{2}} - 2T_{ij}^{n+\frac{1}{2}} + T_{i-1,j}^{n+\frac{1}{2}}}{\Delta x^2} + \frac{1}{Pe}\frac{T_{i,j+1}^{n+1} - 2T_{ij}^{n+1} + T_{i,j-1}^{n+1}}{\Delta y^2}.$$

$$(10.57)$$

If a square grid is chosen, $\Delta x = \Delta y = l$, then Eqs.(10.56) and (10.57) become

$$T_{ij}^{n+\frac{1}{2}}\left(1 + \frac{\Delta t}{Pel^2}\right) + T_{i+1,j}^{n+\frac{1}{2}}\left(\frac{\Delta t}{4l}u_{ij}^{n+\frac{1}{2}} - \frac{\Delta t}{2Pel^2}\right) + T_{i-1,j}^{n+\frac{1}{2}}\left(-\frac{\Delta t}{4l}u_{ij}^{n+\frac{1}{2}} - \frac{\Delta t}{2Pel^2}\right)$$

$$= T_{ij}^{n}\left(1 - \frac{\Delta t}{Pel^2}\right) + T_{i,j-1}^{n}\left(\frac{\Delta t}{4l}v_{ij}^{n} + \frac{\Delta t}{2Pel^2}\right) + T_{i,j+1}^{n}\left(-\frac{\Delta t}{4l}v_{ij}^{n} + \frac{\Delta t}{2Pel^2}\right)$$

$$(10.58)$$

$$T_{ij}^{n+1}\left(1 + \frac{\Delta t}{Pel^2}\right) + T_{i,j+1}^{n+1}\left(\frac{\Delta t}{4l}v_{ij}^{n+1} - \frac{\Delta t}{2Pel^2}\right) + T_{i,j-1}^{n+1}\left(-\frac{\Delta t}{4l}v_{ij}^{n+1} - \frac{\Delta t}{2Pel^2}\right)$$

$$= T_{i,j}^{n+\frac{1}{2}}\left(1 - \frac{\Delta t}{Pel^2}\right) + T_{i-1,j}^{n+\frac{1}{2}}\left(\frac{\Delta t}{4l}u_{ij}^{n+\frac{1}{2}} + \frac{\Delta t}{2Pel^2}\right) + T_{i+1,j}^{n+\frac{1}{2}}\left(-\frac{\Delta t}{4l}u_{ij}^{n+\frac{1}{2}} + \frac{\Delta t}{2Pel^2}\right)$$

$$(10.59)$$

If in addition, it is assumed that slug flow exists, that is, the x-velocity component u' = a constant, u_∞, and the y-velocity component v = 0. Then, Eq. (10.54) is simplified to

$$\frac{\partial T}{\partial t} = -\frac{\partial T}{\partial x} + \frac{1}{Pe}\frac{\partial^2 T}{\partial x^2} + \frac{1}{Pe}\frac{\partial^2 T}{\partial y^2} \qquad (10.60)$$

Let us assume that the initial temperature condition in the porous medium is T(x,y, t = 0) = 0. The boundary conditions shown are as follows:

$T(x = 0, y, t) = 1$
$T(x = 1, y, t) = 0$
$T(x, y = 0, t) = 0$
$T(x, y = 0.5, t) = 0.5.$ (10.61)

If a rectangular grid is chosen with Δx and Δy as the dimensions of the individual rectangles, then the finite difference form of Eq. (10.60) over the time step Δt, using the ADI method, is given in 2 steps by

$$\left(T_{ij}^{n+\frac{1}{2}} - T_{ij}^{n}\right)\frac{2}{\Delta t} = -\frac{T_{i+1,j}^{n+\frac{1}{2}} - T_{i-1,j}^{n+\frac{1}{2}}}{2\Delta x} + \frac{1}{Pe}\frac{T_{i+1,j}^{n+\frac{1}{2}} - 2T_{ij}^{n+\frac{1}{2}} + T_{i-1,j}^{n+\frac{1}{2}}}{\Delta x^2}$$

$$+ \frac{1}{Pe}\frac{T_{i+1,j}^{n} - 2T_{ij}^{n} + T_{i,j+1}^{n}}{\Delta y^2}$$

(10.62)

$$\left(T_{ij}^{n+1} - T_{ij}^{n+\frac{1}{2}}\right)\frac{2}{\Delta t} = -\frac{T_{i+1,j}^{n+\frac{1}{2}} - T_{i-1,j}^{n+\frac{1}{2}}}{2\Delta x} + \frac{1}{Pe}\frac{T_{i+1,j}^{n+\frac{1}{2}} - 2T_{ij}^{n+\frac{1}{2}} + T_{i-1,j}^{n+\frac{1}{2}}}{\Delta x^2}$$

$$+ \frac{1}{Pe}\frac{T_{i+1,j}^{n+1} - 2T_{ij}^{n+1} + T_{i,j+1}^{n+1}}{\Delta y^2}$$

(10.63)

If a square grid is chosen, $\Delta x = \Delta y = l$, then Eqs. (10.62) and (10.63) become

$$T_{ij}^{n+\frac{1}{2}}\left(1 + \frac{\Delta t}{Pel^2}\right) + T_{i+1,j}^{n+\frac{1}{2}}\left(\frac{\Delta t}{4l} - \frac{\Delta t}{2Pel^2}\right) + T_{i-1,j}^{n+\frac{1}{2}}\left(-\frac{\Delta t}{4l} - \frac{\Delta t}{2Pel^2}\right)$$

$$= T_{ij}^{n}\left(1 - \frac{\Delta t}{Pel^2}\right) + T_{i,j-1}^{n}\left(\frac{\Delta t}{2Pel^2}\right) + T_{i,j+1}^{n}\left(\frac{\Delta t}{2Pel^2}\right)$$

(10.64)

$$T_{ij}^{n+1}\left(1+\frac{\Delta t}{Pel^2}\right)+T_{i,j+1}^{n+1}\left(-\frac{\Delta t}{2Pel^2}\right)+T_{i,j-1}^{n+1}\left(-\frac{\Delta t}{2Pel^2}\right)$$

$$=T_{i,j}^{n+\frac{1}{2}}\left(1-\frac{\Delta t}{Pel^2}\right)+T_{i-1,j}^{n+\frac{1}{2}}\left(\frac{\Delta t}{4l}+\frac{\Delta t}{2Pel^2}\right)+T_{i+1,j}^{n+\frac{1}{2}}\left(-\frac{\Delta t}{4l}+\frac{\Delta t}{2Pel^2}\right)$$

$$(10.65)$$

For current consideration, take $\dfrac{\Delta t}{2Pel^2}=\dfrac{0.1}{2(2)(0.1)^2}=2.5$, and

$$\frac{\Delta t}{4l}=\frac{0.1}{4(0.1)}=0.25,$$

so that Eqs. (10.64) and (10.65) become

$$T_{ij}^{n+\frac{1}{2}}(6)+T_{i+1,j}^{n+\frac{1}{2}}(-2.25)+T_{i-1,j}^{n+\frac{1}{2}}(-2.75)=T_{ij}^{n}(-4)+T_{i,j-1}^{n}(2.5)+T_{i,j+1}^{n}(2.5)$$

$$(10.66)$$

$$T_{ij}^{n+1}(6)+T_{i,j+1}^{n+1}(-2.25)+T_{i,j-1}^{n+1}(-2.75)=T_{ij}^{n+\frac{1}{2}}(-4)+T_{i-1,j}^{n+\frac{1}{2}}(2.5)+T_{i+1,j}^{n+\frac{1}{2}}(2.5)$$

$$(10.67)$$

The tridiagonal matrix obtained is shown below:

$$\begin{bmatrix} 6 & -2.25 & 0 & 0 & .. & .. & .. & .. & .. & 0 \\ -2.25 & 6 & -2.25 & 0 & .. & .. & .. & .. & .. & 0 \\ 0 & -2.25 & 6 & -2.25 & .. & .. & .. & .. & .. & 0 \\ 0 & 0 & -2.25 & 6 & .. & .. & .. \\ & & & & & & & & & \\ & & & & & & & & & \\ & & & & & & & & & \\ .. & .. & .. & .. & .. & .. & .. & .. & 6 & -2.25 \\ 0 & 0 & .. & .. & 0 & 0 & .. & .. & .. & .. & -2.25 & 6 \end{bmatrix}\begin{bmatrix} T_{2,2} \\ T_{3,2} \\ T_{4,2} \\ T_{5,2} \\ . \\ . \\ . \\ T_{n-1} \\ T_n \end{bmatrix}=\begin{bmatrix} 2.75 \\ 0 \\ \\ 0 \\ 0 \\ . \\ . \\ 0 \\ 0 \end{bmatrix}$$

$$(10.68)$$

10.8.1 Thomas Algorithm for Tridiagonal Systems

For tridiagonal matrices, the decomposition of the matrix into a product of a lower and an upper diagonal matrix leads to an efficient algorithm known as the Thomas algorithm. For a system of the form

$$a_k x_{k-1} + b_k x_k + c_k x_{k+1} = f_k \qquad\qquad k = 1,\ldots\ldots,N \qquad (10.69)$$

$$\text{with} \quad a_1 = c_N = 0 \qquad\qquad (10.70)$$

the algorithm below is obtained :

Forward step

$$\beta_1 = b_1 \qquad \beta_k = b_k - a_k \frac{c_{k-1}}{\beta_{k-1}} \qquad k = 2,\ldots\ldots, N$$

$$\gamma_1 = \frac{f_1}{\beta_1} \qquad \gamma_k = \frac{(-a_k \gamma_{k-1} + f_k)}{\beta_k} \qquad k = 2,\ldots\ldots, N$$

$$\qquad\qquad (10.71)$$

Backward step

$$x_N = \gamma_N$$

$$x_k = \gamma_k - x_{k+1} \frac{c_k}{\beta_k} \qquad\qquad k = N-1,\ldots\ldots,1. \qquad (10.72)$$

This calculation involves 5N operations.

The Thomas algorithm will always converge if the tridiagonal matrix is diagonally dominant. In other words, the matrix is such that

$$|b_k| \geq |a_k| + |c_k| \qquad\qquad k = 2,\ldots\ldots, N-1$$

$$|b_1| > |c_1| \quad \text{and} \quad |b_N| > |a_N|. \qquad (10.73)$$

A subroutine in FORTRAN code is written below for the Thomas algorithm.

Subroutine THOMAS

        Subroutine THOMAS (PP,QQ,RR,SS,N1,N)

```
c
c       Solution of a tridiagonal system of equations
c
c       PP(K)*X(K-1) + QQ(K)*X(K) + RR(K)*X(K+1) = SS(K)
c
c       range of K from N1 to N
c       Solution X(K) is stored in SS(K)
c
c
        DIMENSION PP(1),QQ(1),RR(1),SS(1)
        QQ(N1)=1./QQ(N1)
        PP(N1)=SS(N1)*QQ(N1)
        N2=N1+1
        JN=N1+N
        DO 15 K=N2,N
        K1=K-1
        RR(K1)=RR(K1)*QQ(K1)
        QQ(K)=QQ(K)-PP(K)*RR(K1)
        QQ(K)=1./QQ(K)
        PP(K)=(SS(K)-PP(K)*PP(K1))*QQ(K)
   15 CONTINUE
c
c       Back Substitution
c
        SS(N)=PP(N)
        DO 33 K1=N2,N
        K=JN-K1
        SS(K)=PP(K)-RR(K)*SS(K+1)
   33 CONTINUE
        RETURN
        END
```

Instead of a computer program, an Excel spreadsheet may be used to solve the Eqs. (10.66) and (10.67) since there are only a small number of grid points in the rectangular region considered.

PROBLEMS

10.1. Use the linear Burgers equation for heat convection in a channel where the water is flowing with uniform velocity of 0.1 m/s across the cross section of the channel (boundary layers are neglected). The water is initially at 25°C throughout. At time t = 0 sec, waste heat is continuously rejected at $x = 0$ m, and the channel is long such that $dT/dx = 0$ for $x \geq 1$ m. The amount of heat rejected is 6.23 W/m^2 for t > 0. Using the MacCormack explicit scheme, calculate the first 9 time steps to show the transient temperature distributions.

10.2. Solve Prob. 10.1 using an implicit scheme. Compare the results with that obtained in Prob. 10.1.

10.3. Air for ventilation purposes flows through a 10 m insulated duct at 0.75 m/s. Initially, the air is at 25°C. A cooling coil at the entrance cools the air to 15°C. At time = 0, the cooling coil is turned on and the temperature there is maintained constant at 15°C. At the duct exit, the temperature gradient of the air may be assumed unchanging. Use the Burgers equation to model this physical problem, and solve it with an appropriate finite difference scheme.

10.4. Gases flow between two insulated parallel plates, 1.5 m long, and the flow may be considered uniform and one-dimensional. Initially, the gases are at 5°C. At the entrance, the gases are maintained at 25°C by using a radiation source. At the exit, the temperature cannot be less than 20°C. Model this practical problem using Burgers equation, and solve it with an efficient finite difference scheme.

10.5. Hot water is flowing through an insulated 3 m pipe. The pipe contains a cooling coil at the entrance. Initially, the water in the pipe is at 90°C, and the flowrate throughout the time period of interest is 0.5 m/s. At time = 0, the cooling coil is turned on and heat is removed at the entrance of the pipe. At the exit of the pipe, the water comes in contact with a large reservoir of water at 30°C. Use the Burgers equation to model this physical problem, and solve it with an appropriate finite difference scheme.

10.6   The water in a 1.2 m insulated pipe is initially at room temperature, 20°C. At time = 0, cooling water at 0°C enters the pipe at 1 m/s. The entrance of the pipe is maintained at 5°C, and the exit cannot be more than 8°C. Model this practical problem using the Burgers equation, and solve it with an efficient finite difference scheme.

10.7   Using Taylor series expansion, find the forward second-order accurate finite difference expansion for the first derivative of the temperature T with respect to x. In other words, find $\dfrac{\partial T}{\partial x}_i$ in terms of T_i, T_{i+1} and T_{i+2}.

10.8   Consider laminar flow of a fluid over a flat plate. Use the fully implicit method of finite differencing to compute the two dimensionless velocity-component distributions within the boundary layer.

10.9   Consider laminar flow of a fluid over a flat plate. Use the Crank-Nicholson method of finite differencing to compute the two dimensionless velocity-component distributions within the boundary layer.

10.10   Consider convection with incompressible, laminar flow of a constant-temperature fluid over a flat plate maintained at a constant temperature. With the velocity distributions found in either Prob. 10.1 or Prob. 10.2, compute the dimensionless temperature distribution within the thermal boundary layer for the Peclet number equal to 0.1,1.0,10.0,100.0. Use the ADI method.

10.11   Figure 10.1 shows slug flow of a fluid through a rectangular porous medium. Compute the temperature distribution with the same boundary conditions except at y = 0, where the condition is

now $\left.\dfrac{\partial T}{\partial y}\right|_{y=0} = 0$. For a 101 x 51 square mesh, program an

Excel spreadsheet to solve the problem.

10.12 Figure 10.1 shows slug flow of a fluid through a rectangular porous medium. Compute the temperature distribution with the same boundary conditions except at y = 0, where the condition is

now $\left.\dfrac{\partial T}{\partial y}\right|_{y=0} = 0$. Use the ADI method.

REFERENCES

1. L Howarth. On the Solution of the Laminar boundary Layer Equations. Proc R Soc (London), A164:546, 1938.

APPENDIX A

```
      implicit double precision (a-h,o-z)
      implicit integer (i-m)
          REAL*8 U(102,202),V(102,202),XX
c         character string*13
      write(*,*)
      write(*,*)
      write(*,*)
      write(*,*) "            Program to Solve for the
      write(*,*)
      write(*,*) "            Laminar Boundary Layer Equations
      write(*,*)
      write(*,*)

c         read(*,*) string

      write(*,*)
```

```
      write(*,*)
      write(*,*)
      write(*,*)
      write(*,*) ' 1. Input Value of Kinematic Viscosity'
      read(*,*) XX

      WRITE(*,*)  XX
      write(*,*)
c     Let delta x = 0.01, delta y = 0.001
c     Stability criterion satisfied for explicit method.
c     pause
      do 100 I=1,101
      do 100 J=2,201
      U(I,J)=1.0D0
      V(I,J)=0.0D0
100   continue

      do 101 I=1,101
      U(I,1)=0.0D0
      U(I,201)=1.0D0
101   continue

      do 102 J=1,201
      V(1,J)=0.0D0
102   continue

      do 201 K=1,3000,1
    do 200 I=2,100,1
      dO 200 J=2,200,1
    U(I+1,J)=U(I,J)-5.0D0*V(I,J)/U(I,J)*(U(I,J+1)-U(I,J-1))-
2.0D3*XX+
    C    (1.0D3*XX)/U(I,J)*(U(I,J+1)+U(I,J-1))
      V(I+1,J)=V(I+1,J-1)-5.0D-2*(U(I+1,J)+U(I+1,J-1)-U(I,J)-U(I,J-
    C 1))
200  continue
201      continue
    write(*,*)'============================='
      do 301 J=2,200
      write(*,*)        (U(I,J),I=2,100)
      write(*,*)
```

```
301     continue
        do 302 J=2,200
        write(*,*)        (V(I,J),I=2,100)
        write(*,*)
302     continue
```

```
cccccccccccccccccccccccccc
cccccccccccccccccccccccccc
```

```
        stop

        END
```

Alternating Direction Implicit Method

The alternating direction implicit method of finite differencing
Is a method of variable direction in finite differencing
Employs splitting of one time step into two to obtain implicit method
Requires only inversion of tridiagonal matrix in this method.

Stability of this two-dimensional method is unconditional
Stability of fully implicit method is also unconditional
Only tridiagonal matrices to be solved for problems with two dimensions
Usually true only in problems with one dimension, not two dimensions.

K.V. Wong

11

Basic Relations of Radiation

11.1 Thermal Radiation

Thermal radiation is the energy emitted by bodies because of their temperature level. Other types of radiation include gamma rays, x-rays, for instance. Radiation is often treated as electromagnetic waves that propagate according to Maxwell's classic electromagnetic theory. Radiation may also be treated as photons as prescribed in Max Planck's concept of the quantum of energy. The electromagnetic theory has been used to predict the radiant properties of materials, while the quantum theory has been used to predict the amount of radiant energy emitted by a body because of its level of temperature.

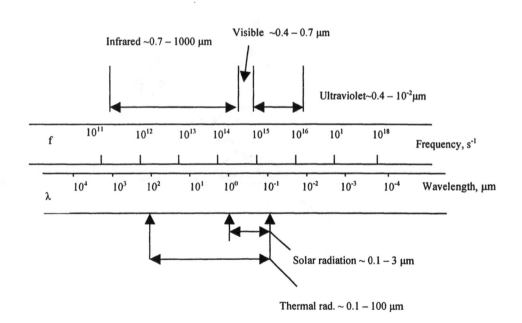

Figure 11.1 Electromagnetic wave spectrum.

In Fig.11.1, is shown a large range of the electromagnetic-wave spectrum. In theory, electromagnetic waves of zero to infinity wavelengths have thermal radiant energy. In practice, a big portion of the thermal radiation lies in the range from about 0.1 to 100 μm. This portion is labeled as such in the figure. The visible range is from 0.4 to 0.7 μm; it is important to the extent that it tells the scholars of heat transfer to use their eyes to obtain insight into the thermal radiation phenomenon. When radiation is considered an electromagnetic wave, its transport in a medium takes place with the speed of light, c. The wavelength λ and the frequency f are related to the speed of light by c $= f$. When thermal radiation travels in a vacuum, for instance, for most of the distance between the sun and the earth, the speed of light is 2.9979 x 10^8 m/s. This speed is attenuated by the atmosphere surrounding the earth.

11.2 Radiation Intensity and Blackbody

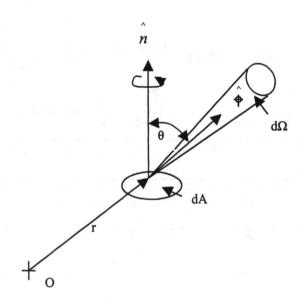

Figure 11.2 Notation for radiation intensity.

Radiation may be conceived as being propagated as a beam (like visible light), as in Fig.11.2. A basic quantity that is used to quantify radiative energy in a given direction $\hat{\Omega}$, at a wavelength λ, at a position r is the spectral radiation intensity $I_\lambda(r,\hat{\Omega})$. This represents the quantity of energy streaming through a unit area perpendicular to the direction $\hat{\Omega}$, per unit time, per unit solid angle about the direction $\hat{\Omega}$ and per unit wavelength about the wavelength λ. The radiation intensity $I(r,\hat{\Omega})$ represents the amount of energy emitted over the entire wavelength spectrum from $\lambda = 0$ to ∞ in a beam and is defined from the spectral radiation intensity $I_\lambda(r,\hat{\Omega})$,

$$I(r,\hat{\Omega}) = \int_{\lambda=0}^{\infty} I_\lambda(r,\hat{\Omega})d\lambda \tag{11.1}$$

The radiation intensity I is the amount of radiant energy passing through a unit area normal to the direction of propagation $\hat{\Omega}$, per unit time, per unit solid angle about the direction $\hat{\Omega}$ and per unit time, per unit solid angle about the direction $\hat{\Omega}$.

Consider the radiation intensity $I(r,\hat{\Omega})$ within a solid angle $d\Omega$ to (or from) a surface element dA, propagating in the direction $\hat{\Omega}$ at an angle θ with the normal n to the surface element, as shown in Fig. 11.2. The quantity

$$dq = I(r,\hat{\Omega})\cos\theta\, d\Omega \tag{11.2}$$

is the amount of radiant energy per unit time, to (or from) per unit area of the surface owing to radiation contained within a solid angle $d\Omega$. The radiative energy flux q to (or from) a surface owing to radiation contained in a solid angle over an entire hemisphere is obtained by the integration of Eq. (11.2) as

$$q = \int_\cap I \cos\theta d\Omega \tag{11.3}$$

where the symbol   indicates the integration with respect to a solid angle over an entire hemisphere. As shown in Fig. 11.2, θ is the polar angle and φ is the azimuthal angle. Since $d\Omega = \sin\theta d\theta d\varphi$, Eq. (11.3) may be written as

$$q = \int_{\varphi=0}^{2\pi} \int_{\vartheta=0}^{\pi/2} I(r,\theta,\varphi)\cos\theta\sin\theta d\varphi \tag{11.4}$$

The dimensions of q are energy per unit time, per unit area of the surface (e.g., $kJ/h.m^2$.)

There is a maximum amount of radiant energy emitted by a body at a given absolute temperature T at a wavelength λ. This maximum amount of radiant emission is the spectral blackbody radiation intensity $I_{\lambda b}(T)$; the emitter of such radiation is named a blackbody. This spectral blackbody radiation intensity is independent of direction. For a blackbody at an absolute temperature T and emitting radiative energy into a vacuum, $I_{\lambda b}(T)$ is calculated from the relation given by Planck, 1959 [1], in the form

$$I_{\lambda b}(T) = \frac{2hc^2}{\lambda^5[\exp(hc/\lambda kT)-1]} \tag{11.5}$$

where h ($= 6.6256 \times 10^{-34}$ J.s) and k ($= 1.38054 \times 10^{-23}$ J.K) are the Planck and Boltzmann constants, respectively, T is the absolute temperature and c is the speed of light in a vacuum.

For engineering practice, the spectral blackbody emissive flux $q_{\lambda b}(T)$ at a surface is defined as

$$q_{\lambda b}(T) = \int_\cap I_{\lambda b}(T)\cos\theta d\Omega. \tag{11.6}$$

As $I_{\lambda b}(T)$ is independent of direction,

$$q_{\lambda b}(T) = I_{\lambda b}(T) \int_{\varphi=0}^{2\pi} \int_{\theta=0}^{\pi/2} \cos\theta \sin\theta d\varphi = \pi I_{\lambda b}(T). \tag{11.7}$$

The quantity in Eq. (11.7) is the amount of radiative energy emitted by a blackbody at temperature T per unit of its surface, per unit time, per unit wavelength in all directions in the hemispherical space. Substituting Eq. (11.5) into Eq. (11.7),

$$q_{\lambda b}(T) = \frac{c_1}{\lambda^5 [\exp(c_2 / \lambda T) - 1]} \tag{11.8}$$

where $q_{\lambda b}(T)$ is the spectral blackbody emissive flux as the surface (W/m^2.μm),

$c_1 = 2\pi hc^2 = 3.743$ x 10^8 W.μm^4/m^2
$c_2 = hc/k = 1.4387$ x 10^4 μm.K.

Figure 11.3 is a plot of the spectral blackbody emissive flux as a function of wavelength at various temperatures. From this figure, it is clear that at any given wavelength, the radiative energy emitted by a blackbody increases as the absolute temperature of the body increases. Each curve displays a peak, and the peaks shift toward smaller wavelengths as the temperature rises. The locus of the peaks calculated analytically by Wien's displacement rule is

$$(\lambda T)_{max} = 0.28976 \text{ cm.K} = 28997.6 \, \mu\text{m.K} \tag{11.9}$$

The blackbody radiation intensity $I_b(T)$ is found by the integration of $I_{\lambda b}(T)$ over the wavelengths ranging from 0 to ∞.

$$I_b(T) = \int_{\lambda=0}^{\infty} I_{\lambda b}(T) d\lambda \tag{11.10}$$

Put Eq.(11.5) into Eq.(11.10) and integrate,

$$I_b(T) = \frac{\sigma T^4}{\pi} \tag{11.11}$$

where the Stefan-Boltzmann constant $\sigma = 5.6697$ x 10^{-8} W/(m^2.K^4).

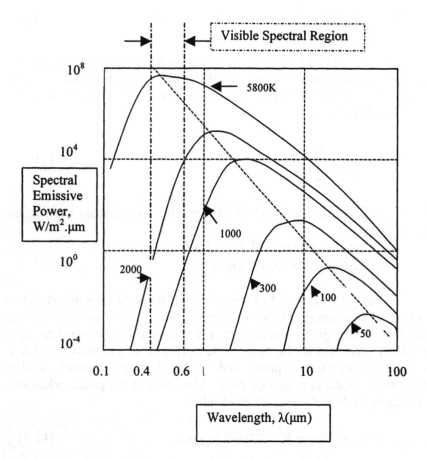

Figure 11.3   Spectral blackbody emissive power.

The blackbody emissive flux $q_b(T)$ at an absolute temperature T is gained by integrating $q_{\lambda b}(T)$ over all wavelengths,

$$q_b(T) = \int_{\lambda=0}^{\infty} q_{\lambda b}(T)d\lambda = \pi \int_{\lambda=0}^{\infty} I_{\lambda b}(T)d\lambda = \pi I_b(T) = \sigma T^4. \quad (11.12)$$

This emissive flux has the dimensions of energy per unit time, per unit area. From Eqs. (11.11) and (11.12), it can be seen that

$$I_b(T) = \frac{q_b(T)}{\pi} \qquad\qquad (11.13)$$

The generalized idea of a blackbody is one that possesses the characteristic of allowing all incident radiation to enter the medium without surface reflection and without allowing it to leave the medium again. A blackbody absorbs all incident radiation from all directions at all frequencies without reflecting, transmitting, or scattering it outwards. The blackbody emits as much radiative energy as it absorbs, if it is at thermal equilibrium with the enclosure walls. For practical purposes, a cavity such as a hollow sphere whose interior surfaces are kept at a uniform temperature T can be used to approximate a blackbody. If a very tiny hole (compared to the cavity) is made, any radiation entering the cavity through the hole is almost entirely absorbed since it has very little possibility to escape through the hole. Such a cavity is considered an approximate blackbody. By a similar argument, radiation leaving the cavity through the hole is considered almost a blackbody radiation at temperature T.

11.3 Reflectivity, Absorptivity, Emissivity and Transmissivity

<u>Real Surfaces</u>

Consider a beam of radiant energy incident on a real surface. Part of this radiation is reflected, part of it is absorbed and the rest is transmitted. Let $I'_\bullet$ be the spectral radiation intensity incident on the surface. The spectral radiant heat flux incident on the surface can be expressed as

$$q'_\lambda = \int_\Omega I'_\lambda \cos\theta' d\Omega' \quad \text{energy/(time x area x wavelength)} \qquad (11.14)$$

where θ' is the polar angle between the direction of the incident radiation and the normal to the surface. The spectral hemispherical reflectivity ρ_λ is defined as

$$\rho_\lambda = \frac{\text{radiant energy reflected}/(\text{ time x area x wavelength })}{q_\lambda}.$$

(11.15)

The spectral hemispherical absorptivity α_λ is defined as

$$\alpha_\lambda = \frac{\text{radiant energy absorbed}/(\text{ time x area x wavelength })}{q_\lambda}.$$

(11.16)

For an opaque surface, the relationship between the spectral hemispherical reflectivity and the spectral hemispherical absorptivity is

$$\rho_\lambda + \alpha_\lambda = 1.$$ (11.17)

For much of engineering practice, the reflectivity and the absorptivity, averaged over the entire wavelengths, is of relevance. When this is done, the resulting hemispherical reflectivity ρ and the hemispherical absorptivity α are defined as follows:-

$$\rho = \frac{\int_0^\infty \rho_\lambda q_\lambda' d\lambda}{\int_0^\infty q_\lambda' d\lambda}$$

(11.18)

$$\alpha = \frac{\int_0^\infty \alpha_\lambda q_\lambda' d\lambda}{\int_0^\infty q_\lambda' d\lambda}.$$

(11.19)

For an opaque surface,

$$\rho + \alpha = 1.$$ (11.20)

The radiant energy emitted by a real surface at an absolute temperature T is always less than that emitted by a blackbody surface at the same temperature. Let $q_\lambda(T)$ be the spectral emissive flux from a real surface at an absolute temperature T and $q_{\lambda b}(T)$ be the spectral blackbody emissive heat flux for a blackbody surface at the same temperature. The hemispherical emissivity ε_λ of the surface is defined as

$$\varepsilon_\lambda = \frac{q_\lambda(T)}{q_{\lambda b}(T)}.$$ (11.21)

The hemispherical emissivity ε over the entire range of wavelengths is found by

$$\varepsilon = \frac{\int_{\lambda=0}^{\infty} \varepsilon_\lambda q_{\lambda b}(T)d\lambda}{\int_{\lambda=0}^{\infty} q_{\lambda b}(T)d\lambda} = \frac{q(T)}{q_b(T)}$$ (11.22)

where q(T) and q$_b$(T) are the emissive fluxes from the real surface at temperature T, and the blackbody at temperature T, respectively.

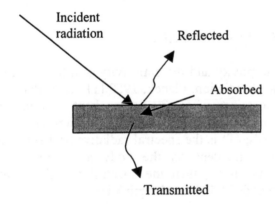

Figure 11.4.   Incident radiation on a translucent body.

When radiation is incident on a translucent body, part of the incident radiation is reflected, part is absorbed, and the remainder is transmitted through the translucent body (Fig. 11.4). An example of a translucent body is a pane of glass. The relationship between the spectral reflectivity ρ_λ, the spectral absorptivity α_λ and the spectral transmissivity τ_λ of the translucent body is

$$\rho_\lambda + \alpha_\lambda + \tau_\lambda = 1.$$ (11.23)

When these radiative properties are averaged over all wavelengths, we get

$$\rho + \alpha + \tau = 1. \tag{11.24}$$

The reflectivity, absorptivity and transmissivity of a translucent body depend in large part on the surface conditions, the wavelength of the radiation, the composition of the material and the thickness of the body. Since the attenuation of radiation within a body should be analyzed as a bulk process, the evaluation of the reflectivity and transmissivity of a translucent object is more involved.

Graybody

For simplicity, the graybody assumption is used in many applications. The radiative properties ρ_λ, α_λ, ε_λ and τ_λ are assumed to be uniform over the entire wavelength spectrum. In other words, graybodies have radiative properties ρ, α, ε and τ that are independent of wavelength.

11.4 Kirchhoff's Law of Radiation

The absorptivity and the emissivity of a body can be related by Kirchhoff's law of radiation, Planck, 1959 [1]. Consider a body inside a black, closed container whose walls are kept at a uniform absolute temperature T and has reached thermal equilibrium with the walls of the container. If flux $q_\lambda(T)$ is the spectral radiative heat flux from the walls at temperature T incident on the body and $\alpha_\lambda(T)$ is the spectral absorptivity of the body, then the spectral radiative heat flux $q_\lambda(T)$ absorbed by the body at the wavelength λ is

$$q_\lambda(T) = \alpha_\lambda(T)q_\lambda^{'}(T). \tag{11.25}$$

Since the body is in radiative equilibrium, $q_\lambda(T)$ also expresses the spectral radiative flux emitted by the body at the wavelength λ. The incident radiation $q'_\lambda(T)$ comes from the black walls of the enclosure at temperature T, and the emission by the walls is not influenced by the body regardless if it is a blackbody or not. Let $q_{\lambda b}(T)$ be the spectral blackbody emissive flux at temperature T. Then,

$$q_{\lambda b}(T) = q'_{\lambda}(T).$$ (11.26)

From Eqs. (11.25) and (11.26),

$$\frac{q_{\lambda}(T)}{q_{\lambda b}(T)} = \alpha_{\lambda}(T).$$ (11.27)

The spectral emissivity $\varepsilon_{\lambda}(T)$ of the body for radiation at temperature T is defined as the ratio of the spectral emissive flux $q_{\lambda}(T)$ of the body to the spectral blackbody emissive flux $q_{\lambda b}(T)$ at the same temperature. Expressed mathematically,

$$\frac{q_{\lambda}(T)}{q_{\lambda b}(T)} = \varepsilon_{\lambda}(T).$$ (11.28)

From Eqs. (11.27) and (11.28), it can be deduced that

$$\varepsilon_{\lambda}(T) = \alpha_{\lambda}(T).$$ (11.29)

Equation (11.29) is Kirchhoff's law of radiation. The law states that the spectral emissivity for the emission of radiation at temperature T is equal to the spectral absorptivity for radiation from a blackbody at the same temperature T. The relation

$$\varepsilon(T) = \alpha(T)$$ (11.30)

holds only if the incident and emitted radiation have the same spectral distribution or when the body is gray. This later characteristic is one where the radiative properties are independent of wavelength.

PROBLEMS

11.1.   The average internal temperature of an oven is 1500°C, and the emissivity of the internal surface is $\varepsilon = 0.9$ at this temperature. Calculate the radiant energy coming from the oven through an opening 10 cm by 10 cm.

11.2.   A blackbody enclosure at 1000°C has a small aperture into the environment. Determine (i) the blackbody radiation intensity emerging from the aperture, and (ii) the blackbody radiation heat flux from the blackbody.

11.3.   The surface of an outer space station receives solar radiation at a rate of 1.2 kW/m^2. The surface has an absorptivity of α = 0.75 for solar radiation and an emissivity of ε = 0.86. There are no heat losses into the space station. However, heat is dissipated by thermal radiation into the space at absolute zero. Determine the equilibrium temperature of the surface.

11.4.   A solar collector surface receives solar radiation at 1 kW/m^2, and its other side is insulated. The absorptivity of the surface to solar radiation is α = 0.8 while its emissivity is ε = 0.6. Assuming the surface loses heat by radiation into a clear sky at an effective temperature of 10°C, calculate the temperature of the surface.

REFERENCES

1. M Planck. The Theory of Heat Radiation. New York: Dover Publications, 1959.

Blackbody and Graybody

A blackbody absorbs all incident radiation
At all frequencies and from all different directions
No phenomena of reflecting, transmitting or scattering
It is emitting as much as it is absorbing.

At any conditions, graybody has uniform properties
They are not dependent on other properties
Radiative properties are uniform over all wavelengths
Graybody has properties independent of wavelength.

K.V. Wong

12

Radiative Exchange in a Non-Participating Medium

When the medium participates in radiation, the analysis becomes more complicated. The discussion in this chapter concentrates on situations where the participation of the medium may be neglected.

12.1 Radiative Exchange Between Two Differential Area Elements

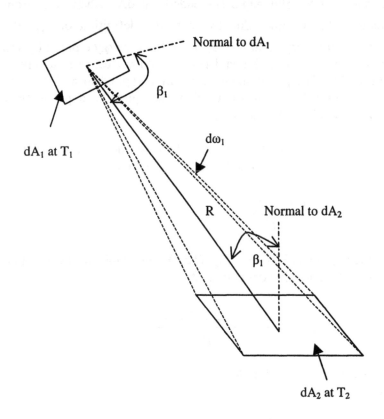

Figure 12.1 Radiative exchange between two differential area elements.

We first look at the radiative exchange between differential elements. Then the relations will be developed for exchange between areas of finite size. Two differential black elements are shown in Fig. 12.1. The elements dA_1 and dA_2 are at temperatures T_1 and T_2 respectively; their normals are at angles β_1 and β_2 to the line of length R joining them.

The total energy per unit time leaving dA_1 and incident upon dA_2 is

$$d^2Q'_{d1-d2} = i'_{b,1}dA_1 \cos\beta_1 d\omega_1, \tag{12.1}$$

and $d\omega_1$ is the solid angle subtended by dA_2 when seen from dA_1. Equation (12.1) comes directly from the definition of $i'_{b,1}$, the total blackbody intensity of surface 1, as the total energy emitted by surface 1 per unit time, per unit of area dA_1 projected normal to R, and per unit of solid angle. The prime shows a quantity applied in a single direction. The second differential shows that the quantity depends upon two differential values, dA_1 and $d\omega_1$.

The solid angle $d\omega_1$ is linked to the distance between the differential elements R, and the projected area dA_2 by

$$d\omega_1 = \frac{dA_2 \cos\beta_2}{R^2}. \tag{12.2}$$

Substituting Eq.(12.2) into Eq. (12.1), the total energy per unit time leaving dA_1 that is incident upon dA_2 is

$$d^2Q'_{d1-d2} = \frac{i'_{b,1}dA_1 \cos\beta_1 dA_2 \cos\beta_2}{R^2}. \tag{12.3}$$

A similar derivation for the radiation leaving dA_2 that arrives at dA_1 gives

$$d^2Q'_{d2-d1} = \frac{i'_{b,2}dA_2 \cos\beta_2 dA_1 \cos\beta_1}{R^2}. \tag{12.4}$$

Equations (12.3) and (12.4) provide the expressions for the energy emitted by one element that is incident upon the second element. If the elements are black, all incident energy is absorbed. For this special case, Eqs. (12.3) and (12.4) provide the expressions for the energy from one element that is absorbed by the second.

The net energy per unit time $d^2Q'_{d1 \leftrightarrow d2}$ exchanged from black element dA_1 to dA_2 along path R is then the difference of $d^2Q'_{d1-d2}$ and $d^2Q'_{d2-d1}$. From Eqs. (12.3) and (12.4),

$$d^2Q'_{d1 \leftrightarrow d2} \equiv d^2Q'_{d1-d2} - d^2Q'_{d2-d1} = (i'_{b,1} - i'_{b,2})\frac{\cos\beta_1 \cos\beta_2}{R^2}dA_1 dA_2.$$

$$(12.5)$$

From the previous chapter, the blackbody total intensity is related to the blackbody total hemispherical emissive power by

$$i'_b = \frac{e_b}{\pi} = \frac{\sigma T^4}{\pi}.$$

$$(12.6)$$

Equation (12.5) may be written as

$$d^2Q'_{d1 \leftrightarrow d2} = \sigma(T_1^4 - T_2^4)\frac{\cos\beta_1 \cos\beta_2}{\pi R^2}dA_1 dA_2.$$

$$(12.7)$$

Example 12.1

Problem: A black element 0.5 cm by 0.5 cm, is at a temperature of 800°C and is near a tube of 2 cm diameter. The opening of the tube may be approximated as a black surface, and is at 400°C. Calculate the net radiation exchange along the connecting path R between the square element and the tube opening.

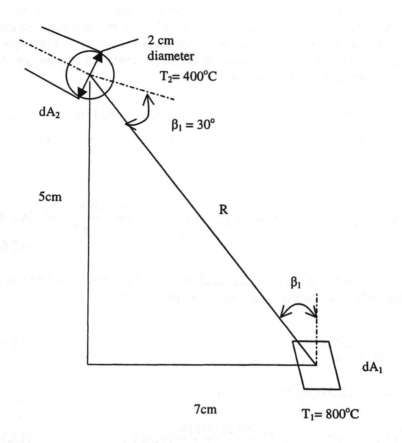

Figure 12.2   Sketch for Example 12.1

<u>Solution</u>
From Eq. (12.7),

$$d^2Q'_{d1\leftrightarrow d2} = \sigma(T_1^4 - T_2^4)\frac{\cos\beta_1 \cos\beta_2}{\pi R^2}dA_1 dA_2.$$

From the figure, $\cos\beta_1 = \dfrac{5}{\sqrt{5^2+7^2}} = \dfrac{5}{\sqrt{74}}$

$$d^2 Q'_{d1 \leftrightarrow d2} = 5.669 \times 10^{-8} \frac{W}{m^2 K^4} (1073.15^4 - 673.15^4) K^4 \times$$

$$\frac{5}{\sqrt{74}} \frac{\cos 30^\circ}{\pi (74 \times 10^{-4})} \frac{1}{m^2} (0.5^2 m^2)(\pi 1^2 m^2) \times 10^{-8} = 0.01W$$

12.2 Concept of View Factor

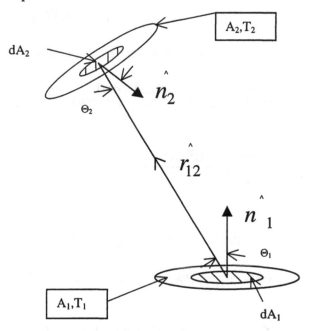

Figure 12.3   Coordinates for the definition of diffuse view factor.

The physical significance of view factor is the fraction of the radiative energy leaving one surface element that strikes the other surface directly.

(a)   Diffuse view factor between two elemental surfaces.

Using the notation in Fig. (12.3), the diffuse view factor between two elemental surfaces is given by

$$dF_{dA_1-dA_2} = \frac{dq_1}{q_1} = \frac{\cos\theta_1\,\cos\theta_2 dA_2}{\pi r^2} \tag{12.8}$$

$$dF_{dA_2-dA_1} = \frac{\cos\theta_1\,\cos\theta_2 dA_1}{\pi r^2}. \tag{12.9}$$

Reciprocity theorem gives $\qquad dA_1 dF_{dA_1-dA_2} = dA_2 dF_{dA_2-dA_1} \tag{12.10}$

(b)   Diffuse view factor between surfaces dA_1 and A_2.

$$F_{dA_1-A_2} = \int_{A_2} dF_{dA_1-dA_2} = \int_{A_2} \frac{\cos\theta_1\,\cos\theta_2}{\pi r^2} dA_2 \tag{12.11}$$

$$F_{A_2-dA_1} = \frac{dA_1}{A_2} \int_{A_2} \frac{\cos\theta_1\,\cos\theta_2}{\pi r^2} dA_2 \tag{12.12}$$

Reciprocity theorem gives $\qquad dA_1 F_{dA_1-A_2} = A_2 F_{A_2-dA_1}. \tag{12.13}$

(c)   Diffuse view factor between two finite surfaces A_1 and A_2.

$$F_{A_1-A_2} = \frac{[\text{Radiative energy leaving surface } A_1 \text{ that strikes } A_2 \text{ directly}]}{[\text{Radiative energy leaving surface } A_1 \text{ in all directions in the hemispherical space}]}$$

$$F_{A_1-A_2} = \frac{1}{A_1} \int_{A_1} \int_{A_2} \frac{\cos\theta_1\,\cos\theta_2}{\pi r^2} dA_2\, dA_1 \tag{12.14}$$

$$F_{A_2-A_1} = \frac{1}{A_2} \int_{A_1} \int_{A_2} \frac{\cos\theta_1\,\cos\theta_2}{\pi r^2} dA_2\, dA_1 \tag{12.15}$$

Reciprocity theorem gives $A_1 F_{A_1-A_2} = A_2 F_{A_2-A_1}$.   (12.16)

12.3 Properties of Diffuse View Factors

Reciprocity theorem: $A_i F_{A_i-A_j} = A_j F_{A_j-A_i}$

          or $A_i F_{i-j} = A_j F_{j-i}$ (12.17)

Summation: $\displaystyle\sum_{k=1}^{N} F_{i-k} = 1$ (12.18)

$$F_{ii} = 0 \quad \text{for plane or convex surfaces.}$$
$$F_{ii} \neq 0 \quad \text{for concave surfaces.}$$

There is a reciprocity relationship that can be derived from the symmetry of a geometry. Consider the areas in Fig. 12.4(i). It is clear from symmetry that $A_2 = A_4$ and $F_{2-3} = F_{4-1}$. Hence, $A_2 F_{2-3} = A_4 F_{4-1}$. From reciprocity, $A_4 F_{4-1} = A_1 F_{1-4}$. Therefore, the following relationship holds:

$$A_2 F_{2-3} = A_1 F_{1-4}$$

The arrows in the figure show the diagonal directions.

Analogously, the symmetry of Fig.12.4(ii) gives

$$A_2 F_{2-8} = A_3 F_{3-7}.$$

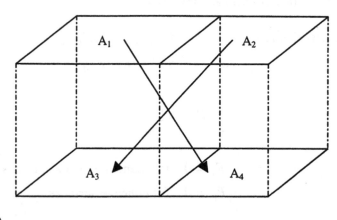

(i)

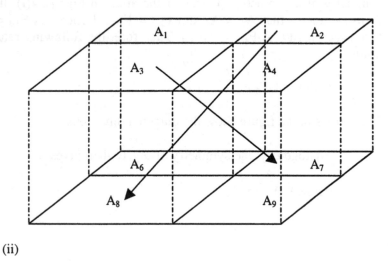

(ii)

Figure 12.4   Reciprocity between diagonally opposing rectangles. (i)
Two pairs of opposing rectangles. (ii) Four pairs of opposing rectangles.

12.4   Determination of Diffuse View Factor by Contour Integration

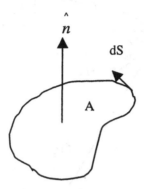

Figure 12.5   Convention for the direction of circulation in Stokes Theorem.

Stokes theorem states that the circulation of a vector $\vec{v}$ around the boundary s of a closed surface A is equal to the flux of the curl of the vector $\vec{v}$ over the surface A; it is given as:

$$\int_{surface\ A} \hat{n} \bullet \left(\nabla \times \vec{v} \right) dA = \oint_{contour\ of\ A} \vec{v} \bullet \vec{ds} \qquad (12.19)$$

where $\vec{v} = \hat{i} v_x + \hat{j} v_y + \hat{k} v_z$

and $\vec{n} = \hat{i} l + \hat{j} m + \hat{k} n$

$$\iint_{surface\ A} \left[\left(\frac{\partial v_z}{\partial y} - \frac{\partial v_y}{\partial z} \right)_l + \left(\frac{\partial v_x}{\partial z} - \frac{\partial v_z}{\partial x} \right)_m + \left(\frac{\partial v_y}{\partial x} - \frac{\partial v_x}{\partial y} \right)_n \right] dA$$
$$= \oint_{contour\ of\ A} (v_x dx + v_y dy + v_z dz). \qquad (12.20)$$

(a) Diffuse View Factor between Surfaces dA_1 and A_2.

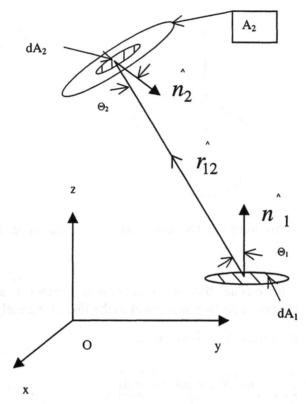

Figure 12.6   Application of Stokes Theorem to determine diffuse view factor $F_{dA_1-dA_2}$.

The diffuse view factor from dA_1 to A_2 is defined as

$$F_{dA_1-A_2} = \int_{A_2} \frac{\cos\theta_1 \cos\theta_2}{\pi r^2} dA_2 \qquad (12.21)$$

where $\cos\theta_1 = \dfrac{\hat{n}_1 \bullet \vec{r}_{12}}{r}$ and $r = \left|\vec{r}_{12}\right|$ \qquad (12.22a)

and
$$\cos\theta_2 = \frac{\hat{n}_2 \bullet \vec{r}_{21}}{r}.$$
(12.22b)

Substituting Eqs. (12.21a and b) in Eq.(12.21),

$$F_{dA_1-A_2} = -\frac{1}{\pi}\int_{A_2}\left(\frac{\hat{n}_1 \bullet \vec{r}_{12}}{r^2}\right)\left(\frac{\hat{n}_2 \bullet \vec{r}_{12}}{r^2}\right)dA_2$$

$$F_{dA_1-A_2} = -\frac{1}{\pi}\int_{A_2}\hat{n}_2 \bullet \left(\frac{\hat{n}_1 \bullet \vec{r}_{12}}{r^2}\right)\left(\frac{\vec{r}_{12}}{r^2}\right)dA_2$$
(12.23)

But
$$-\frac{\vec{r}_{12}}{r^2}\left(\frac{\hat{n}_1 \bullet \vec{r}_{12}}{r^2}\right) = \frac{1}{2}\nabla\times\left(\frac{\vec{r}_{12}\times\hat{n}_1}{r^2}\right), \text{ hence}$$

$$F_{dA_1-A_2} = \frac{1}{2\pi}\int_{A_2}\hat{n}_2 \bullet \left[\nabla\times\left(\frac{\vec{r}_{12}\times\hat{n}_1}{r^2}\right)\right]dA_2$$

Using Stokes Theorem,

$$F_{dA_1-A_2} = \frac{1}{2\pi}\oint_{contour\ of A_2}\left(\frac{\vec{r}_{12}\times\hat{n}_1}{r^2}\right)\bullet \vec{ds}$$
(12.24)

In the x, y, z rectangular coordinate system,
$$\vec{r}_{12} = (x_2 - x_1)\hat{i} + (y_2 - y_1)\hat{j} + (z_2 - z_1)\hat{k} \quad (12.25)$$
$$\hat{n}_1 = l_1\,\hat{i} + m_1\,\hat{j} + n_1\,\hat{k}$$
$$\vec{ds} = dx_2\,\hat{i} + dy_2\,\hat{j} + dz_2\,\hat{k}$$

$$F_{dA_1-A_2} = \frac{l_1}{2\pi} \oint_{contour\ of\ A_2} \frac{(z_2-z_1)dy_2 - (y_2-y_1)dz_2}{r^2}$$

$$+ \frac{m_1}{2\pi} \oint_{contour\ of\ A_2} \frac{(x_2-x_1)dz_2 - (z_2-z_1)dx_2}{r^2}$$

$$+ \frac{n_1}{2\pi} \oint_{contour\ of\ A_2} \frac{(y_2-y_1)dx_2 - (x_2-x_1)dy_2}{r^2}$$

$$(12.26)$$

where $r^2 = (x_2-x_1)^2 + (y_2-y_1)^2 + (z_2-z_1)^2$ and l_1, m_1,n_1 are the

direction cosines. If the unit normal vector $\hat{n}_1$ to dA_1 lies along one of

the coordinate axes, the direction cosines of $\hat{n}_1$ with respect to the other
two axes become zero; two integrals of Eq. (12.26) vanish. If any one of
the boundaries of A_2 is parallel to a coordinate axis, the integration is
simplified.

(b) Diffuse View Factor Between A_1 and A_2

The diffuse view factor $F_{A_1-A_2}$ from surface A_1 to surface A_2 is

$$A_1 F_{A_1-A_2} = \int_{A_1} F_{dA_1-A_2} dA_1 \qquad (12.27)$$

Substituting $F_{dA_1-A_2}$ from Eq. (12.26) into Eq. (12.27) and rearranging,

$$A_1 F_{A_1-A_2} = \frac{1}{2\pi} \oint_{contour\ of\ A_2} \left[\int_{A_1} \frac{(y_2-y_1)n_1 - (z_2-z_1)m_1}{r^2} dA_1 \right] dx_2$$

$$+ \frac{1}{2\pi} \oint_{contour\ of\ A_2} \left[\int_{A_1} \frac{(z_2-z_1)l_1 - (x_2-x_1)n_1}{r^2} dA_1 \right] dy_2$$

$$+\frac{1}{2\pi}\oint_{contour\ of\ A_2}\left[\int_{A_1}\frac{(x_2-x_1)m_1-(y_2-y_1)l_1}{r^2}dA_1\right]dz_2$$

$$(12.28)$$

The surface integrals in Eq. (12.28) will be changed into contour integrals. The first surface integral on the right-hand side can be written as

$$\int_{A_1}\frac{(y_2-y_1)n_1-(z_2-z_1)m_1}{r^2}dA_1=\int_{A_1}\hat{n}_1\bullet\left(\nabla\times\vec{v}_1\right)dA_1$$

where $\quad\hat{n}_1=l_1\,\hat{i}+m_1\,\hat{j}+n_1\,\hat{k}$

$\qquad\vec{v}_1\equiv\hat{i}\ln r$

$$\int_{A_1}\frac{(y_2-y_1)n_1-(z_2-z_1)m_1}{r^2}dA_1=\oint_{contour\ of\ A_1}\vec{v}_1\bullet\vec{ds}_1=\oint_{contour\ of\ A_1}\ln r\,dx_1$$

$$(12.29)$$

since $\quad\vec{ds}=dx_2\,\hat{i}+dy_2\,\hat{j}+dz_2\,\hat{k}$

Similarly,

$$\int_{A_1}\frac{(z_2-z_1)l_1-(x_2-x_1)n_1}{r^2}dA_1=\oint_{contour\ of\ A_1}\vec{v}_2\bullet\vec{ds}_1=\oint_{contour\ of\ A_1}\ln r\,dy_1$$

$$(12.30)$$

$$\int_{A_1}\frac{(x_2-x_1)n_1-(y_2-y_1)l_1}{r^2}dA_1=\oint_{contour\ of\ A_1}\vec{v}_3\bullet\vec{ds}_1=\oint_{contour\ of\ A_1}\ln r\,dz_1$$

$$(12.31)$$

where $\quad\vec{v}_2=\hat{j}\ln r\quad$ and $\quad\vec{v}_3=\hat{k}\ln r$

Substituting Eqs. (12.29)-(12.31) in Eq. (21),

$$A_1 F_{A_1-A_2} = \frac{1}{2\pi} \oint_{contour\ of\ A_2} \left[\oint_{contour\ of\ A_1} \ln r\ dx_1 \right] dx_2$$

$$+ \frac{1}{2\pi} \oint_{contour\ of\ A_2} \left[\oint_{contour\ of\ A_1} \ln r\ dy_1 \right] dy_2$$

$$+ \frac{1}{2\pi} \oint_{contour\ of\ A_2} \left[\oint_{contour\ of\ A_1} \ln r\ dz_1 \right] dz_2$$

$$A_1 F_{A_1-A_2} = \frac{1}{2\pi} \oint_{contour\ of\ A_1} \oint_{contour\ of\ A_2} (\ln r\,dx_2 dx_1 + \ln r\,dy_2 dy_1 + \ln r\,dz_2 dz_1)$$

$$(12.32)$$

where $r = \sqrt{(x_2 - x_1)^2 + (y_2 - y_1)^2 + (z_2 - z_1)^2}$.

Example 12.2

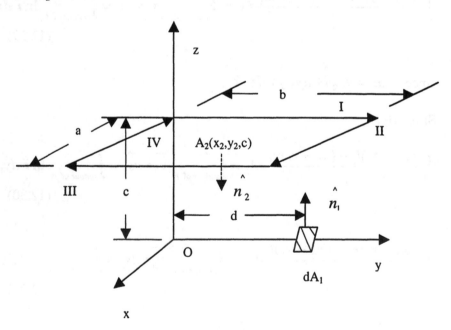

Figure 12.7   Evaluation of $F_{dA_1-A_2}$ by contour integration.

Problem: Consider an elemental surface dA_1 and a finite rectangular surface A_2 which are parallel to each other and positioned as shown in the figure above. Surface dA_1 is parallel to the xy plane and positioned on the oy axis at a distance d from the origin. Surface A_2 has one corner at the oz axis, and its sides a and b are parallel to the ox and oy axes, respectively. Find the diffuse view factor $F_{dA_1-A_2}$.

Solution

The coordinates of dA_1 are $x_1 = 0$, $y_1 = d$, $z_1 = 0$. The direction cosines of $\hat{n}_1$ to surface dA_1 are $l_1 = 0$, $m_1 = 0$, $n_1 = 1$. Substituting into Eq. (12.26),

$$F_{dA_1-A_2} = \frac{1}{2\pi} \oint_{\text{contour of } A_2} \frac{(y_2 - d)dx_2 - x_2 dy_2}{x_2^2 + (y_2 - d)^2 + c^2} \qquad (12.33)$$

where x_2, y_2, c are the coordinates of any point on surface A_2. We can divide contour A_2 into 4 segments, I, II, III, and IV as shown in Fig. 12.7.

Segment I: $x_2 = 0$, $dx_2 = 0$; then integral vanishes.
Segment II: $y_2 = b$, $dy_2 = 0$; x_2 varies from 0 to a.
Segment III: $x_2 = a$, $dx_2 = 0$; y_2 varies from b to 0.
Segment IV: $y_2 = 0$, $dy_2 = 0$; x_2 varies from a to 0.

Therefore,

$F_{dA_1-A_2}$

$$= \frac{1}{2\pi} \left[0 + \int_{x_2=0}^{a} \frac{b-d}{x_2^2 + (b-d)^2 + c^2} - \int_{y_2=b}^{0} \frac{a}{a^2 + (y_2-d)^2 + c^2} dy_2 - \int_{x_2=a}^{0} \frac{d}{x_2^2 + d^2 + c^2} dx_2 \right]$$

$$(12.34)$$

The integrals may be obtained using standard integral tables.

Example 12.3

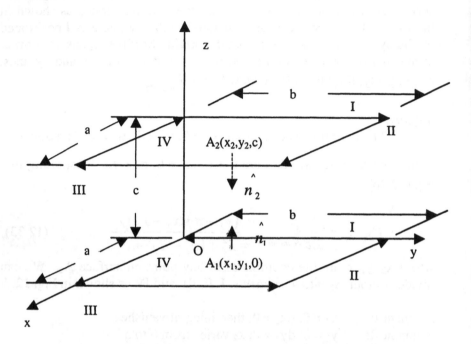

Figure 12.8   Evaluation of $F_{A_1-A_2}$ by contour integration.

Problem: Determine the diffuse view factor $F_{A_1-A_2}$ between the two parallel rectangular finite surfaces A_1 and A_2, separated by a distance c as illustrated above.

Solution
    $z_1 = 0$, $z_2 = c$ for surfaces A_1 and A_2, respectively. Also, $dz_1 = dz_2 = 0$. Equation (12.32) becomes

$$2\pi A_1 F_{A_1-A_2} = \oint_{contour\ of\ A_1} \oint_{contour\ of\ A_2} (\ln r\, dx_2\, dx_1 + \ln r\, dy_2\, dy_1). \quad (12.35)$$

Following the path around A_2, $dx_2 = 0$ for segments I and III, and $dy_2 = 0$ for segments II and IV.

$$2\pi ab F_{A_1-A_2} = \oint_{\text{contour of } A_1} \left\{ \int_{y_2=0}^{b} \ln\left[x_1^2 + (y_2 - y_1)^2 + c^2\right]^{\frac{1}{2}} dy_2 \right\} dy_1$$

$$+ \oint_{\text{contour of } A_1} \left\{ \int_{x_2=0}^{a} \ln\left[(x_2 - x_1)^2 + (b - y_1)^2 + c^2\right]^{\frac{1}{2}} dx_2 \right\} dx_1$$

$$+ \oint_{\text{contour of } A_1} \left\{ \int_{y_2=b}^{0} \ln\left[(a - x_1)^2 + (y_2 - y_1)^2 + c^2\right]^{\frac{1}{2}} dy_2 \right\} dy_1$$

$$+ \oint_{\text{contour of } A_1} \left\{ \int_{x_2=a}^{0} \ln\left[(x_2 - x_1)^2 + y_1^2 + c^2\right]^{\frac{1}{2}} dx_2 \right\} dx_1$$

$$(12.36)$$

Similarly, we follow the path of integration around A_1,

$$2\pi ab F_{A_1-A_2} = \int_{y_1=b}^{0} \int_{y_2=0}^{b} \ln\left[(y_2 - y_1)^2 + c^2\right]^{\frac{1}{2}} dy_2 dy_1$$

$$+ \int_{y_1=b}^{0} \int_{y_2=0}^{b} \ln\left[a^2 + (y_2 - y_1)^2 + c^2\right]^{\frac{1}{2}} dy_2 dy_1$$

+ Integrals for segments II, III and IV.

The resulting integrals can be found from standard integral tables.

$$2\pi ab F_{A_1-A_2} = \int_{y_1=0}^{b} \int_{y_2=0}^{b} \ln\left[\frac{a^2 + (y_2 - y_1)^2 + c^2}{(y_2 - y_1)^2 + c^2}\right] dy_2 dy_1$$

$$+ \int_{x_1=0}^{a} \int_{x_2=0}^{a} \ln\left[\frac{(x_2 - x_1)^2 + b^2 + c^2}{(x_2 - x_1)^2 + c^2}\right] dx_2 dx_1 \qquad (12.37)$$

$$F_{A_1-A_2} = \frac{2}{\pi AB}\left\{\ln\left[\frac{(1+A^2)(1+B^2)}{1+A^2+B^2}\right]^{\frac{1}{2}} + A\sqrt{1+B^2}\,\tan^{-1}\left[\frac{A}{\sqrt{1+B^2}}\right]\right.$$

$$\left. + B\sqrt{1+A^2}\,\tan^{-1}\left[\frac{B}{\sqrt{1+A^2}}\right] - A\tan^{-1}A - B\tan^{-1}B\right\}$$

$$(12.38)$$

where $A = a/c$, $B = b/c$.

12.5   Relations Between View Factors

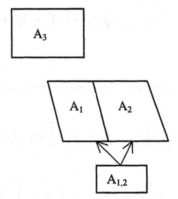

Figure 12.9   Relations between view factors.

The view factor from surface A_3 to surfaces A_1 and A_2 together may be written as $F_{3-1,2}$. This may be expressed in terms of the view factor from surface A_3 to surface A_1 and the view factor from surface A_3 to surface A_2. The total view factor is the sum of its parts.

$$F_{3-1,2} = F_{3-1} + F_{3-2} \qquad (12.39)$$
$$A_3\,F_{3-1,2} = A_3\,F_{3-1} + A_3\,F_{3-2}. \qquad (12.40)$$

Making use of the reciprocity relations

$$A_3\,F_{3-1,2} = A_{1,2}\,F_{1,2-3}. \qquad (12.41)$$

Also, $A_3\,F_{3-1} = A_1\,F_{1-3}$ $\qquad (12.42)$

$$A_3 F_{3\text{-}2} = A_2 F_{2\text{-}3}. \tag{12.43}$$

The expression then becomes

$$A_{1,2} F_{1,2\text{-}3} = A_1 F_{1\text{-}3} + A_2 F_{2\text{-}3}. \tag{12.44}$$

This states that the total radiative energy reaching surface A_3 is the sum of the radiative energies from surfaces A_1 and A_2.

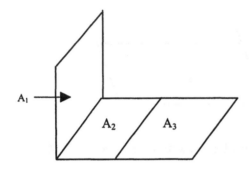

Figure 12.10   View factor of perpendicular rectangles with a common edge.

For perpendicular rectangles as shown in Fig. 12.10, one method of finding the view factors is as described below. These view factors have to be expressed in terms of perpendicular rectangles with a common edge, because charts exist only for such cases (Incropera and DeWitt, 2002 [1], Kreith and Bohn, 2001 [2], Ozisik, 1985 [3]). The reason is that an infinite number of charts cannot be prepared for an infinite number of combinations of shapes and configurations. The exercise is to express view factors between perpendicular rectangles in terms of view factors between perpendicular rectangles with a common edge. In Fig. 12.10,

$$F_{1\text{-}2,3} = F_{1\text{-}2} + F_{1\text{-}3} \tag{12.45}$$

$F_{1\text{-}2,3}$ and $F_{1\text{-}2}$ may be determined from the charts. Hence, $F_{1\text{-}3}$ can be determined.

$$F_{1\text{-}3} = F_{1\text{-}2,3} - F_{1\text{-}2} \tag{12.46}$$

Example 12.4

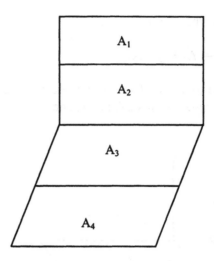

Figure 12.11   Sketch for Example 12.4.

Problem: Determine the view factor F_{1-4} for the geometry shown in Fig. 12.11.   The expression has to be in terms of view factors for perpendicular rectangles with a common edge.

Solution
In accordance with Eq. (12.40),

$$A_{1,2}\, F_{1,2\text{-}3,4} = A_1\, F_{1\text{-}3,4} + A_2\, F_{2\text{-}3,4} \qquad (12.47)$$

$F_{1,2\text{-}3,4}$ and $F_{2\text{-}3,4}$ can be obtained from the charts.

$$A_1\, F_{1\text{-}3,4} = A_1\, F_{1\text{-}3} + A_1\, F_{1\text{-}4} \qquad (12.48)$$
$$A_{1,2}\, F_{1\text{-}2,3} = A_1\, F_{1\text{-}3} + A_2\, F_{2\text{-}3} \qquad (12.49)$$

Eq. (12.48)-Eq. (12.49), $A_1\, F_{1\text{-}3,4} = A_{1,2}\, F_{1,2\text{-}3} + A_1\, F_{1\text{-}4} - A_2\, F_{2\text{-}3}$ (12.50)
Substituting Eq.(12.50) in Eq.(12.47),

$$A_{1,2}\, F_{1,2\text{-}3,4} = A_{1,2}\, F_{1,2\text{-}3} - A_2 F_{2\text{-}3} + A_1\, F_{1\text{-}4} + A_2\, F_{2\text{-}3,4} \qquad (12.51)$$

All view factors in Eq.(12.51) except F_{1-4} may be determined from a chart, [1]-[3]. Hence,

$$F_{1-4} = \quad (A_{1,2}\,F_{1,2-3} - A_2 F_{2-3} + A_1\,F_{1-4} + A_2\,F_{2-3,4}\,)/A_1. \qquad (12.52)$$

Since the surfaces were flat, $F_{11} = F_{22} = F_{33} = 0$.

12.6   Diffuse View Factor Between an Elemental Surface and an Infinitely Long Strip

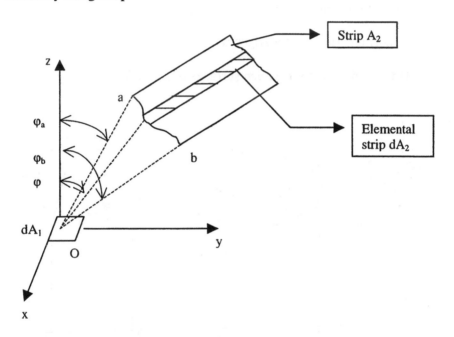

Figure 12.15   An elemental surface and an infinitely long strip.

Consider an elemental surface dA_1 lying on the xy plane at the origin and an infintely long strip A_2 whose generating lines are parallel to the x-axis as shown in the figure. Let ab be the contor of intersection of strip A_2 with the yz plane. Let φ_a, φ_b be the angles between the oz axis and the lines oa and ob. The elemental view factor between the elemental area dA_1 and the elemental strip dA_2 is given as

$$dF_{dA_1-strip\ dA_2} = \frac{1}{2}\cos\varphi d\varphi = \frac{1}{2}d(\sin\varphi) \tag{12.53}$$

Integrating,

$$F_{dA_1-strip\ A_2} = \int_{\varphi_1}^{\varphi_2} \frac{1}{2}\cos\varphi d\varphi = \frac{1}{2}(\sin\varphi_b - \sin\varphi_a) \tag{12.54}$$

This relationship is applicable also when dA_1 is an elemental strip on the xy plane parallel to the ox axis. The original derivation of this relationship was done by Jacob in 1957 [4].

12.7   Diffuse View Factor Algebra

(a) Diffuse view factor between surfaces dA_1 and A_2.

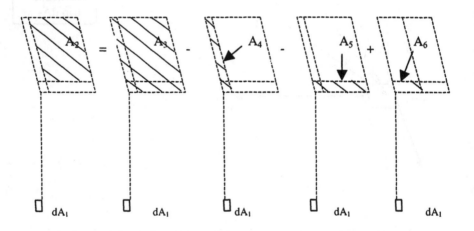

Figure 12.16   View factor between surfaces dA_1 and A_2.

Since $A_2 = A_3 - A_4 - A_5 + A_6$,

$$F_{dA_1-A_2} = F_{dA_1-A_3} - F_{dA_1-A_4} - F_{dA_1-A_5} + F_{dA_1-A_6}. \tag{12.55}$$

The relationship is true because dA_1 is an elemental surface and the cross-effects are eliminated.

To use diffuse view factor algebra effectively in a complex problem, one has to recognize the corresponding simple case as a first step in solving the problem. In the next subsection, the view factor between two arbitrary rectangles are expressed in terms of view factors of perpendicular rectangles that share a common edge (the view factors of which are easily available in formulae or charts [1-3]). Then, Example 14.5 is presented as a more complex case from that discussed in subsection (b). One uses the results of the simple case and applies them to the more complex case.

(b) Diffuse view factor between surfaces A_1 and A_2'.

We define $G_{\alpha-\beta} \equiv A_\alpha F_{\alpha-\beta}$. The reciprocity principle gives us

$G_{\alpha-\beta} = G_{\beta-\alpha}$.

In general, if the surfaces A_1 and A_2 are subdivided according to the following:

$$A_1 = A_i + A_j$$
$$A_2 = A_k + A_l$$

The diffuse view factor between the surfaces of this composite system obeys the following laws of arithmetic:

$$G_{1-2} \equiv G_{ij-kl} = G_{ij-k} + G_{ij-l}$$
$$= G_{i-kl} + G_{j-kl}$$
$$= G_{i-k} + G_{i-l} + G_{j-k} + G_{j-l}$$

where $G_{ij-kl} \equiv \left(A_i + A_j\right) F_{(A_i + A_j)-(A_k - A_l)}$

$G_{i-kl} = A_i F_{A_i - (A_k - A_l)}$, etc.

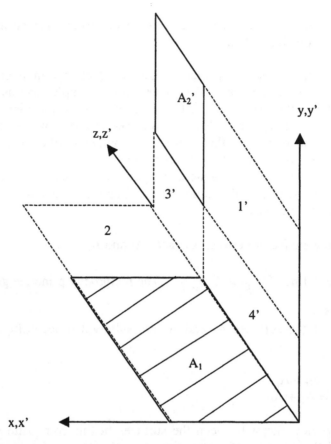

Figure 12.17   Arrangement of surfaces for diffuse view factor $F_{A_1-A_2}$.

Expand the view factor $G_{12-1'2'3'4'}$ according to the above laws of arithmetic,

$$G_{12-1'2'3'4'} = G_{12-1'2'} + G_{12-3'4'} = (G_{1-1'} + G_{1-2'} + G_{2-1'} + G_{2-2'}) + G_{12-3'4'}.$$

(12.56)

In Eq. (12.56), $G_{1-2'}$ is the diffuse view factor required. $G_{12-1'2'3'4'}, G_{12-3'4'}$ can be evaluated from a chart for view factors between perpendicular rectangles with a common edge [1-3]. View factors $G_{1-1'}, G_{2-1'}, G_{2-2'}$ are obtained from a chart.   Use the following relationships:

$$G_{1-1'} = G_{1-1'4'} - G_{1-4'} \qquad (12.57a)$$
$$G_{2-2'} = G_{2-2'3'} - G_{2-3'} \qquad (12.57b)$$

The right hand side of Eqs. (12.57a-b) can be obtained from the chart. It can be shown that

$$G_{2-1'} = G_{1-2'} \qquad (12.58)$$

The relationship, Eq. (12.58), has been shown by Frank Kreith in 1962 [5]. Substituting Eqs. (12.57a,b) and (12.58) into Eq. (12.56), we obtain

$$2G_{1-2'} = G_{12-1'2'3'4'} + G_{1-4'} + G_{2-3'} - G_{1-1'4'} - G_{2-2'3'} - G_{12-3'4'} \qquad (12.59)$$

All terms on the right side of Eq. (12.59) are obtainable from a chart [1]-[3].

Example 12.5

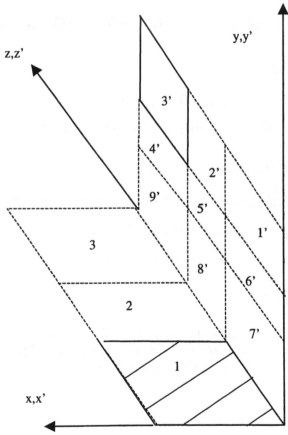

Figure 12.18 Sketch for Example 12.5.

Problem: Find $G_{1-3'}$.

Solution
       From Eq. (12.59) and Fig. 12.18, we can deduce that

$G_{12-3'} = 0.5[G_{123-1'2'3'4'5'6'7'8'9'} + G_{12-5'6'7'8'} + G_{3-4'9'} - G_{12-1'2'5'6'7'8'} - G_{3-3'4'9'}$
$- G_{123-4'5'6'7'8'9'}]$ (12.60)

But $G_{12-3'} = G_{1-3'} + G_{2-3'}$

and

$$G_{2-3'} = \frac{1}{2}[G_{23-2'3'4'5'8'9'} + G_{2-5'8'} + G_{3-4'9'} - G_{2-2'5'8'} - G_{3-3'4'9'} - G_{23-4'5'8'9'}]$$

$G_{1-3'} = G_{12-3'} - G_{2-3'}$
$$= \frac{1}{2}[G_{123-1'2'3'4'5'6'7'8'9'} + G_{12-5'6'7'8'} + G_{3-4'9'} - G_{12-1'2'5'6'7'8'} - G_{3-3'4'9'} - G_{123-4'5'6'7'8'9'}$$
$$- G_{23-2'3'4'5'8'9'} - G_{2-5'8'} - G_{3-4'9'} + G_{2-2'5'8'} + G_{3-3'4'9'} + G_{23-4'5'8'9'}]$$

$G_{1-3'} = G_{12-3'} - G_{2-3'}$
$$= \frac{1}{2}[G_{123-1'2'3'4'5'6'7'8'9'} + G_{12-5'6'7'8'} - G_{12-1'2'5'6'7'8'} - G_{123-4'5'6'7'8'9'}$$
$$- G_{23-2'3'4'5'8'9'} - G_{2-5'8'} + G_{2-2'5'8'} + G_{23-4'5'8'9'}].$$
                                                                    (12.61)

PROBLEMS

12.1.   Determine the view factor F_{d1-2} from an element dA_1 to a right
         triangle BCD as shown in the figure. The sides of the triangle
         are BC = a, CD = b and DB = c.

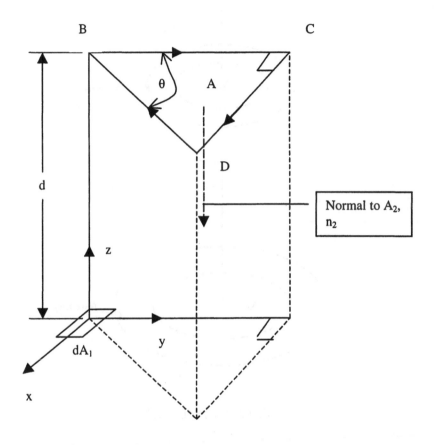

Figure for Prob. 12.1.

12.2.   The view factor is known between two parallel disks of any finite size whose centers lie on the same axis. From this, find the view factor between the two rings A_2 and A_3 of the figure. Supply the result in terms of disk-to-disk factors from disk areas on the lower level to disk areas on the upper level.

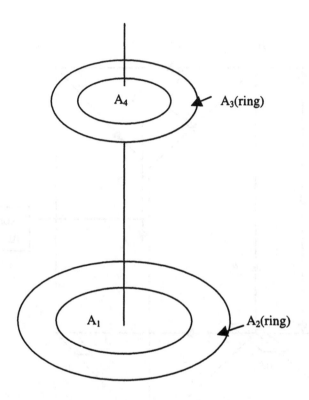

Figure for Prob. 12.2.

12.3.   The curved internal surface of a hollow circular cylinder of
          radius α is radiating to a disk A_4 of radius β. Find the view
          factor from the curved side A_2 to the disk in terms of disk-to-disk
          factors for the case of β less than α.

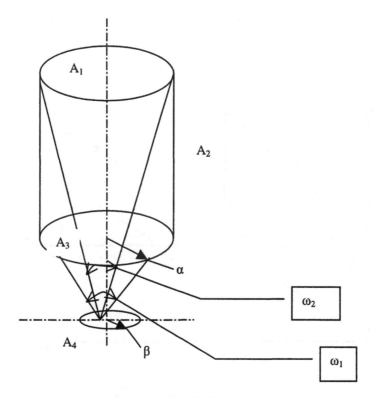

Figure for Prob. 12.3.

12.4.   The figure shows four areas on two perpendicular rectangles having a common edge. Show the validity of the relation $A_1F_{1\text{-}2} = A_3F_{3\text{-}4}$.

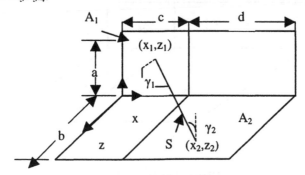

Figure for Prob. 12.4.

12.5.   Find the expression for the view factor product $G_{2-1'}$.

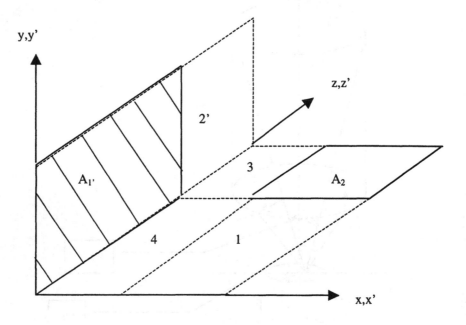

Figure for Prob. 12.5.

12.6.   If the view factor is known for two perpendicular rectangles with a common edge, derive the view factor F_{7-6} for the figure shown.

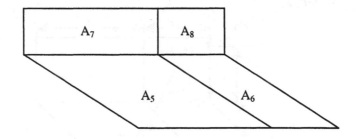

Figure for Prob. 12.6.

12.7.   If the view factor is known for two perpendicular rectangles with a common edge, derive the view factor F_{1-6} for the figure shown. Use the results from Prob. 12.6.

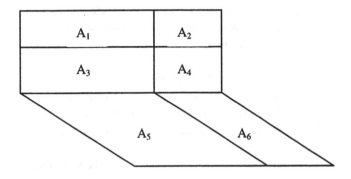

Figure for Prob. 12.7.

REFERENCES

1. F P Incropera and D P DeWitt. Fundamentals of Heat and Mass Transfer, 5th ed. New York: John Wiley & Sons, 2002.

2. F Kreith and M S Bohn. Principles of Heat Transfer, 6th ed. Pacific Grove, CA: Brooks/Cole, 2001.

3. M N Ozisik. Heat Transfer, A Basic Approach, 1st ed. New York: McGraw-Hill, 1985.

4. M. Jacob. Heat Transfer, Vol.1. New York: John Wiley & Sons, 1957.

5. F Kreith. Radiation Heat Transfer for Spacecraft and Solar Power Design. Scranton, PA: International Textbook Company, 1962.

View Factor

Thermal radiation includes the visible spectrum
Use your eye to help with concept of view factor
If an object cannot be seen by a heat source,
The object cannot receive radiation from that source.

There is the reciprocity theorem for the view factor
For any object, sum to unity for all view factors
View factor to itself is zero for plane surfaces
View factor to itself is zero for convex surfaces.

K.V. Wong

13

Radiation Exchange in Long Enclosures

In long enclosures, the radiation problem essentially reduces to a two-dimensional problem. Under these conditions, a particular useful theorem, Hottel's theorem, applies. This chapter presents many practical examples.

13.1 Diffuse View Factor in Long Enclosures

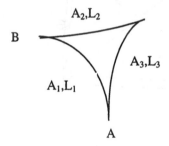

Figure 13.1   Diffuse view factor between the surfaces of a long enclosure.

Consider an enclosure shown above, consisting of three surfaces which are very long in the direction perpendicular to the plane of the figure. The summation rule for view factors gives

$$\sum_{j=1}^{3} F_{i-j} = 1, \quad i = 1,2,3 \tag{13.1a}$$

with $F_{jj} = 0.$ (13.1b)

There is the assumption that the surfaces are flat or convex. The reciprocity rule gives

$$A_i F_{i-j} = A_j F_{j-i}, \quad i, j = 1,2,3. \tag{13.2}$$

319

The objective is to find F_{1-2}. Solving Eqs. (13.5) and (13.6), we obtain

$$A_1 F_{1-2} = \frac{A_1 + A_2 - A_3}{2} \qquad (13.3a)$$

which can be written as $L_1 F_{1-2} = \dfrac{L_1 + L_2 - L_3}{2}.$ $\qquad$ (13.3b)

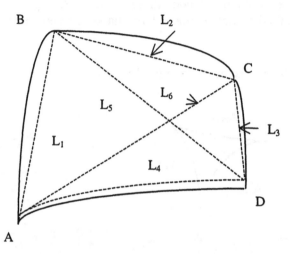

Figure 13.2   Diffuse view factor between the surfaces of a long enclosure.

Consider an enclosure consisting of 4 surfaces, very long in the direction perpendicular to the plane of the figure, Fig. 13.2. The surfaces of the enclosure can be flat, convex or concave (restriction of $F_{jj} = 0$ is removed). Assume that imaginary strings (shown by dotted lines) are tightly stretched among the corners A, B, C and D.

Let L_i, i = 1,2,3,4,5,6 denote the lengths of strings joing the corners A-B, B-C, C-D, D-A, D-B, and A-C, respectively. The objective is to find F_{AB-CD} from the surface AB to the surface CD.

Consider the imaginary enclosures ABC and ABD formed by imaginary strings. By application of Eq. (13.3b), we can determine $L_1 F_{1-2}$ and $L_1 F_{1-4}$ for the imaginary enclosures ABC and ABD, respectively.

The summation rule gives

$$F_{1-2} + F_{1-3} + F_{1-4} = 1 \qquad (F_{1-1} = 0) \qquad (13.4)$$

Substitution of F_{1-2} and F_{1-4} evaluated from Eq. (13.3b) into the summation rule gives

$$L_1 F_{1-3} = \frac{(L_5 + L_6) - (L_2 + L_4)}{2}. \qquad (13.5)$$

Also, $L_4 F_{4-2} = \frac{1}{2}[(L_5 + L_6) - (L_1 + L_3)]$. It can be shown that $L_1 F_{1-3} =$ AB $F_{AB\text{-}CD}$ where AB and CD characterize the curved surfaces. This is the Hottel's crossed-string theorem. Note that $(L_5 + L_6)$ in Eq.(13.5) is the sum of the crossed strings, and ($L_2 + L_4$) is the sum of the uncrossed strings. In other words, the right-hand side of the equation is equal to one-half of the total quantity formed by the sum of the lengths of crossed strings connecting the outer edges of areas A_1 and A_2 minus the sum of the lengths of the uncrossed strings.

The rest of this chapter discusses practical examples where Hottel's crossed-stringed theorem may be used effectively.

Example 13.1

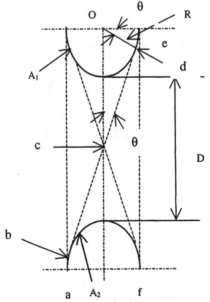

Figure 13.3   Sketch for Ex.13.1.

Problem: Two infinitely long semicylindrical surfaces of radius R are separated by a minimum distance D as shown in the figure. Derive the view factor F_{1-2} for this case.

Solution
        Let the length of the crossed string abcde = L_1, and the length of the uncrossed string ef = L_2 = D + 2R. From symmetry,

$$F_{1-2} = \frac{2L_1 - 2L_2}{2A_1} = \frac{L_1 - L_2}{\pi R}.$$

The segment of L_1 from c to d is found from right-angled triangle Ocd,

$$L_{1,c-d} = \left[\left(\frac{D}{2} + R \right)^2 - R^2 \right]^{\frac{1}{2}} = \left[D \left(\frac{D}{4} + R \right) \right]^{\frac{1}{2}}$$

$$L_{1,d-c} = R\theta.$$

From triangle Ocd, $\theta = \sin^{-1} \dfrac{R}{D/2 + R}$

$$F_{1-2} = \frac{L_1 - L_2}{\pi R} = \frac{2(L_{1,c-d} + L_{1,d-e}) - L_2}{\pi R}$$

$$= \frac{\left[4D \left(\frac{D}{4} + R \right) \right]^{\frac{1}{2}} + 2R \sin^{-1} \left[\frac{R}{(D/2+R)} \right] - D - 2R}{\pi R}.$$

Letting X = 1+D/(2R),

$$F_{1-2} = \frac{2}{\pi} \left[(x^2 - 1)^{\frac{1}{2}} + \sin^{-1} \left(\frac{1}{X} \right) - X \right]$$

$$F_{1-2} = \frac{2}{\pi} \left[(x^2 - 1)^{\frac{1}{2}} + \frac{\pi}{2} - \cos^{-1} \left(\frac{1}{X} \right) - X \right].$$

Example 13.2

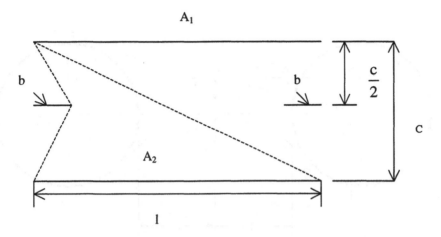

Figure 13.4 Partially blocked view between parallel strips.

Problem: The view between two infinitely long parallel strips of width l is partially blocked by opaque strips of width b as shown in the figure above. Find $F_{1\text{-}2}$.

Solution

$$\text{Length of each crossed string} = \sqrt{l^2 + c^2}$$

$$\text{Length of each uncrossed string} = 2\sqrt{b^2 + \left(\frac{c}{2}\right)^2}$$

From Hottel's crossed-string method,

$$F_{1\text{-}2} = \frac{\sqrt{l^2 + c^2} - 2\sqrt{b^2 + (c/2)^2}}{l} = \sqrt{1 + \left(\frac{c}{l}\right)^2} - \sqrt{\left(\frac{2b}{l}\right)^2 + \left(\frac{c}{l}\right)^2}$$

It is a good practice to check the solution in the limits if the answer is known at the limits. In this case, as b tends to 0.5l, $F_{1\text{-}2}$ tends to zero. This limit makes sense because the view factor becomes zero as the partial blockage becomes full blockage.

Example 13.3

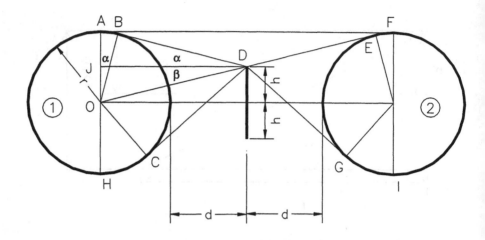

Figure 13.5   Figure for Example 13.3.

Problem: As shown in the figure above, two long cylinder pipes, with different temperatures, run horizontally parallel to each other. Both have radius r and the distance is $2d$ between them. A long opaque plate is placed at the middle and it is parallel to the pipes. Study the relationship between the value of the view factor F_{1-2} and the height of the plate.

<u>*Solution*</u>

In this case, both pipes and plate are long, therefore it is a two-dimensional long enclosure problem. Hottel's cross-string method can be used.

Study the geometry in the figure above. It is symmetric in both vertical and horizontal direction. The view factor can be computed by calculating the upper and lower paths respectively and adding them. Because of the symmetry, only the upper path need to be considered.

The sum of the length of crossed strings:

$$L_{A-B-D-G-I} + L_{H-C-D-E-F}$$

The sum of the length of uncrossed strings:

$$L_{A-F} + L_{H-C-D-G-I}$$

Therefore, the view factor F_{1-2} is calculated as following,

$$L_{A-B-C-H} F_{1-2} = 2\frac{L_{A-B-D-G-I} + L_{H-C-D-E-F} - L_{A-F} - L_{H-C-D-G-I}}{2}$$

The factor '2' in the numerator comes from 2 similar paths. After rearrangement,

$$F_{1-2} = \frac{L_{A-B-D-G-I} + L_{H-C-D-E-F} - L_{A-F} - L_{H-C-D-G-I}}{L_{A-B-C-H}}$$

Here,

$$L_{A-B-C-H} = \pi r$$

Therefore, after simplification,

$$F_{1-2} = \frac{L_{A-B} + L_{B-D} + L_{D-E} + L_{E-F} - L_{A-F}}{\pi r}$$

Because of the symmetry about the vertical axis,

$$F_{1-2} = \frac{2(L_{A-B} + L_{B-D}) - L_{A-F}}{\pi r} = \frac{2(L_{A-B} + L_{B-D}) - 2(r+d)}{\pi r} \qquad (i)$$

Study each segment in the equation above,

(i) L_{B-D}

$$L_{B-D} = \sqrt{OD^2 - OB^2}$$

where,

$$OD^2 = (r+d)^2 + h^2$$

Therefore,

$$L_{B-D} = \sqrt{(r+d)^2 + h^2 - r^2} = \sqrt{d^2 + 2dr + h^2} \qquad (ii)$$

(ii) L_{A-B}

$$L_{A-B} = r\alpha \qquad (iii)$$

Study the geometry relationship in the figure above,

$$\alpha + \beta = \angle BDO = \tan^{-1}\left(\frac{OB}{BD}\right) = \tan^{-1}\left(\frac{r}{\sqrt{d^2 + 2dr + h^2}}\right) \qquad \text{(iv)}$$

$$\beta = \angle JDO = \tan^{-1}\left(\frac{JO}{JD}\right) = \tan^{-1}\left(\frac{h}{r+d}\right) \qquad \text{(v)}$$

Therefore,

$$\alpha = (\alpha + \beta) - \beta = \tan^{-1}\left(\frac{r}{\sqrt{d^2 + 2dr + h^2}}\right) - \tan^{-1}\left(\frac{h}{r+d}\right) \qquad \text{(vi)}$$

Substitute Eq. (vi) into Eq. (iii),

$$L_{A-B} = r\left[\tan^{-1}\left(\frac{r}{\sqrt{d^2 + 2dr + h^2}}\right) - \tan^{-1}\left(\frac{h}{r+d}\right)\right] \qquad \text{(vii)}$$

Substitute the expressions Eqs. (vii) and (ii) into Eq. (i),

$$F_{1-2} = 2\frac{r\left[\tan^{-1}\left(\frac{r}{\sqrt{d^2 + 2dr + h^2}}\right) - \tan^{-1}\left(\frac{h}{r+d}\right)\right] + \sqrt{d^2 + 2dr + h^2} - (r+d)}{\pi r}$$

$$\text{(viii)}$$

When h decreases to zero, the value of F_{1-2} becomes maximum.

$$F_{1-2} = 2\frac{r\left[\tan^{-1}\left(\frac{r}{\sqrt{d^2 + 2dr}}\right)\right] + \sqrt{d^2 + 2dr} - (r+d)}{\pi r} \qquad \text{(ix)}$$

F_{1-2} decreases as h increases. It reaches its minimum value 0 at $h = r$.

Example 13.4

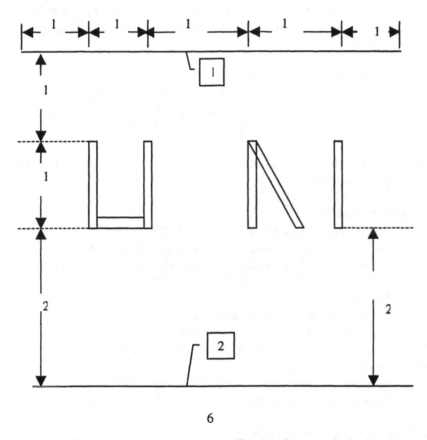

Figure 13.6 Damaged sign. Dimensions are in inches.

Problem: The sign above is drawn to scale in inches. The "M' letter in the sign (shown in the figure) is defective because one of the "slanted" parts has fallen off. Find the view factor between surfaces 1 and 2 before and after the defect takes place, and hence the change in the value of the view factor because of the defect. Assume that the letters are opaque to radiation.

Solution

      The paths of radiation from surface 1 to 2 may be identified as the left-most path, the middle path and the right-most path. Consider the left-most path.

      Sum of crossed strings = 7.6875" + 8.125" = 15.8125"
      Sum of uncrossed strings = 4" + 11.4375" = 15.4375"
By Hottel's crossed-string method, denoting the view factors by a '.
      $A_1F_{1-2}' = 0.5(15.8125" - 15.4375") = 0.1875$
      $F_{1-2}' = 0.03125$
Owing to symmetry, the right-most path is the same as the left-most path. Denoting the view factors by a ''',
      $F_{1-2}''' = F_{1-2}' = 0.03125$
Consider the middle path.
      Sum of crossed strings = 7.236" + 7.303" = 14.539"
      Sum of uncrossed strings = 6.064" + 6.895" = 12.959"
By Hottel's crossed-string method, denoting the view factors by a ''.
      $A_1F_{1-2}'' = 0.5(14.539" - 12.959") = 0.79$
      $F_{1-2}'' = 0.132$
Hence, the view-factor before the defect is
      $F_{1-2} = F_{1-2}' + F_{1-2}'' + F_{1-2}''' = 0.198$.
After the defect, there are 4 paths. For the path due to the defect, we denote the view factors by a *.
      Sum of crossed strings = 7.625" + 7.375" = 15"
      Sum of uncrossed strings = 9.8125" + 5" = 14.8125"

By Hottel's crossed-string method,
      $A_1F_{1-2}* = 0.5(15" - 14.8125") = 0.09375$
      $F_{1-2}* = 0.0156$, which is the change of view factor due to the defect.
$(F_{1-2})_{after\ defect} = (F_{1-2})_{before\ defect} + F_{1-2}* = 0.214$.

      The discussion is how would one modify Hottel's crossed-string method when the obstructing objects are not completely opaque, i.e. $\tau \neq 0$. The procedure is a two-step one. First, the view-factor between surface 1 and surface 2 that is attributable to the obstructing objects (i.e., when $\tau = 0$) have to be found. The second step is to multiply this view factor by the non zero τ. Example 13.5 below shows the procedure.

Example 13.5

Problem: For the figure shown, the objects A and B have a nonzero transmissivity of τ. Find the view factor between surface 1 and surface 2.

Solution
The paths of radiation from surface 1 to 2 may be identified as the left-most path, the middle path and the right-most path. Consider the left-most path.

Sum of crossed strings =
$$\sqrt{a^2 + \frac{9a^2}{4}} + \sqrt{25a^2 + \frac{9a^2}{4}} + \sqrt{a^2 + \frac{25a^2}{4}} + \sqrt{25a^2 + \frac{a^2}{4}}$$

Sum of uncrossed strings $= 3a + \left[a + \sqrt{25a^2 + \frac{a^2}{4}} + \sqrt{25a^2 + \frac{9a^2}{4}} \right]$

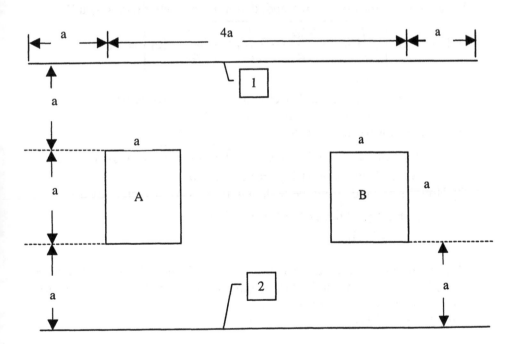

6
Figure 13.7   Sketch for Example 13.5

By Hottel's crossed-string method, denoting the view factors by a '.

$$A_1F_{1\text{-}2}' = 0.5\left[\sqrt{a^2 + \frac{9a^2}{4}} + \sqrt{a^2 + \frac{25a^2}{4}} - 4a\right] = 0.298a$$

$$F_{1\text{-}2}' = 0.0413.$$

Owing to symmetry, the right-most path is the same as the left-most path. Denoting the view factors by a ''',

$$F_{1\text{-}2}''' = F_{1\text{-}2}' = 0.0413$$

Consider the middle path.

$$\text{Sum of crossed strings} = 2\sqrt{4a^2 + \frac{a^2}{4}} + 2\sqrt{16a^2 + \frac{25a^2}{4}}$$

$$\text{Sum of uncrossed strings} = 2a + 2\sqrt{4a^2 + \frac{a^2}{4}} + 2\sqrt{4a^2 + \frac{9a^2}{4}}$$

By Hottel's crossed-string method, denoting the view factors by a '',

$$A_1F_{1\text{-}2}'' = 0.5\left[2\sqrt{16a^2 + \frac{25a^2}{4}} - 2\sqrt{4a^2 + \frac{9a^2}{4}} - 2a\right] = 1.217a$$

$$F_{1\text{-}2}'' = 0.203$$

Hence, if the objects A and B were opaque, the view-factor is

$$F_{1\text{-}2} = F_{1\text{-}2}' + F_{1\text{-}2}'' + F_{1\text{-}2}''' = 0.2856.$$

If the objects were not present,

$$\text{Sum of the crossed strings} = 2\sqrt{9a^2 + 36a^2} = 2a\sqrt{45}$$

$$\text{Sum of uncrossed strings} = 3a + 3a = 6a$$

By Hottel's crossed-string method, denoting the view factors by a *.

$$A_1F_{1\text{-}2}^* = 0.5\left[2a\sqrt{45} - 6a\right] = 3.708a$$

$$F_{1\text{-}2}^* = 0.618$$

Hence, the view factor attributable to the objects is $0.618 - 0.2856 = 0.3324$. If the transmissivity through the objects is τ, then the view factor with the translucent objects is

$$= 0.2856 + 0.3324\tau.$$

Example 13.6

Problem: Most of the Earth's thermal energy is received from the short wave radiation of the sun. Although it receives radiation from other bodies in space, it is negligible compared to with the solar energy. Incoming solar energy is at approximately at the same intensity as when it left the surface of the sun, before it enters the earth's atmosphere. However once it enters the atmosphere approximately 6% is reflected by particles in the atmosphere, 16% is absorbed by the atmosphere, 20-30% is reflected by the clouds, and 3% is absorbed by the clouds. On any given day all of these factors can limit the amount of net solar radiation received by a solar panel.

The objective of this problem is to calculate the view factor of a solar field, taking into account all of the above-mentioned facts.

Solution

First assume that the clouds and the atmosphere act as translucent bodies absorbing or blocking the radiation coming from the sun and calculate their transmissivity (τ) accordingly. For example, let us calculate the transmissivity of the atmosphere on a clear day with no clouds. Since there are no clouds the radiation reaching the earth's surface is: 100%(from sun) − 16% (absorbed by atmosphere) − 6% (reflected by atmosphere).

$$\tau_{atm} = 1 - (0.16 + 0.06) = 0.78$$

For upper level clouds assume they are less dense lower level clouds and only reflect 15% and absorb 2% of the radiation. Therefore

$$\tau_{up\text{-}cloud} = 1 - (0.15 + 0.06) = 0.83$$

For lower level clouds assume 30% reflected and 3% absorbed.

$$\tau_{low\text{-}cloud} = 1 - (0.3 + 0.03) = 0.67$$

Here is the problem layout assuming the surface of the sun is flat because its radius is so large compared with the earth's.

Calculate the view factor using Hottel'string theorem assuming the clouds to be opaque. (The view factor should be calculated assuming no radiation passes through clouds; then calculate the view factor assuming open air gaps are opaque and clouds are translucent, and add them. The atmosphere is also translucent.)

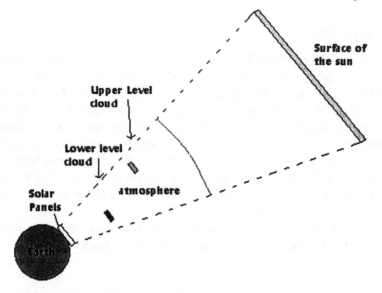

Figure 13.8 and 13.9   First and second figures for Example 13.6

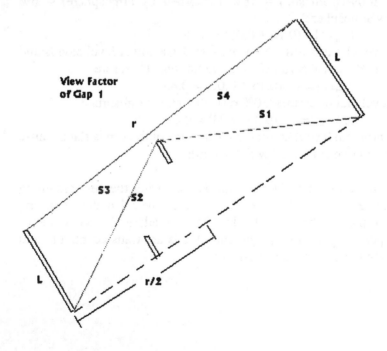

The gap spacing and cloud lengths are as follows:

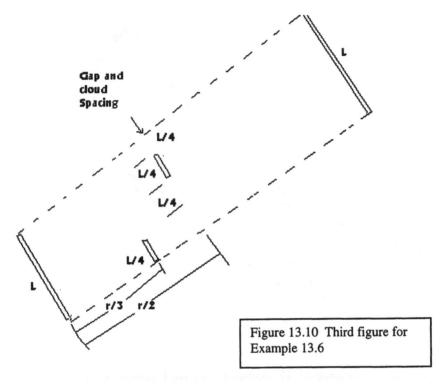

Figure 13.10 Third figure for Example 13.6

The view factor from Hottel's theorem of the first gap shown in Figure 13.9 is

$$F_{gap} = \frac{1}{L} \left\{ \left[\sqrt{\left(\frac{3L}{4}\right)^2 + \left(\frac{r}{2}\right)^2} + \sqrt{\left(\frac{r}{2}\right)^2 + \left(\frac{L}{4}\right)^2} \right] - \left[r + 2\sqrt{r^2 + \left(\frac{L}{4}\right)^2} \right] \right\}$$

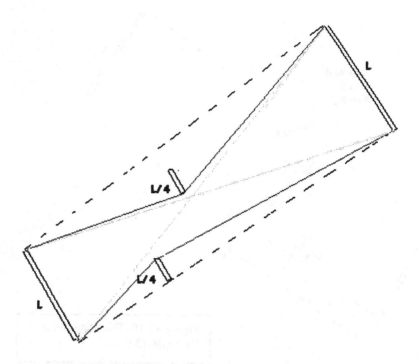

Figure 13.11 Fourth figure for Example 13.6

For the second gap the schematic is shown in Fig. 13.11. From Hottel's theorem for gap 2,

$$F_{gap-2} = \frac{1}{2L}\left\{\left[2\sqrt{r^2+L^2}\right]-\left[2\sqrt{\left(\frac{L}{2}\right)^2+\left(\frac{r}{2}\right)^2}+\sqrt{\left(\frac{r}{3}\right)^2+\left(\frac{L}{4}\right)^2}+\sqrt{\left(\frac{2r}{3}\right)^2+\left(\frac{L}{4}\right)^2}\right]\right\}$$

The view factors for each gap must be multiplied by the transmissivity of the atmosphere to get the actual view factors from the sun to the solar panel assuming the clouds are completely opaque.

$$F_{\text{sun-panel-opaque}} = \tau_{\text{atm}}\left(F_{\text{gap-1}} + F_{\text{gap-2}}\right) = 0.78\left(F_{\text{gap-1}} + F_{\text{gap-2}}\right)$$

Now the view factors for the two translucent clouds must be added to the view factor between the sun and the panel assuming the clouds where opaque.

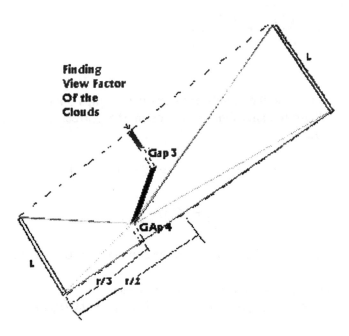

Figure 13.12 Fifth figure for Example 13.6

The view factor for the cloud can be found by making the previous gaps 1 and 2 closed off and finding view factors for gaps 3 and 4 and then multiplying by their respective transmissivity.

The view factors for gaps 3 and 4 are as follows:

$$F_{gap-3} = \frac{1}{2L}\left\{ 2\left[\sqrt{\left(\frac{L}{4}\right)^2 + \left(\frac{r}{2}\right)^2} + 2\sqrt{\left(\frac{r}{2}\right)^2 + \left(\frac{L}{2}\right)^2}\right] - \left[2\sqrt{\left(\frac{L}{4}\right)^2 + \left(\frac{r}{2}\right)^2} + 2\sqrt{\left(\frac{r}{2}\right)^2 + \left(\frac{L}{2}\right)^2}\right.$$

$$F_{gap-4} = \frac{1}{2L}\left\{ \sqrt{\left(\frac{3L}{4}\right)^2 + \left(\frac{2r}{3}\right)^2} + \sqrt{\left(\frac{L}{4}\right)^2 + \left(\frac{r}{3}\right)^2} + \sqrt{\left(\frac{2r}{3}\right)^2 + \left(\frac{L}{4}\right)^2} + \sqrt{\left(\frac{3L}{4}\right)^2 + \left(\frac{r}{3}\right)^2} - \right.$$

$$\left[\sqrt{\left(\frac{3L}{4}\right)^2 + \left(\frac{2r}{2}\right)^2} + \sqrt{\left(\frac{3L}{4}\right)^2 + \left(\frac{r}{3}\right)^2} + r\right]\right\}.$$

However, since the clouds are translucent, gaps 3 and 4 must be multiplied by their transmissivities to obtain the correct view factors through the clouds.

$$F_{up-cloud} = \tau_{up-cloud} F_{gap-3} = .83\left(F_{gap-3}\right)$$

and

$$F_{low-cloud} = \tau_{low-cloud} F_{gap-4} = .67\left(F_{gap-4}\right)$$

Therefore, the total view factor between the sun and the solar panels is

$$F_{sun-panel} = F_{sun-panel-opaque} + F_{up-cloud} + F_{low-cloud}$$
$$= 0.78(F_{gap-1} + F_{gap-2}) + 0.83F_{gap-3} + 0.67F_{gap-4}.$$

Example 13.7

Problem: Below is a cross-sectional view of a heat exchanger consisting of a multi-pass configuration with baffles and five pipes running down the middle. The pipe diameter is small compared to the combined width of the baffles (of dimension l each) and the distance between them. The baffles are spaced at a distance 4d apart with the five pipes positioned as shown in the diagram. The central cooling pipe is at distance 0.5l from the heat exchanger wall. The other four pipes are positioned at a distance of l/3 from the nearest heat exchanger wall. The temperature at area A_1 (of width l) is significantly larger than that of A_2 (of width l) such that finding the view factor F_{1-2} would be beneficial.

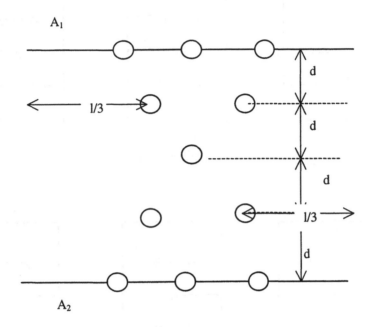

Figure 13.13 Sketch for Example 13.7

<u>Solution</u>
The pipe diameters are small with respect to d and l, but not small enough for the obstruction to radiation to be neglected.

There are 18 different radiation pathways that can be identified. Nine of them are shown in Fig. 13.14, and the other nine are mirror-images of those shown, using the right vertical border as the position of the mirror. For each of the radiation pathways, Hottel cross-string theorem is applied. The final view factor F_{1-2} is the sum of the eighteen view factors corresponding to the eighteen radiation paths.

The view factor for path 1, F'_{1-2}, is given by

$$F'_{1-2} = \frac{1}{l}\sqrt{9d^2 + {l^2}\Big/{9}} - \frac{3d}{l}. \tag{i}$$

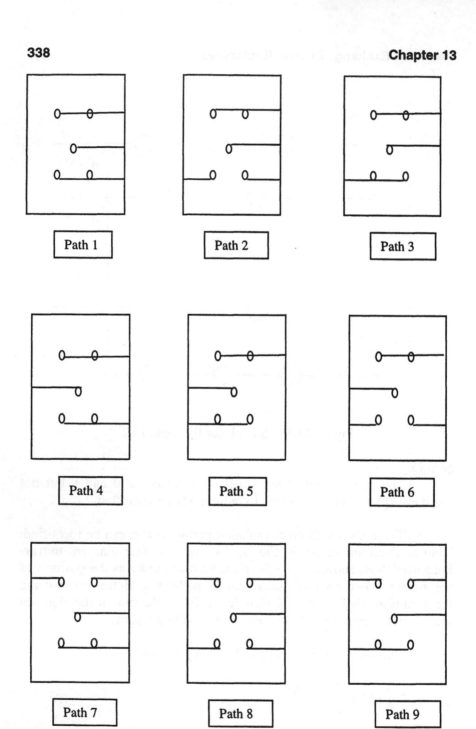

Figure 13.14   The nine radiation pathways.

$$F_{1-2}^2 = \frac{1}{2l}\sqrt{9d^2 + 4l^2/9} + \frac{d}{l} - \frac{1}{2l}\sqrt{9d^2 + l^2/9} - \frac{1}{2l}\sqrt{4d^2 + l^2/9} \qquad \text{(ii)}$$

$$F_{1-2}^3 = \frac{1}{2l}\sqrt{4d^2 + l^2/9} + \frac{1}{2l}\sqrt{16d^2 + l^2} - \frac{1}{2l}\sqrt{9d^2 + 4l^2/9} - \frac{1}{2l}\sqrt{d^2 + l^2/36}$$

$$- \frac{1}{2l}\sqrt{4d^2 + l^2/4}$$

(iii)

$$2l.F_{1-2}^4 = \sqrt{4d^2 + l^2/4} + \sqrt{d^2 + l^2/36} + \sqrt{d^2 + 4l^2/9} + \sqrt{d^2 + 4l^2/9}$$

$$+ \sqrt{d^2 + l^2/36} + \sqrt{4d^2 + l^2/4}$$

$$- 2\sqrt{4d^2 + l^2/4} - 2\sqrt{d^2 + 4l^2/9} - 2\sqrt{d^2 + l^2/36}$$

$$\mathbf{F^4}_{1-2} = 0 \qquad \text{(iv)}$$

$$F_{1-2}^5 = \frac{1}{2l}\sqrt{4d^2 + l^2/9} + \frac{1}{2l}\sqrt{16d^2 + l^2} - \frac{1}{2l}\sqrt{4d^2 + l^2/4} - \frac{1}{2l}\sqrt{d^2 + l^2/36}$$

$$- \frac{1}{2l}\sqrt{9d^2 + 4l^2/9}$$

(v)

$$F_{1-2}^6 = \frac{1}{l}\sqrt{d^2 + l^2/36} - \frac{1}{2l}\sqrt{4d^2 + l^2/9} \qquad \text{(vi)}$$

$$F_{1-2}^7 = \frac{d}{l} + \frac{1}{2l}\sqrt{9d^2 + 4l^2/9} - \frac{1}{2l}\sqrt{9d^2 + l^2/9} - \frac{1}{2l}\sqrt{4d^2 + l^2/9} \qquad \text{(vii)}$$

$$F_{1-2}^8 = \frac{1}{l}\sqrt{4d^2 + l^2/9} - \frac{d}{l} - \frac{1}{l}\sqrt{d^2 + l^2/36} \qquad \text{(viii)}$$

$$F_{1-2}^9 = \frac{1}{2l}\sqrt{4d^2 + l^2/4} + \frac{1}{2l}\sqrt{d^2 + l^2/36} - \frac{1}{2l}\sqrt{4d^2 + l^2/9} - \frac{1}{2l}\sqrt{d^2 + l^2/9}$$

(ix)

Adding the nine different components, and multiplying by two,

$$F_{1-2} = -\frac{4d}{l} + \frac{2}{l}\sqrt{16d^2 + l^2} - \frac{1}{l}\sqrt{4d^2 + l^2/4} - \frac{1}{l}\sqrt{d^2 + l^2/36} - \frac{1}{l}\sqrt{d^2 + l^2/9}$$

Hence, the view factor

$$F_{1-2} = -\frac{4d}{l} + \frac{3}{l}\sqrt{4d^2 + l^2/4} - \frac{1}{l}\sqrt{d^2 + l^2/36} - \frac{1}{l}\sqrt{d^2 + l^2/9}.$$

(x)

Depending on the relative sizes of d and l, the expression for F_{1-2}, may be slightly different from that given by equation (x) above.

In one limit when $l \ll d$, equation (x) gives $F_{1-2} = -\frac{4d}{l} + \frac{6d}{l} - \frac{d}{l} - \frac{d}{l} = 0$. For two parallel plates of width l and at a distance 4d apart without any obstructions in between, the view factor is $= \frac{1}{l}\sqrt{16d^2 + l^2} - \frac{4d}{l}$. In the same limit when $l \ll d$, this view factor $= 0$ also.

PROBLEMS

13.1.   Find an expression for the view factor between two parallel plates, F_{1-2}, of width L and a distance d apart. Intuitively, explain what would happen to F_{1-2} when the plates become very wide, that is, $L \rightarrow \infty$, and (ii) when the plates are very far apart, that is, $d \rightarrow \infty$. Show that the expression obtained provides the mathematical basis for your intuition.

13.2.

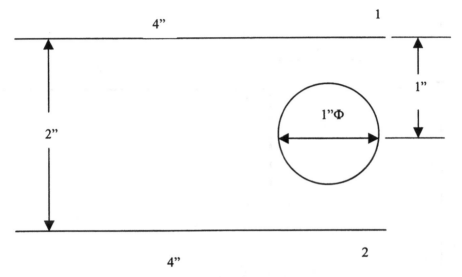

Figure for Prob.13.2.

For the system shown, the view between surface 1 and surface 2 is partially blocked by an intervening circle. Determine the view-factor F_{1-2}.

13.3. For the 2-D geometry shown, the view between A_1 and A_2 is partially blocked by an intervening structure of negligible thickness. Determine the view factor F_{1-2}.

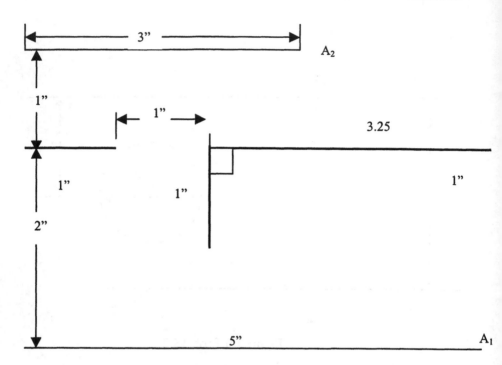

Figure for Prob.13.3.

13.4.   A tube bundle is as configured in the figure. The center long
         tube is surrounded by 6 other identical equally spaced tubes.
         What is the view factor from the central tube to each of the
         surrounding tubes?

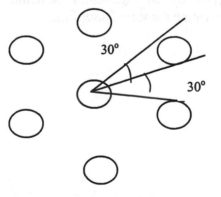

Figure for Prob. 13.4.

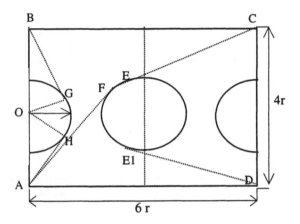

Figure for Prob. 13.5.

13.5. Consider an enclosure consisting of 4 surfaces, AD, BC, AB, CD, very long in the direction perpendicular to the plane of the figure. Two infinitely long semi-cylindrical surfaces of radius R are put both sides, AB and CD, and one infinitely long cylindrical surface of radius R in the central of the enclosure. The lengths of AB, BC are 4r, 6r, respectively. The objective is to find F_{AB-CD} from the surface AB to the surface CD.

13.6. Find an expression for the view factor between two parallel plates, F_{1-2}, of width L and a distance d apart, with an intervening plate placed at a distance d/2 from each of them as shown. There are 2-D openings spaced uniformly in the intervening plate that allows radiation from the one plate to the other, such that the openings and obstructions are of width L/9 each. Intuitively, explain what would happen to F_{1-2} when the holes become very small, and show that the expression obtained provides the mathematical basis for your intuition.

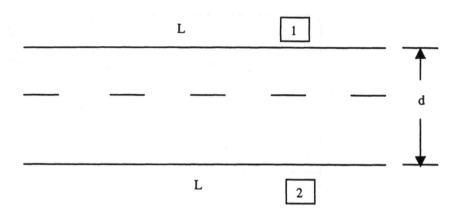

Figure for Prob. 13.6.

13.7.   Consider radiation between two parallel surfaces, with two
        translucent obstructions A and B of transmissivity τ as shown.
        Indicate the method of finding the view factor between the two
        parallel surfaces.

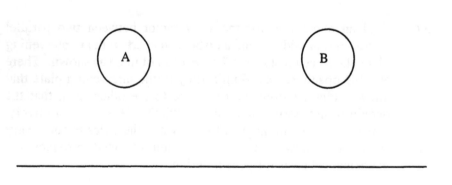

Figure for Prob.13.7.

13.8.   Find the view factor between the two parallel surfaces.

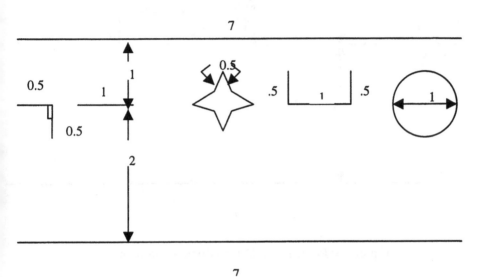

Figure for Prob. 13.8. Dimensions are in inches.

13.9.   Find the view factor between the two parallel surfaces, with the 4
         opaque objects in between.

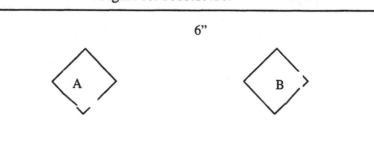

7

0.5   1   1   0.5   0.5   1   0.5   1.5

0.5   0.5   2

7

Figure for Prob. 13.9. Dimensions are in inches.

13.10. Find the view factor between the parallel surfaces, if A and B are opaque.   The parallel surfaces are 6 inches in width each, and 3 inches apart. A and B are squares with 1-inch diagonals, and set back an inch each from the edges. The straight line connecting the centers of A and B is 1 inch from the upper plate. What is the view factor if A and B are translucent with transmissivity τ?

Figure for Prob.13.10.

6"

A   B

6"

13.11. In an annealing process, a steel sheet is passed under an electric heater in order to raise its temperature and increase its hardness. The sheet is passed through a large oven on rollers. The floor of the oven is a slab of concrete thick enough to absorb the radiation from the extremely hot sheet. Engineers must know the view factor from the sheet to the floor in order to calculate the thickness. The length of the oven is L, the diameter of the rollers is d, the height from floor to sheet is h, and the pitch of the rollers is p. Calculate the view factor from the sheet to the floor. Assume that d is small compared to h and p.

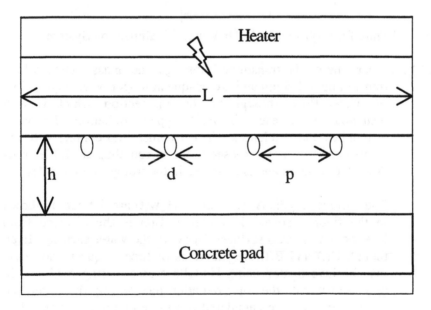

Figure for Prob. 13.11.

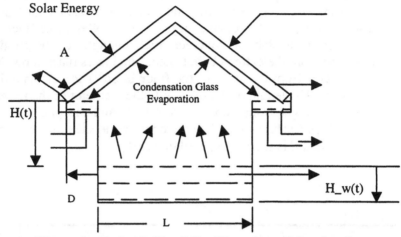

Figure for Prob.13.12.   Solar Water Desalinization System.

13.12.   Solar energy is transmitted through the glass plates which transmit up to 2.5 μm and are opaque at longer wavelengths. The sea water flows through the bottom section, which is well insulated. As a result of the "trapped" radiation, the water evaporates and condenses when it comes into contact with the slanted glass surfaces exposed to the surrounding air. It then runs down the surfaces and is collected in the troughs on the sides.

The various lengths A, H(t), D, L, H_w(t) are defined as shown in the figure above.   In addition, B(t) is the distance from intersection of glass surfaces (apex) to the water surface. Even though H(t) and B(t) are functions of time, a quasi-static case may be investigated using Hottel's crossed string method. In order to estimate the solar radiation received by the water, the view factor is to be calculated from one of the two slant glass surfaces to the water surface.

13.13.   The Hottel crossed-string method allows us to calculate the view factor of a surface that is very long in the direction perpendicular to the cross section of the objects. This problem stretches this assumption to use the method for a space heater in a room that has a shape of a cylinder and a person at some distance away.

Figure 1 for Prob. 13.13 shows a side view of the heater and the person. This figure is only used to understand the problem. Since the heater and the person's height are about the same and they are relatively tall compared with their cross sections, we will be able to use Hottel's crossed-string method.

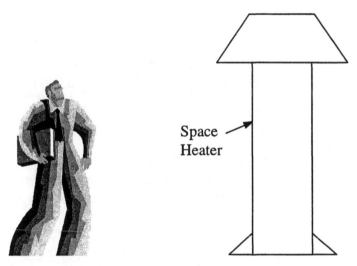

Figure 1 for Prob. 13.13. Side view of the heater and the person.

The assumptions made are that the person's cross sections are considered an ellipse and that the heights relative to the cross sections are very large. Due to the complexity of ellipses, actual angles and numbers are to be used.

Two cases are described by the two figures below. The first one is with the ellipse's longer side facing the heater. The second is with the shorter side facing the heater. Figure 2 for Prob. 13.13 gives the dimensions (in feet) required to find the view factor for orientation 1.

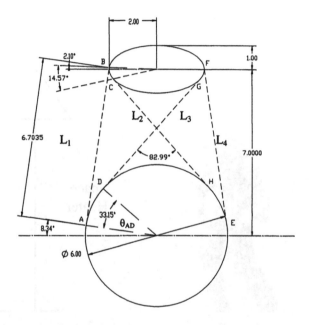

Figure 2 for Prob.13.13. Dimensions of configuration 1 with the long side of the ellipse facing the heater.

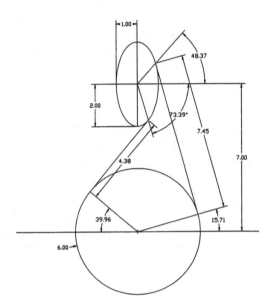

Figure 3 for Prob. 13.13. Dimensions of configuration with the short side of the ellipse facing the heater.

13.14. (a) A patient is receiving radiation therapy to diminish a certain group of cells which lies one inch beneath the surface of his skin. The source of radiation is positioned one inch above and parallel to the top of the skin surface. Using ultrasonic technology, a thin growth of tissue 1" wide is noticed in the path of radiation. The tissue has a transmissivity of 0.5. You are called to inspect the situation to determine the view factor for the cells. Assume that the tissue is negligibly thin and that it is at the surface of the skin. The location of the tissue can be seen in the accompanying figure.

(b) The above found view factor was acceptable and therapy was done upon the patient. After a certain period of time, the ultrasonic technology noticed that the tissue has grown to 1.75" and thickened to an amount of 0.25", thus it cannot be neglected. The tissue now has a transmissivity of 0.3. You are again called to inspect the situation to determine the new view factor. Assume that the tissue's thickness starts at the surface of the skin. The location of the tissue can be seen in the accompanying figure.

After you solve the problem, biomedical engineers will determine if the radiation experienced by the cells will be sufficient at the same dosage of radiation therapy.

Part (a)

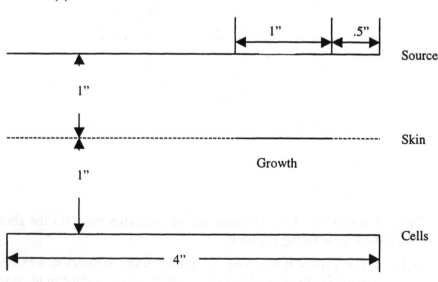

Part (b)

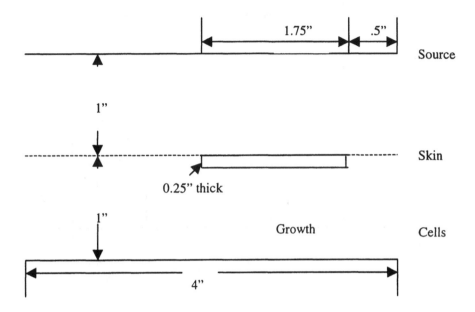

1.75"   .5"
Source

1"

Skin

0.25" thick

1"   Growth   Cells

4"

Figure for Prob.13.14.

13.15. The inlet manifold depicted is used to supply air to a large displacement rotary engine (10 liters). The layout of the intake trumpet is long and ovular in shape, making it well suitable for analysis by Hottel's theorem. Find the radiation view factor from interior wall CD to AB in the given intake manifold geometry. Assume negligible radiation from the side walls AC and BD. Assume also that all parts of wall CD radiate equally and evenly. The intake trumpet of the port is the only obstruction between the two walls. Assume that the trumpet itself does not radiate heat nor does it "see" itself in any manner. Assume the temperature of walls CD and AB are 400 and 40 degrees F respectively. Assume also that the ends of the trumpet and manifold have no significant contribution to the amount of heat radiated. In the figure, the dimensions are in inches.

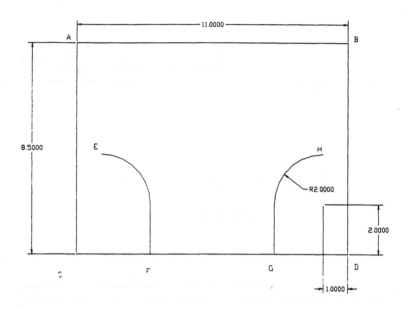

Figure for Prob. 13.15.

13.16. Below is a cross-sectional view of a heat exchanger consisting of
        a multi-pass configuration with baffles and five square pipes
        running down the middle. Each side of the square pipe is 2r, and
        the combined width of the baffles is l each. The baffles are
        spaced at a distance 4d apart with the five pipes positioned as
        shown in the diagram. The central cooling pipe is at distance 0.5l
        from the heat exchanger wall. The centers of the other four
        pipes are positioned at a distance of l/3 from the nearest heat
        exchanger wall. The temperature at area A_1 (of width l) is
        significantly larger than that of A_2 (of width l). Find the view
        factor F_{1-2}. Determine the view factor as r becomes small
        compared to d and l.

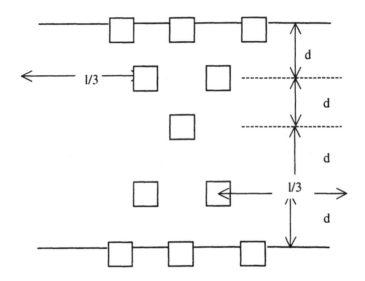

Figure for Problem 13.16.

Hottel's Theorem

In an enclosure, Hottel's theorem is very useful
It is only for two dimensions, not all that wonderful
However, convenient to evaluate the view factor
Between two surfaces that make up an enclosure.

Study the value obtained by taking the sum of the crossed strings
And subtracting away the sum of the uncrossed strings
Half of this calculated value is the required product
Of the length of one area and its view factor to the other.

K.V. Wong

14

Radiation with Other Modes of Heat Transfer

14.1   Introduction

Physical situations that involve radiation with other modes of heat transfer are fairly common. If conduction enters the problem, the Fourier conduction law states that the heat flow depends upon the temperature gradient, thus introducing derivatives of the first power of the temperature. If convection matters, the heat flow depends roughly on the first power of the temperatures, the exact power depends on the type of flow. For instance, natural convection depends on a temperature difference between the 1.25 and 1.4 power. Physical properties that are temperature dependent introduce more temperature dependencies. This all means that the governing equations are highly nonlinear.

Since the radiative terms are usually in the form of integrals and the conductive terms involve derivatives, the energy balance equations are in the form of nonlinear integrodifferential equations. These equations are not solved readily with currently available mathematical techniques. Numerical techniques have to be used to solve such equations. The scholar is referred to the extensive mathematical literature on numerical methods for these techniques, as they are not discussed here. This chapter focuses on the setting up of the energy balance equations and obtaining physical insight into practical problems. In addition, the assumption in this chapter is that the medium is not participating in the radiative heat transfer.

14.2   Radiation with Conduction

Physical situations that involve radiation with conduction are fairly common indeed. Examples include heat transfer through "superinsulation" made up of separated layers of very reflective material, heat transfer and temperature distributions in satellite and spacecraft structures, and heat transfer through the walls of a vacuum flask.

14.2.1 Radiating Longitudinal Plate Fins

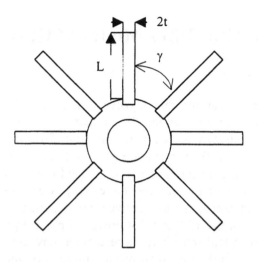

Figure 14.1   A radiator for a space vehicle.

The following assumptions are made.

(1)   Dimension of the plate normal to the plane of the figure is sufficiently long so that there is no temperature variation in that direction.

(2)   L>>t, so that the temperature in the plate is a function of the axial coordinate only, i.e., $T_1(x_1)$ and $T_2(x_2)$.

(3)   The radiative energy incident on the fin surface from the external environment is negligible.

(4)   The temperature at the fin base is uniform throughout, i.e., $T_1(0) = T_2(0) = T_b$.

(5)   The heat loss from the fin tips is negligible, i.e., $\dfrac{dT_1(L)}{dx_1} = 0$ and $\dfrac{dT_2(L)}{dx_2} = 0$.

(6)   The surfaces are opaque, gray, diffuse emitters and have uniform emissivity, ε. The thermal conductivity k is uniform through the plates.

(7)   Kirchhoff's law is applicable, i.e., $\varepsilon_\lambda(T) = \alpha_\lambda(T)$ or $\varepsilon(T) = \alpha(T), \rho_\lambda(T) = 1 - \varepsilon_\lambda(T)$.

(8)   The surfaces are diffuse reflectors.

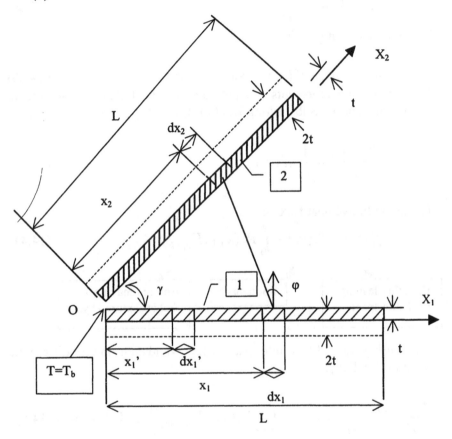

Figure 14.2 Longitudinal plate fin.

The steady-state balance equation for a differential volume element of the fin is

(Net conductive heat gains) + (net radiative heat gains) = 0

or $dQ^c + dQ^r = 0$ (14.1)

Let w = width of plates normal to the plane of the figure.   The conductive net gain in plate 1 for volume element twdx is:

$$dQ^c = ktwdx_1 \frac{d^2 T_1}{dx_1^2}$$
(14.2a)

The radiative heat gain is:

$$dQ^r = -dx_1 wq_1^r(x_1) \quad \text{since} \quad w \gg t$$
(14.2b)

The minus sign represents the net radiative heat flux leaving the fin surface into space. Substituting Eq. (14.2) into Eq. (14.1),

$$\frac{d^2 T_1(x_1)}{dx_1^2} = \frac{1}{kt} q_1^r(x_1)$$
(14.3)

The net radiative heat flux is:

$$q_1^r(x_1) = R_1(x_1) - \int_{x_2}^{L} R_2(x_2) dF_{dx_1 - dx_2}$$
(14.4)

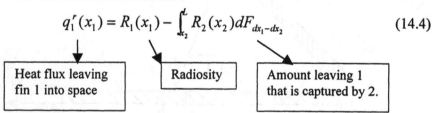

| Heat flux leaving fin 1 into space | Radiosity | Amount leaving 1 that is captured by 2. |

Substituting Eq. (14.4) into Eq. (14.3) yields an integrodifferential equation for $T_1(x_1)$.

$$\frac{d^2 T_1(x_1)}{dx_1^2} = \frac{1}{kt} \left[R_1(x_1) - \int_{x_2=0}^{L} R_2(x_2) dF_{dx_1 - dx_2} \right] \quad \text{in } 0 \le x_1 \le L \quad (14.5)$$

The boundary conditions obtained from assumptions (14.4) and (14.5) above are as follows:

At $x_1 = 0$, $T_1(x_1) = T_b$
(14.6a)

At $x_1 = L$, $\frac{dT_1(x_1)}{dx_1} = 0$
(14.6b)

The equation for the radiosity is given below, where ρ is replaced by $(1-\varepsilon)$ according to assumption (7).

$$R_1(x_1) = \varepsilon\sigma T_1^4(x_1) + (1-\varepsilon)\int_{x_2=0}^{L} R_2(x_2)dF_{dx_1-dx_2} \qquad (14.7)$$

$\downarrow$ $\downarrow$

| Emitted | Amount of energy from 2 reflected by 1. |

A set of relations similar to Eqs. (14.5)-(14.7) can be written for temperature distribution $T_2(x_2)$ and the radiosity $R_2(x_2)$ at plate 2. However, this is not necessary since symmetry exists. Because of this symmetry between the fins, $R_1(x_1) = R_2(x_2)$ and $T_1(x_1) = T_2(x_2)$ for $x_1 = x_2$. Removing subscripts 1 and 2 from Eqs. (14.5)-(14.7), and writing in dimensionless form,

$$\frac{d^2\theta_1(x_1)}{d\xi_1^2} = \frac{1}{N_c}\left[\beta(\xi_1) - \int_{\xi_2=0}^{L}\beta_2(\xi_2)dF_{d\xi_1-d\xi_2}\right] \text{ in } 0 \le \xi_1 \le L$$

$$(14.8)$$

At $\xi_1 = 0$, $\theta(\xi_1) = 1$ (14.9)

At $\xi_1 = 1$, $\dfrac{d\theta(\xi_1)}{d\xi_1} = 0$ (14.10)

and $\beta_1(\xi_1) = \varepsilon\sigma\theta^4(\xi_1) + (1-\varepsilon)\int_{\xi_2=0}^{L}\beta(\xi_2)dF_{d\xi_1-d\xi_2}.$ (14.11)

The dimensionless quantities have been defined as

$$\theta \equiv \frac{T}{T_b}, \ \beta = \frac{R}{\sigma T_b^4}, \ N_c = \frac{kt}{L^2\sigma T_b^3} = \frac{conduction}{radiation}, \ \xi_1 \equiv \frac{x_1}{L}, \ \xi_2 \equiv \frac{x_2}{L}.$$

$$(14.12)$$

When N_c is large, conduction $>>$ radiation.
When N_c is small, conduction $<<$ radiation.

When $N_c \rightarrow \infty$,   Eq. (14.8) becomes $\dfrac{d^2\theta(\xi_1)}{d\xi_1} = 0,$ and the situation

reduces to the pure conduction case.

Recall that the elemental diffuse view factor between strips $d\xi_1$ and $d\xi_2$ is as follows:

$$dF_{d\xi_1-d\xi_2} = \frac{1}{2}d(\sin\varphi) \qquad (14.13)$$

where φ is the angle between the normal to the strip $d\xi_1$ and the straight line joining strips $d\xi_1$ and $d\xi_2$.

$$\sin\varphi = \frac{x_1 - x_2\cos\gamma}{\left[(x_1 - x_2\cos\gamma)^2 + (x_2\sin\gamma)^2\right]^{\frac{1}{2}}}$$

$$\sin\varphi = \frac{x_1 - x_2\cos\gamma}{\left(x_1^2 - 2x_1x_2\cos\gamma + x_2^2\right)^{\frac{1}{2}}}. \qquad (14.14)$$

Then $\quad dF_{dx_1-dx_2} = \dfrac{1}{2}\dfrac{x_1x_2\sin^2\gamma}{\left(x_1^2 - 2x_1x_2\cos\gamma + x_2^2\right)^{\frac{3}{2}}}dx_2 \qquad (14.15\text{a})$

or $\quad dF_{d\xi_1-d\xi_2} = \dfrac{1}{2}\dfrac{\xi_1\xi_2\sin^2\gamma}{\left(\xi_1^2 - 2\xi_1\xi_2\cos\gamma + \xi_2^2\right)^{\frac{3}{2}}}d\xi_2. \qquad (14.15\text{b})$

Once Eqs.(14.8)-(14.11) have been solved, $\beta(\xi_1)$ can be determined. The distribution of net radiative heat flux on the surface of the fin is

$$\frac{q^r(\xi_1)}{\sigma T_b^4} = \beta(\xi_1) - \int_{\xi_L=0}\beta(\xi_2)dF_{d\xi_1-d\xi_2}. \qquad (14.16)$$

The net rate of heat dissipation by radiation Q^r, from one fin surface per unit width normal to the plane, is

$$Q^r = \int_{x_1=0}^{L} q^r(x_1)dx_1$$

or $\qquad \dfrac{Q^r}{\sigma T_b^4} = L \displaystyle\int_{\xi_1=0}^{1} \left[\beta(\xi_1) - \int_{\xi_2=0}^{1} \beta(\xi_2) dF_{d\xi_1-d\xi_2} \right] d\xi_1$ $\qquad$ (14.17)

Consider black surfaces, that is, $\varepsilon = 1$. There is uniform temperature T_b everywhere. The net rate of heat dissipation by radiation of an ideal fin is given below.

$$Q^r_{ideal} = \sigma T_b^4 [L \sin \dfrac{\gamma}{2}].$$ $\qquad$ (14.18)

The radiative effectiveness is defined as:

$$\eta \equiv \dfrac{Q^r}{Q^r_{ideal}} = \dfrac{1}{\sin\left(\dfrac{\gamma}{2}\right)} \int_{\xi_1=0}^{1} \left[\beta(\xi_1) - \int_{\xi_2=0}^{1} \beta(\xi_2) dF_{d\xi_1-d\xi_2} \right] d\xi_1.$$ $\qquad$ (14.19)

Discussion of Results

Equations (14.8) and (14.11) are two coupled integrodifferential equations which must be solved simultaneously for unknowns, $\theta(\xi_1)$ and $\beta(\xi_1)$. Analytical solutions are unlikely, but we can solve numerically for prescribed values of ε, γ and N_c.

The results calculated by Sparrow, Eckert, and Irvine, 1967 [1], are presented in Fig. 14.3. The plots are made as a function of the conduction to radiation parameter N_c for two values of emissivity, and for several values of the opening angle. The curves for emissivity are equal to one, and converge at the highest heat loss value as $N_c \to \infty$. When emissivity is 0.5, this ideal case is not reached as the surfaces are nonblack. From the figure, it is clear that as Nc is decreased, the fin effectiveness is decreased. In addition, as would be expected intuitively, the smaller opening angles result in greater fin effectiveness.

Figure 14.3   Radiation and Conduction in Non-participating Media

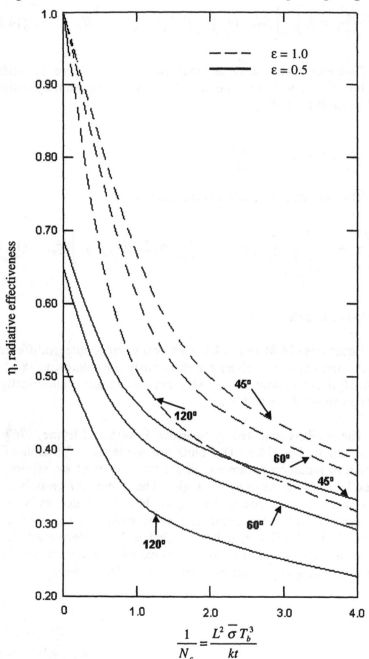

Example 14.1

Problem: Two black infinite parallel plates separated by a transparent medium of thickness b and thermal conductivity k. Plate 2 is at temperature T_2, and a known amount of energy Q_1/A is added per unit area to plate 1 and removed at plate 2. What is the temperature T_1 of plate 1?

Solution

The energy transfer per unit time and area by radiation between two infinite parallel black plates is

$$\frac{Q_R}{A} = \sigma\left(T_1^4 - T_2^4\right)$$

and by conduction is

$$\frac{Q_C}{A} = \frac{k}{b}(T_1 - T_2)$$

Hence,

$$\frac{Q_1}{A} = \frac{Q_R}{A} + \frac{Q_C}{A} = \sigma\left(T_1^4 - T_2^4\right) + \frac{k}{b}(T_1 - T_2)$$

$$\sigma T_1^4 + \frac{k}{b}T_1 = \sigma T_2^4 + \frac{k}{b}T_2 + \frac{Q_1}{A}$$

Solve iteratively.

Example 14.2

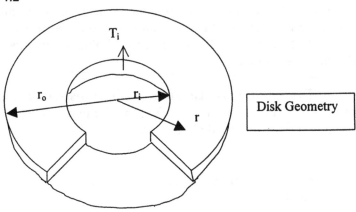

Figure 14.4a   Sketch for Example 14.2.

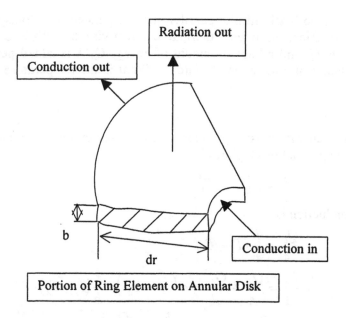

Portion of Ring Element on Annular Disk

Figure 14.4b   Sketch for Example 14.2.

Problem: A thin annular fin in a vacuum is enclosed in insulation so that there is no heat transfer on one face and around its outside edge. The disk is of thickness b, has an inner radius r_i, an outer radius r_o, and a thermal conductivity k. Energy is supplied to the inner edge, say from a solid rod of radius r_i that fits the central hole, and this keeps the inner edge at a temperature T_i.

The exposed annular surface, with emissivity ε, radiates to the environment at temperature $T_e = 0$. Find the temperature distribution of the fin as a function of radial position.

Solution:
Assumption: Disk is thin, so that the local temperature is constant across the thickness b.

For any ring element of width dr,

Conduction in = Conduction out + Radiation out

$$-k2\pi rb\frac{dT}{dr} = \varepsilon\sigma T^4 2\pi r dr - k2\pi rb\frac{dT}{dr} + \frac{d}{dr}\left(-k2\pi rb\frac{dT}{dr}\right)dr.$$

(i)

If b and k are constants,

$$kb\frac{1}{r}\frac{d}{dr}\left(r\frac{dT}{dr}\right) - \varepsilon\sigma T^4 = 0.$$

(ii)

The boundary conditions are as given below.

At $r = r_i$, $T = T_i$

At $r = r_o$, $\dfrac{dT}{dr} = 0$ (insulated at the outer edge)

Using $\theta = \dfrac{T}{T_i}$, $R = \dfrac{(r - r_i)}{(r_o - r_i)}$

$$\frac{d^2\theta}{dR^2} + \frac{1}{R + \dfrac{r_i}{(r_o - r_i)}}\frac{d\theta}{dR} - \frac{(r_o = r_i)^2 \varepsilon\sigma T_i^2}{kb}\theta^4 = 0.$$

(iii)

Using $\delta = \dfrac{r_o}{r_i}$, $\gamma = \dfrac{(r_o - r_i)^2 \varepsilon\sigma T_i^3}{kb}$

$$\frac{d^2\theta}{dR^2} + \frac{1}{R + \dfrac{1}{(\delta - 1)}}\frac{d\theta}{dR} - \gamma\theta^4 = 0.$$

(iv)

The boundary conditions become: at $R = 0$, $\theta = 1$

at $R = 1$, $\dfrac{d\theta}{dR} = 0.$

The solution may be obtained by numerical methods.

The fin efficiency is defined as

$$\eta = \frac{\text{Energy actually radiated by fin}}{\text{Energy radiated if entire fin at temp } T_i}.$$

Hence,

$$\eta = \frac{2\pi\varepsilon\sigma \int_{r_i}^{r_o} T^4 dr}{\pi\left(r_o^2 - r_i^2\right)\varepsilon\sigma T_i^4} \qquad (va)$$

$$\eta = \frac{2\int_0^1 [R(\delta+1)+1]\theta^4 dR}{\delta+1} \qquad (vb)$$

14.3   Radiation with Convection

Physical situations that involve radiation with convection are fairly common. Examples include solar radiation interacting with the earth's environment to produce complex natural convection, water environmental studies for predicting natural convection patterns in lakes, seas and oceans, and heat transfer along copper tubes in the furnace of a boiler.

14.3.1   Laminar Boundary Layer Flow Along a Flat Plate with Radiation Boundary Condition

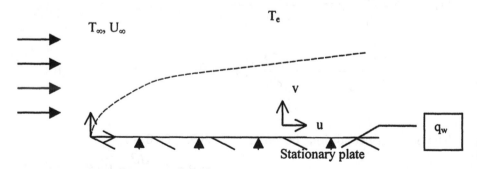

Figure 14.5 Flow along a flat plate with radiation boundary condition.

Consider a steady, laminar boundary layer flow of incompressible, transparent fluid along a flat plate, with a constant applied heat flux q_w Btu/(hr ft^2) at the wall surface. The properties of the fluid are assumed constant. The main considerations are conduction to the fluid, and radiation from the plate to the environment at T_e. Surface of the plate is opaque and gray, and the uniform emissivity is ε. The fluid which is at a temperature of T_∞, flows at a uniform velocity of U_∞. Flow velocities are sufficiently small so that viscous dissipation may be neglected.

Continuity

For the incompressible fluid, continuity equation is

$$\frac{\partial u}{\partial x} + \frac{\partial v}{\partial y} = 0 \tag{14.20}$$

Velocity Distribution

Since the y-component of the velocity, v, is small compared to u, the y-momentum equation yields no useful information. The x-momentum equation is

$$u\frac{\partial u}{\partial x} + v\frac{\partial v}{\partial y} = \upsilon\frac{\partial^2 u}{\partial y^2} \tag{14.21}$$

The boundary conditions are the no slip boundary conditions at y = 0, and the free-stream velocity, that is,

$$u = v = 0 \text{ at } y = 0 \tag{14.22a}$$
$$u = u_\infty \text{ at } y \to \infty \tag{14.22b}$$

Temperature Distribution

The energy conservation equation is in the form

$$\left(u\frac{\partial T}{\partial x}+v\frac{\partial T}{\partial y}\right)=\alpha\frac{\partial^2 T}{\partial y^2}. \tag{14.23}$$

The boundary conditions are as follows:

$$q_w = -k\frac{\partial T}{\partial y}+\varepsilon\sigma\left(T^4-T_e^4\right) \qquad \text{at } y=0 \tag{14.24a}$$

or $\qquad \dfrac{\partial T}{\partial y}=-\dfrac{q_w}{k}+\dfrac{\phi\sigma}{k}\left(T^4-T_e^4\right) \qquad \text{at } y=0.$ (14.24b)

Since the momentum equation is not coupled to the energy equation, they may be solved independently of each other.

We define the stream function such that

$$u = \frac{\partial \psi(x,y)}{\partial y} \text{ and } v = -\frac{\partial \psi(x,y)}{\partial x}. \tag{14.25}$$

Introduce similarity variables f(η) and η,

$$\eta = y\sqrt{\frac{u_\infty}{\upsilon x}} \tag{14.26a}$$

$$f(\eta) = \frac{\psi(x,y)}{\sqrt{x\upsilon u_\infty}} \tag{14.26b}$$

Momentum equation becomes

$$2f''' + ff''=0 \tag{14.26a}$$

with boundary conditions

f = 0, f' = 0 at η = 0   (14.27b)
f' = 1 at η → ∞.   (14.27c)

The solution of Eq. (14.27a) was given by Blasius,

$$u = u_\infty f'$$ (14.28a)

$$v = 0.5\sqrt{\frac{\upsilon u_\infty}{x}}(\eta f' - f).$$

(14.28b)

To solve the energy equation, we define the similarity variable, ξ, such that

$$\xi = \frac{\varepsilon \sigma T_\infty^3}{k}\sqrt{\frac{\upsilon x}{u_\infty}}$$ (14.29)

When the transformation Eqs. (14.29) and (14.26a), the velocity components Eqs. (14.28a) and (14.28b), are introduced into the energy equation (14.23), the later becomes

$$\frac{\partial^2 T}{\partial \eta^2} + \frac{1}{2}\mathrm{Pr}\, f\, \frac{\partial T}{\partial \eta} - \frac{1}{2}\mathrm{Pr}\, \frac{df}{d\eta}\xi\frac{\partial T}{\partial \xi} = 0.$$ (14.30)

Equation (14.30) is solved using a power series technique.

$$T(\xi,\eta) - T_\infty = T_\infty \sum_{n=1}^{\infty} a_n \theta_n(\eta)\xi^n$$ (14.31)

with the requirement that

$$\theta_1(0) = \theta_2(0) = \theta_3(0) = \ldots\ldots\ldots = 1$$ (14.32)

where the coefficients a_n and functions $\theta_n(\eta)$ are unknowns.

Substituting Eq. (14.31) into Eq. (14.30), and equating coefficients of ξ^n to zero (for $a_n \neq 0$), it is found that the functions $\theta_n(\eta)$ constitute the solution of the following ordinary differential equation.

$$\theta_n'' + \frac{1}{2}\mathrm{Pr}\,f\theta_n' - \frac{n}{2}\mathrm{Pr}\,f'\theta_n = 0 \qquad\qquad (14.33)$$

where the prime denotes differentiation with respect to η. The boundary conditions are

$$\theta_n = 1 \text{ at } \eta = 0 \qquad\qquad (14.34a)$$

$$\theta_n = 0 \text{ at } \eta \rightarrow \infty. \qquad\qquad (14.34b)$$

Equation (14.33) with boundary conditions (14.34a) and (14.34b) can be solved numerically because functions f and f' have been found.

When functions $\theta_n(\eta)$ are known, the problem of determining the temperature distribution $T(\xi,\eta)$ in the boundary layer reduces to evaluating unknown expansion coefficients an in Eq. (14.31). The boundary condition Eq. (14.34a) is now utilized to determine these coefficients. From Eq. (14.31),

$$T(\xi,0) = T_\infty\left(1 + \sum_{n=1}^{\infty} a_n \xi^n\right). \qquad\qquad (14.35)$$

The gradient $\dfrac{\partial T}{\partial y}$ at the wall can be evaluated as

$$\left.\frac{\partial T(x,y)}{\partial y}\right|_{y=0} = \left.\frac{\partial T}{\partial \eta}\right|_{\eta=0}\frac{\partial \eta}{\partial y}$$

$$= \left[T_\infty \sum_{n=1}^{\infty} a_n \theta_n'(0)\xi^n\right]\sqrt{\frac{u_\infty}{\upsilon x}}$$

$$= \left[T_\infty \sum_{n=1}^{\infty} a_n \theta_n'(0)\xi^n\right]\frac{1}{\xi}\frac{\varepsilon\sigma T_\infty^3}{k}$$

$$= \frac{\varepsilon \sigma T_\infty^4}{k} \sum_{n=1}^{\infty} a_n \theta_n'(0) \xi^{n-1}. \tag{14.36}$$

Substituting Eq. (14.35) and Eq.(14.36) into the boundary condition Eq. (14.34a) and equating like powers of ξ, the desired relation for the determination of coefficients an is obtained. Equating the constant terms (i.e., the coefficients of ξ^0 for instance),

$$a_1 = \frac{-1}{\theta_1'(o)} \left[\frac{q_w}{\varepsilon \sigma T_\infty^4} + \left(\frac{T_e}{T_\infty}\right)^4 - 1 \right] \tag{14.37a}$$

Equating coefficients of ξ',

$$\frac{a_2}{a_1} = \frac{4}{\theta_2'(0)}. \tag{14.37b}$$

Other coefficients are determined in a similar manner.

Knowing the functions $\theta_n(\eta)$ and the coefficients a_n, the distribution of temperature in the boundary layer can be evaluated from Eq. (14.31).

The local Nusselt number is defined as

$$Nu = -\frac{x}{T_w - T_\infty} \frac{\partial T(y)}{\partial y}\bigg|_{y=0} \tag{14.38}$$

Substituting Eqs. (14.35) and (14.36) into Eq. (14.38) and in the resulting expression replacing $\frac{\varepsilon \sigma T_\infty^4}{k}$ by $\xi \sqrt{\frac{u_\infty}{\upsilon x}}$ according to Eq. (14.29a),

$$\frac{Nu}{Re^{0.5}} = -\frac{\sum_{n=1}^{\infty} a_n \theta_n'(0) \xi^n}{\sum_{n=1}^{\infty} a_n \xi^n} \qquad (14.39a)$$

where

$$Re = \frac{u_\infty x}{\upsilon}. \qquad (14.39b)$$

Dividing one series into the other,

$$\frac{Nu}{Re^{0.5}} = -\theta_1'(0) - \frac{a_2}{a_1}\left[\theta_2'(0) - \theta_1'(0)\right]\xi \qquad (14.40)$$

Substituting a_2/a_1 from Eq. (14.37b) into Eq. (14.40),

$$\frac{Nu}{Re^{0.5}} = -\theta_1'(0) - 4\left[1 - \frac{\theta_1'(0)}{\theta_2'(0)}\right]\xi. \qquad (14.41)$$

The reference for the above solution is P.L. Donoughe and N.B. Livingood, 1958 [2]. Donoughe and Livingood found that

$$\theta_1'(0) = -0.4059, \quad \theta_2'(0) = -0.4803. \qquad (14.42)$$

The local Nusselt number becomes

$$\frac{Nu}{Re^{0.5}} = 0.4059 - 0.620\xi. \qquad (14.43)$$

The first term on the right in Eq. (14.43) is the conductive or convective term, and the second term is the first-order effect of radiation. Cess, 1962 [3], showed that higher-order terms are small.

Example 14.3

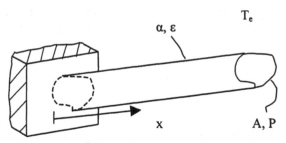

Figure 14.6 Fin of constant cross-sectional area transferring energy by radiation and convection.

Problem: A gas at T_e is flowing over the fin and removing heat by convection. The environment is at T_e. The cross-area of the fin is A, its perimeter is P, and its radiative properties are α,ε. Find x in terms of the heat transfer properties and geometry of the fin.

Solution

An energy balance in an element of length dx yields

$$kA\frac{d^2T}{dx^2}\,dx = \sigma\left(\varepsilon T^4 - \alpha T_e^4\right)Pdx + hPdx(T-T_e) \qquad \text{(i)}$$

Conductive   Radiative   Convective

We neglect radiative exchange between the fin and its base.

$$\frac{d^2T}{dx^2}\frac{dT}{dx} = \frac{\varepsilon\sigma P}{kA}\left(T^4 - \frac{\alpha}{\varepsilon}T_e^4\right)\frac{dT}{dx} + \frac{hP}{kA}(T-T_e)\frac{dT}{dx}$$

Integrating once,

$$\frac{1}{2}\left(\frac{dT}{dx}\right)^2 = \frac{\varepsilon\sigma P}{kA}\left(\frac{T^5}{5} - \frac{\alpha}{\varepsilon}TT_e^4\right) + \frac{hP}{kA}\left(\frac{T^2}{2} - TT_e\right) + C. \qquad \text{(ii)}$$

Now, we use the simplification of letting $T_e = 0$ and the fin be very long. As $x \rightarrow \infty$ (large), $T(x) \rightarrow 0$ and $dT/dx \rightarrow 0$. Therefore, $C = 0$.

$$\frac{dT}{dx} = -\left(\frac{2}{5}\frac{P\varepsilon\sigma}{kA}T^5 + \frac{hP}{kA}T^2\right)^{\frac{1}{2}} \qquad \text{(iii)}$$

The minus sign is taken because T decreases as x increases. We separate the variables in Eq. (iii) and using $T(x) = T_b$ at $x = 0$,

$$\int_0^x dx = -\int_{T_0}^T \frac{dT}{T\left[\frac{2}{5}\left(\frac{P\varepsilon\sigma}{kA}\right)T^3 + \frac{hP}{kA}\right]^{\frac{1}{2}}}.$$

Integration yields

$$x = \frac{1}{3}M^{-\frac{1}{2}}\left[\ln\frac{\left(GT_b^3 + M\right)^{\frac{1}{2}} - M^{\frac{1}{2}}}{\left(GT_b^3 + M\right)^{\frac{1}{2}} + M^{\frac{1}{2}}} - \ln\frac{\left(GT^3 + M\right)^{\frac{1}{2}} - M^{\frac{1}{2}}}{\left(GT^3 + M\right)^{\frac{1}{2}} + M^{\frac{1}{2}}}\right]$$

$$\text{(iv)}$$

where $G = \frac{2}{5}\frac{P\varepsilon\sigma}{kA}$, $M = \frac{hP}{kA}$.

The reference for the above discussion is Shouman, 1967 [4].

14.4   Radiation with Conduction and Convection

Physical situations that involve radiation with conduction and convection are fairly common. Examples include automobile radiators and heat transfer in the furnace of boilers and incinerators. The energy equations become more complex as they comprise both temperature differences coming from convection and temperature derivatives coming from conduction. Hence, there are no classical methods of solution, but numerical methods and specific methods for particular problems.

This section provides the solution of several examples where radiation is combined with the other modes of heat transfer.

Example 14.4

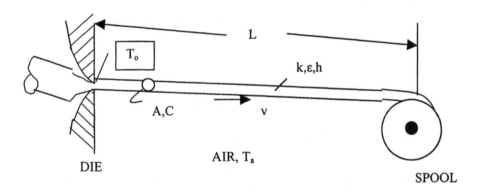

Figure 14.7   Extrusion of a thin wire.

Problem: A thin wire is extruded at a fixed velocity, v, through a die at a temperature of T_0. The wire then passes through air at T_a until its temperature is reduced to T_L. The heat transfer coefficient to the air is h, and the wire emissivity is ε. Find T as a function of wire velocity v and distance L. Derive the differential equation for the wire temperature as a function of the distance from the die.

Solution

Consider the heat balance for flow in and out of a control volume fixed in space, as shown below.

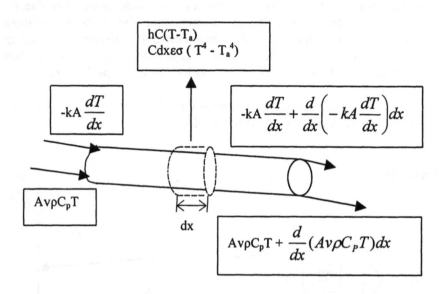

Figure 14.8   Sketch for Example 14.4.

For steady-state conditions, energy in = energy out. Hence,

$$- kA\frac{dT}{dx} + Av\rho C_p T = -kA\frac{dT}{dx} + \frac{d}{dx}\left(-kA\frac{dT}{dx}\right)dx + Av\rho C_p T$$

$$+ \frac{d}{dx}(Av\rho C_p T)dx + hCdx(T-T_a) + Cdx\varepsilon\sigma\left(T^4 - T_a^4\right) \qquad (i)$$

For constant k, the equation reduces to

$$- kA\frac{d^2T}{dx^2} + Av\rho C_p \frac{dT}{dx} + hC(T-T_a) + C\varepsilon\sigma\left(T^4 - T_a^4\right) = 0$$

or

$$\frac{d^2T}{dx^2} - \frac{v\rho C_P}{k}\frac{dT}{dx} - \frac{hC}{kA}(T - T_a) - \frac{C\varepsilon\sigma}{kA}(T^4 - T_a^4) = 0.$$
(ii)

The boundary conditions are as follows:

At $x = 0$, $T = T_o$

At $x = L$, we assume that there is no heat lost after the wire is wound, that is, $\dfrac{dT}{dx}\bigg|_L = 0$.

Example 14.5

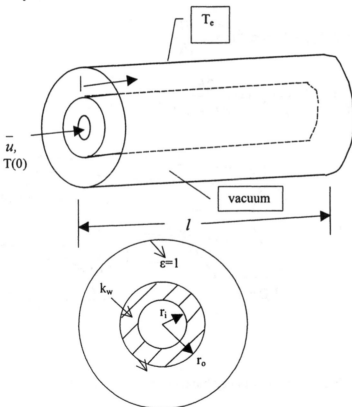

Figure 14.9   Sketch for Example 14.5.

Problem: Opaque liquid at temperature T(0) and mean velocity $\bar{u}$ enters a long vacuum jacket and a concentric electric heater that is kept at a uniform temperature T_e along its length.   The heater can be considered black, and the tube exterior is diffuse-gray with an emissivity ε.   The convective heat transfer coefficient between the liquid and the tube wall is h, and the tube wall conductivity is k_w.   Derive the relations to determine the mean liquid temperature as a function of distance x along the tube (assume that the liquid properties are constant).

Solution
        Heat balance on the liquid where the liquid temperature is T(x), gives

$$2\pi r_i h\left(T_{w,i} - T\right) = \pi r_i^2 \bar{u}\rho C_P \frac{dT}{dx}.$$

Hence, $\dfrac{dT}{dx} + \dfrac{2h}{r_i \pi \rho C_P} T(x) = \dfrac{2h}{r_i \bar{u}\rho C_P} T_{w,i}(x).$ (ia)

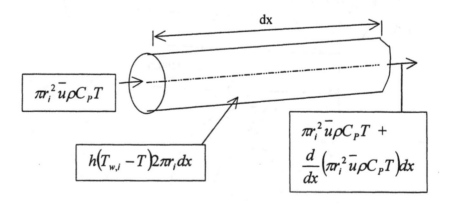

Figure 14.10 Energy balance for Example 14.5.

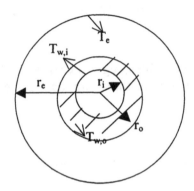

Figure 14.11   Sketch showing temperature boundary conditions for Example 14.5.

The boundary condition given is $T = T(0)$ at $x = 0$. Locally, through the tube wall, neglecting the axial heat conduction,

$$2\pi r_i h(T_{w,i} - T) = (T_{w,o} - T_{w,i}) \frac{2\pi k_w}{\ln\left(\dfrac{r_o}{r_i}\right)} \qquad \text{(ib)}$$

The radiation from element dx = $2\pi r_o \, dx \, \varepsilon \sigma T_{w,o}^4$

The surroundings are black, so no reflected radiation returns to dx. Neglecting radiation from the end planes at $x = 0$ and $x = 1$, radiation absorbed by the element dx is equal to

$$2\pi r_e l \sigma T_e^4 dF_{e-dx} \alpha.$$

For a gray tube $\alpha = \varepsilon$, and by reciprocity,

$$\text{Gain} = 2\pi r_e l \sigma T_e^4 F_{dx-e} \frac{2\pi r_o dx}{2\pi r_e l} \varepsilon = \sigma T_e^4 F_{dx-e} 2\pi r_o dx \varepsilon$$

Then, $$\frac{(T_{w,o} - T_{w,i})2\pi k_w}{\ln\left(\dfrac{r_o}{r_i}\right)} = 2\pi r_o \sigma \varepsilon \left(T_e^4 F_{dx-e} - T_{w,o}^4\right) \qquad \text{(ic)}$$

F_{dx-e} can be found from tables, eg., Siegel and Howell, 1992 [5]. Equations (ia), (ib) and (ic) are three simultaneous equations for $T(x)$, $T_{w,i}(x)$ and $T_{w,o}(x)$.

Example 14.6

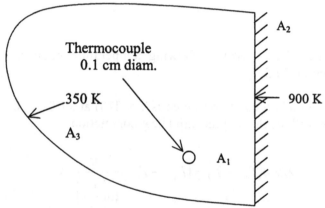

Figure 14.12   Sketch for Example 14.6.

Problem: A copper-constantan thermocouple is in an inert-gas stream at 350 K adjacent to a blackbody surface at 900 K. The heat transfer coefficient from the gas to the thermocouple is 25 W/(m²K). Estimate the temperature of the bare thermocouple. ($\varepsilon = 0.15$ for copper-constantan.)

Solution
For A_1 near A_2, $(A_1 \ll A_2)$,

$F_{1-2} = 0.5$
$F_{1-3} = 0.5$

Thus, $F_{2-1} = \dfrac{1}{2}\dfrac{A_1}{A_2}$ and $F_{3-1} = \dfrac{1}{2}\dfrac{A_1}{A_3}$.

Convective heat loss from the thermocouple = $hA_1 (T_1 - T_g)$.

Radiative loss from thermocouple $= \varepsilon\sigma T_1^4 A_1$.

For the thermocouple assumed gray, the radiative gain

$$= \alpha\left[A_2\sigma T_2^4 F_{2-1} + A_3\sigma T_3^4 F_{3-1}\right]$$
$$= \varepsilon\left[A_2\sigma T_2^4 F_{2-1} + A_3\sigma T_3^4 F_{3-1}\right]$$
$$= \varepsilon\frac{1}{2}A_1\sigma\left(T_2^4 + T_3^4\right)$$

The energy balance gives

$$h\left(T_1 - T_g\right) + \varepsilon\sigma T_1^4 = \frac{1}{2}\varepsilon\sigma\left(T_2^4 + T_3^4\right)$$

$$25\left(T_1 - 350\right) = 0.15 \times 5.729 \times 10^{-8}\left[\frac{900^4 + 350^4}{2} - T_1^4\right]$$

Try $T_1 = 450$ K, $2500 = 2531$

Try $T_1 = 452$ K, $2550 = 2505$

The bare thermocouple is at $T_1 = 451$ K.

PROBLEMS

14.1.   Very long, thin fins of thickness b, width W are attached to a black base that is maintained at a constant temperature T_b, as shown in the figure. There is a larger number of fins. The fin surface is diffuse-gray, and they are in a vacuum at temperature, $T_e = 0$ K. Write the equation that describes the local fin temperature.

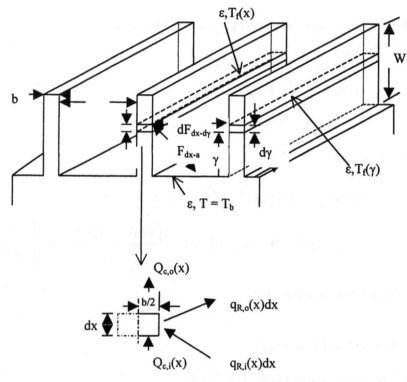

Figure for Prob. 14.1.

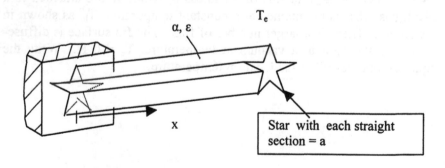

Figure for Prob. 14.2.

14.2.   A gas at T_e is flowing over the fin and removing heat by convection. The environment is at T_e. The cross-area of the fin is a star with each straight section $= a$. Its radiative properties are α, ε. Find x in terms of the heat transfer properties and geometry of the fin.

14.3.   Consider a thin two-dimensional fin in vacuum radiating to outer space at temperature $T_\infty = 0$. Heat loss from the end of the fin can be neglected, and the base of the fin is at T_b. Any radiant exchange with the base surface is negligible also. The fin surface can be considered gray with emissivity ε. Derive the governing equation in dimensionless form for the temperature distribution along the fin. Indicate the boundary conditions used. State the integration to be done to arrive at the temperature distribution $T(x)$.

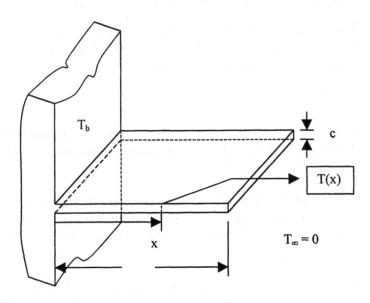

Figure for Prob.14.3.

14.4. A transparent gas flows into and out of a black circular tube of length L and diameter D. The gas has a mean velocity u_m, specific heat at constant pressure c_p and density ρ. The wall of the tube is thin, and the outer surface is insulated. The tube wall is heated electrically and a uniform input of heat is provided per unit area, per unit time. Determine the local wall temperature distribution along the tube length. Assume that the convective heat transfer coefficient h between the gas and the inside of the tube is constant.

14.5. A product is heated in an oven to an equilibrium temperature T_e, which is less than T_o, the temperature of the interior oven walls. The product has the shape of a cylinder with length greater than the diameter, D = 4 cm. The emissivity of the product is ε = 0.75 and T_o = 700 K. Nitrogen at atmospheric pressure flows over the product with a velocity V = 2 m/s, and at a temperature Tg = 300K. Calculate the equilbrium temperature T_e of the product. Use the correlation $\bar{h} = \dfrac{k}{D} 0.26 \, Re_D^{0.6} \, Pr^{0.37}$ for the forced convection of the nitrogen over the product; $v = 20.78 \times 10^{-6}$ m^2/s, k = 0.0293 W/m.K and Pr = 0.711 for nitrogen at the temperature and pressure considered.

REFERENCES

1. E M Sparrow, E R G Eckert, T F Irvine. The Effectiveness of Radiating Fins with Mutual Irradiation. J Aerospace Sc, 28: 763-772, 1967.

2. P L Donoughe and N B Livingood. Exact Solutions of Laminar Boundary-layer Equations with Constant Property Values for Porous Wall with Variable Temperature. NASA Technical Report 1229, 1958.

3. R D Cess. The Effect of Radiation upon Forced Convection Heat Transfer. Applied Science Research, Section A,10, pp. 430-438, 1962.

4. A R Shouman. An Exact General Solution for the Temperature Distribution and the Radiation Heat Transfer Along a Constant Cross-Sectional Area Fin. Paper No. 67-WA/HT-27, ASME, Nov 1967.

5. R Siegel and J R Howell. Radiation Heat Transfer, 3rd ed. Washington D. C.: Taylor and Francis, 1992.

Multimode Heat Transfer

Situations where radiation and conduction occurred,
Integrals and derivatives of temperature are involved
Heat transfer in satellite and spacecraft structures
Heat transfer through walls of a vacuum flask structure.

Situations where radiation and convection occurred,
Integrals and differences of temperature are involved
Heat transfer along the copper tubes in a boiler,
Heat transfer in lakes, seas and environmental waters.

Situations where radiation, conduction and convection occurred,
Integrals, derivatives and differences of temperature
Heat transfer in the extrusion of commercial wires,
Heat transfer of moving fluid with concentric electric heater.

K.V. Wong

Appendix A

Bessel Functions

New functions are sometimes defined as a solution to differential equation, and simply named after the differential equation itself. It is the purview of the mathematician to understand the properties of these functions so that they can be used confidently in numerous other applications. The Bessel function is of this kind, the solution of a differential equation that occurs in many applications of engineering and physics, including heat transfer.

Bessel functions are defined as functions that produce solutions to the class of nonlinear differential equations represented by:

$$x^2 y'' + xy' + (x^2 - n^2)\, y = 0$$

Generally, solutions to differential equations only arise after lengthy calculations using infinite series to find recursions and patterns in the solution. The clever mathematician therefore avoids the lengthy calculation by depending on methods that not only give the solution to the differential equation, but also aid in the understanding of its properties. This way the differential equation can provide essential information about the system in question without actually being solved. One procedure to accomplish this is to find an integral that gives the function; Bessel used this procedure for his functions. A second procedure is to use recurrence formulas that relate functions belonging to different parameters.

The functions $D_n(x)$, which are written as D_n for simplicity, may be defined for any real n by these recurrence relations:

$$D_{n-1} + D_{n+1} = \frac{2n}{x}\, D_n$$

$$D_{n-1} - D_{n+1} = 2\, \frac{dD_n}{dx}$$

From these relations, D_n satisfies the differential equation

$$x^2 D_n'' + x D_n' + (x^2 - n^2) D_n = 0,$$

which is Bessel's equation. Relton, 1946 [1], called the D_n Cylinder functions, but they are also Bessel functions because they satisfy Bessel's equation (Calvert, 2001[2]). Relton, 1946 [1], pointed out that the coefficient of D_n'' shows that the function can touch (i.e., be tangential to) the x-axis only at $x = 0$, because this is the only zero of the coefficient of the second derivative.

The differential equation is the same for -n as for n, so D_{-n} is also a solution, and is generally different from D_n. Thus, a general solution of Bessel's equation with two arbitrary constants is

$$y = A \, D_n(x) + B \, D_{-n}(x).$$

However, when n is integral, from the recurrence relations: $D_{-n} = (-1)^n D_n$. This implies that D_{-n} is linearly dependent of D_n, and a second linearly independent solution to Bessel's equation must still be found.

A second-order differential equation may be changed to normal form by the substitution $y = qw$, selecting $q(x)$ so that the y' term disappears. Starting from

$$y'' + a(x)y' + h(x)y = 0,$$

one obtains

$$w'' + H(x)w = 0,$$

where $q = \exp\{-(1/2)I[a(x)]\}$ (I is the indefinite integral), and

$$H(x) = h(x) - \frac{1}{2}\frac{da}{dx} - \left(\frac{a}{2}\right)^2 .$$

Using this in Bessel's equation,

$$w'' + \left[1 + \frac{1 - 4n^2}{4x^2} \right] w = 0, \text{ where, } D_n(x) = w(x)/x^{1/2}.$$

$w(x)$ is a function similar to a sine or cosine, with a period that slowly shortens, eventually becoming 2π. Except for regions close to the origin, this oscillatory characteristic provides a description of the general behavior of Bessel functions. When n = 1/2, the familiar linear second order differential equation satisfied by the sine and cosine is obtained.

The series solution of Bessel's differential equation will provide facts about the Bessel function's behavior near the origin. The series solution is also used to generate the standard function, and tabulated values of Bessel functions. The resulting series solution is

$$J_n = \left(\frac{x^n}{2^{2n} \Gamma(n+1)} \right) \left\{ 1 - \frac{x^2}{8(2n+2)} + \frac{x^4}{128(2n+2)(2n+4)} - \cdots \right\}.$$

This function is called the Bessel function of the first kind of order n. $\Gamma(n+1)$ is the gamma function of n +1. From this, it can be seen that when n is a positive integer, $J_n(x)$ starts off as x^n. When n = 0, $J_0(0) = 1$. When n is an integer, $J_n(0) = 0$. In all other cases, J_n is infinite at the origin. In many physical problems the solution to Bessel's equation must be defined (finite) and well-behaved at the origin, which eliminates all solutions except for those with integer values of n. It can also be shown that J_n satisfies the same recurrence relations as D_n, verifying that the functions are the same.

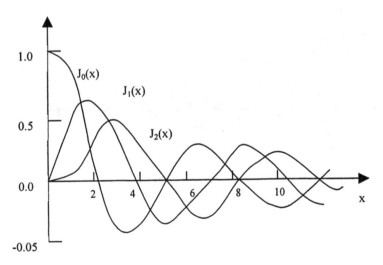

Figure A.1 Bessel functions of the first kind.

When n = 1/2, the result is $J_{1/2}(x) = (2/\pi x)^{1/2}\sin x$. From the recurrence relations, it can be found that $J_{-1/2} = (2/\pi x)^{1/2}\cos x$. The recurrence relation $J_{n+1} = (2n/x)J_n - J_{n-1}$ can the be employed to discover all the other functions of half-integral index. Numerical calculations using recurrence equations are easily impaired by roundoff error, since the error can propagate through successive recurrences.

Some Bessel functions of the first kind are shown in Fig. A.1 to illustrate their behavior. The first five zeros of J_0 are 2.4048, 5.5201, 8.6537, 11.7915, and 14.9309. The interval between the last two is 3.1394, a value near π. Note that as x increases, the absolute value of the maxima and minima decrease. The larger roots are approximately (m - 1/4)π, where m is the number of the root. For n > 1/2, the roots approach π from above instead of from below. The first positive zero of J_n is greater than n, and increase steadily with n. The first zeros are 2.405, 3.832, 5.136, 6.380, 7.588, and 8.771, for n = 0 to 5. The zeros must be found by calculation.

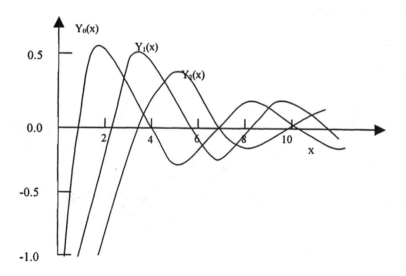

Figure A.2   Bessel functions of the second kind.

When a and b are two separate roots of J_n, the functions $J_n(ax)$ and $J_n(bx)$ are orthogonal to each other over the interval x = 0,1 with weight function x. The implication is that when their product is multiplied by x and integrated from 0 to 1, the result is zero. If b = a, the result is not zero, but $J_n'^2(a)/2 = J_{n+1}^2$ (a)/2. For instance, when n = 1/2, the orthogonality of the functions sin (nπx) in the interval (0,1) is proved. A function can be expanded in a series of $J_n(a_i x)$ corresponding to the zeros of J_n in the same way as a Fourier series is created, using orthogonality to find the coefficients one at a time.

In the box, two additional methods to obtain Bessel functions are summarized. The generating function relates Bessel functions to the exponential, Spiegel, 1971[3]. This relation is useful for obtaining properties of the Bessel function for integral n. Recursions of the Bessel functions are generally derived this way. Bessel's integral relates Bessel and trigonometric function.

Generating Function

$$\exp[\frac{x}{2}\left(t - \frac{1}{t}\right)] = \sum_{-\infty}^{\infty} t^n J_n(x)$$

Bessel's Integral

$$J_n(x) = \frac{1}{\pi} \int_0^\pi \cos(n\theta - x\sin\theta)d\theta$$

When n is an integer, a second solution linearly independent of J_n has to be found. For n = 0, such a function is Neumann's,

$$Y_0(x) = J_0(x) \log x + \{(x/2)^2 - (3/2)(x/2)^2/(2!)^2 + ...\}.$$

This function is tabulated, just like $J_0(x)$, and has zeros that interlace with those of $J_0(x)$. At x = 0, it goes to negative infinity. The general solution of order zero is then

$$y = A J_0(x) + B Y_0 (x).$$

In practice, the important fact is that there are two independent solutions, J and Y, and Y is infinite at the origin. These Y's are called Bessel functions of the second kind. Some of these functions are shown in Figure A.2.

An imaginary argument is also possible for Bessel functions. When this occurs, they become the modified Bessel functions I and K. This substitution changes them from oscillatory to monotonic, as in the analogous case of the trigonometric functions. The modified Bessel function of the first is defined as

$$I_n(x) = (i)^{-n} J_n(ix) = e^{-n\pi i/2} J_n(ix).$$

With proper adjustments due to the factor i, these functions follow recurrence relations similar to those for J_n. $I_0(0) = 1$, $I_n(0) = 0$, and for n >0 the modified Bessel function is monotonically increasing. The second solution, K, does not follow the same recurrence relations as I. Macdonald's definition of the modified Bessel function of the second kind is

$$K_n = \frac{\pi}{2} \left[\frac{I_{-n} - I_n}{\sin(n\pi)} \right].$$

K(x) is infinite at x = 0, and decreases similar to a rectangular hyperbola, approaching the x-axis as an asymptote. In fact,

$$K_{1/2}(x) = K_{-1/2}(x) = (\pi/2x)^2 e^{-x}.$$

The corresponding relations for I, give sinh for n = 0.5, and cosh for n = -0.5. The I_n are the coefficients in the Fourier expansion of $e^{-kr \cos\theta}$, which is $\Sigma\, I_n(kr) \cos n\theta$, where the summation sign on n goes from minus infinity to plus infinity.

The function $y = x^\alpha J_n(\beta x^\gamma)$ is a solution of the equation

$$y'' + \{(1 - 2\alpha)/2\}y' + \{(\beta\gamma x^{\gamma-1})^2 + (\alpha^2 - n^2\gamma^2)/x^2\}y = 0.$$

If the first term in the face brackets is negative, the solution has I_n instead of J_n. This will aid in recognizing equations whose solution can be expressed in terms of Bessel functions that is found in applications.

In this appendix, the essential properties of Bessel functions that are required in physical applications have been discussed. There are many books and articles on Bessel functions, and tables and graphs of their values and properties. There are also several good books giving the essentials of Bessel functions for scientists and engineers. Every textbook on hydrodynamics, elasticity, electromagnetism and vibrations will have examples of the use of these functions. Bowman, 1958 [4], is recommended.

The next table lists the asymptotic formulas for Bessel functions for large values of x. These are useful for problems involving cylindrical geometry in heat transfer.

List A.1   Asymptotic formulas for Bessel functions.

When the values of x are large, the following asymptotic formulas for the Bessel functions apply:

$$J_v(x) \approx \sqrt{\frac{2}{\pi x}} \cos\left(x - \frac{\pi}{4} - \frac{v\pi}{2}\right)$$

$$Y_v(x) \approx \sqrt{\frac{2}{\pi x}} \sin\left(x - \frac{\pi}{4} - \frac{v\pi}{2}\right)$$

$$I_v(x) \approx \frac{e}{\sqrt{2\pi x}}$$

$$K_v(x) \approx \sqrt{\frac{\pi}{2x}} e^{-x}$$

REFERENCES

1. F E Relton. Applied Bessel Functions. London: Blackie and Son, 1946.

2. J B Calvert. Essentials of Bessel Functions, Jan. 2001, www.du.edu.

3. M R Spiegel. Advanced Mathematics for Engineers and Scientists. Schaum's Outline Series. New York: McGraw-Hill, 1971.

4. F Bowman. Introduction to Bessel Functions. New York: Dover, 1958.

Appendix B

Physical Constants and Thermophysical Properties

Universal Gas Constant:

$$R = 8.314 \times 10^{-2} \ m^3.bar/(kmol.K)$$

$$= 8.315 \ kJ/(kmol.K)$$

Stefan-Boltzmann Constant:

$$\sigma = 5.670 \times 10^{-8} \ W/m^2.K^4$$

Standard Atmospheric Pressure:

$$P = 101{,}325 \ N/m^2 = 0.1013 \ MPa$$

Speed of Light in Vacuum:

$$c_o = 2.998 \times 10^5 \ km/s$$

Gravitational Acceleration at Sea Level:

$$g = 9.807 \ m/s^2$$

Table B.1　Thermophysical properties of metals.

Metal	Temp. range in K	Density kg/m²	Thermal cond. W/(m.K)	Sp Heat, kJ/(kg.K)	Emissivity
Aluminum	100-600	2702	240	0.903	0.04-0.06(polished) 0.07-0.09 (foil) 0.76-0.82 (anoxidized)
Brass (70% Cu, 30% Zn)	373-573	8530	110	0.38	0.03-0.07(polished) 0.2-0.25 (foil) 0.45-0.55(oxidized)
Copper	300	8933	401	0.38	0.03-0.04(polished) 0.5-0.8(oxidized)
Gold	100-1000	19300	317	0.129	0.01-0.06(polished) 0.06-0.07(foil)
Iron (4% C, cast)	273-1273	7260	35-52	0.42	0.2-0.25(polished) 0.55-0.65(oxidized)
Lead	273-573	11370	30-35	0.13	0.05-0.08(polished) 0.3-0.6(oxidized)
Mercury	273-573	13400	8-10	0.125	0.1-0.12
Nickel	600-1200	8900	59-93	0.45	0.09-0.17(polished) 0.4-0.57(oxidized)
Platinum	273-1273	21450	72	0.133	0.03-0.05(polished) 0.07-0.11(oxidized)
Silver	273-673	10520	360-410	0.23	0.01-0.03(polished) 0.02-0.04(oxidized)
Steel (1% Cr)	273-1273	7860	33-62	0.46	0.17-0.3(polished)
Steel (1% C)	273-1273	7800	28-43	0.47	0.17-0.3(polished)
Tin	273-473	7300	57-65	0.23	0.04-0.07(polished)
Tungsten	273-1273	19350	76-166	0.13	0.04-0.08(polished) 0.2-0.4(filament)
Zinc	273-673	7140	116	0.39	0.04-0.05(polished) 0.2-0.3 (galvanized)

Table B.2   Thermophysical properties of common materials.

Material	Temp. range in K	Density kg/m^3	Thermal conductivity W/(m.K)	Specific Heat,kJ/(kg.K)	Emissivity
Asbestos	373-1273	470-570	0.15-0.22	0.816	0.93-0.97
Asphalt	273-300	2115	0.062	0.92	0.85-0.93
Brick,common	373-1273	1920	0.72	0.84	0.90-0.95
Clay	273-473	1.46	1.3	0.88	0.91
Concrete	273-473		0.6-1.1		0.94
Cotton	273-300	80	0.06	1.30	
Fiberglass	273-300	32	0.038	0.7	
Glass (pane)	273-873	2200	1.4	0.75	0.90-0.94
Granite	273-300	2630	2.79	0.78	0.80-0.95
Ice	273	920	1.88	2.04	0.95-0.98
Paper	273-300	930	0.18	1.34	0.92-0.97
Wood (pine)	273-300	440	0.11	2.8	0.90

Table B.7 Thermophysical properties of various materials

Index

Absorptivity
  radiation, 281–284
ADI methods, 261–262, 273
Alternating direction implicit method,
      261–262, 273
Aluminum
  thermophysical properties, 398
Annular disk
  portion of ring element, 366
Asbestos
  thermophysical properties, 399
Asphalt
  thermophysical properties, 399
Asymptotic formulas
  Bessel functions, 395
Azimuthal angle, 278

Backward-difference form
  first derivative, 84
Backward finite difference form, 108
Backward time central space (BTCS)
      implicit scheme
  numerical analysis in convection,
      251
Bernoulli's equation
  steady flow over flat plate, 253–254
Bessel functions, 389–395, 390
  asymptotic formulas, 395
  definition, 389
Bessel functions of first kind, 391
Bessel functions of second kind, 392
Bessel's equation, 390
  series solution, 391
Bessel's integral, 393
Bidiagonal systems
  numerical analysis in convection,
      251

Biot number, 22–23, 25
Blackbody, 276–281, 278, 286
  Boltzmann constants, 278
  emissive flux, 279
  emissive power, 280
  incident radiation, 281
  radiation intensity, 278, 279
Black surfaces, 363
Boltzmann constants
  blackbody, 278
Boundary
  with convective heat transfer, 88–89
  insulated, 90–91
Boundary conditions
  constant surface temperature
    method of separation of
        variables, 207–208
  eigenfunctions and eigenvalues, 62
  finite difference formulation, 92–
      101
Boundary layer concept
  for free convection, 221–224
Boundary layer equations, 143–146
  numerical analysis in convection,
      253–259
  example, 256
Boundary layer momentum equation
  laminar flow, 179
Boundary layer with uniform heat flux
  free convection
    similarity solution, 230–233
Boundary layer with uniform
      temperature
  natural convection, 225–230
Bouyancy-driven motion, 5
Bouyancy force
  free convection, 224

[Bouyancy force]
per unit element
momentum integral equation for
natural convection, 234
Brass
thermophysical properties, 398
Brick
thermophysical properties, 399
BTCS implicit scheme
numerical analysis in convection,
251
Burgers equation, 244
viscous, 243, 244

Central-difference approximation
second derivative, 85
Central-difference form
first derivative, 85
Circular tube thermal-entry-length
with hydrodynamically fully
developed laminar flow, 205–
211
Clay
thermophysical properties, 399
Composite medium, 33–35
parallel slabs, 33–34
Concrete
thermophysical properties, 399
Conduction, 1–4
example, 4
in nonparticipating media, 364
numerical analysis, 83–124
in plane wall, 2
and radiation, 357–368, 377–383
through elemental volume, 13
through slab
one fluid to another fluid, 32–
33
Conductive heat flux
for isotropic and homogeneous
media, 13
Conservation of mass equation
continuity equation, 125–126

Constant surface temperature
boundary condition
method of separation of
variables, 207–208
infinite-service solution function,
208, 210
Constant wall heat flux
energy integral equation, 212–215
Constant wall temperature
laminar fluid flow, 215–216
Continuity
convection equations, 245
incompressible fluid, 369
Continuity equation
conservation of mass equation,
125–126
control volume, 125
in different coordinate systems, 137
for two-dimensional steady
incompressible flow
rectangular coordinates, 126
Contour integrals, 299
Control mass system, 104
Control volume
for continuity equation, 125
for integral energy analysis
of laminar boundary layer, 155
Convection, 1, 4–6
classification, 4–5
equations, 125–148, 245–247
continuity, 245
example, 6
forced and free, 5
FTBCS explicit scheme, 248
FTCS explicit scheme, 248
with incompressible flow, 259–261
integral forced, 191–220
numerical algorithms, 247–253
DuFort-Frankel explicit, 248–
249
Nusselt number, 237
vertical surface, 229
and radiation, 9–11, 368–376, 377–

383
example, 10
in two-dimensional porous medium, 243, 262–269
Convective
heat transfer, 9–10
Convergence
finite difference numerical method, 117–119
Copper
thermophysical properties, 398
Copper-constant thermocouple
temperature, 382–383
Cotton
thermophysical properties, 399
Couette flow, 191–197
Crank-Nicholson method
finite difference equation, 108–109
incompressible laminar boundary layer equations, 257–259
FORTRAN code, 271–273
Critical radius of insulation, 43–44
example, 44
Cubic representation
of velocity profile, 157
Curves for emissivity, 363
Cylinder (one-dimensional cylindrical coordinates), 28–29
Cylindrical
coordinate system
continuity equations, 137
energy equation, 139
momentum equations, 138

Differential equation
finite difference formulation, 92–101
Differential Equation of Heat Conduction, 13–15
Diffuse reflectors, 359
Diffuse view factor
algebra, 308
arrangement of surfaces, 310

[Diffuse view factor]
coordinates, 291–293
determination by contour integration, 295–304
example, 301–304
between elemental surface and infinitely long strip, 307–308
long enclosures, 319–340
examples, 321–340
properties, 293
between surfaces, 296–300, 308–312
elemental, 291–292
of long enclosure, 320
Dimensionless adiabatic wall temperature, 186
Dimensionless boundary-layer equations, 146
Dimensionless coefficient of heat transfer, 186
Dimensionless conduction numbers, 25
Dimensionless continuity equation, 141
Dimensionless convection numbers, 148
Dimensionless heat conduction numbers, 21–23, 25
Dimensionless heat generation, 22
Dimensionless numbers, 140–143
Dimensionless parameters, 22
Dimensionless temperature distribution
for laminar free convection, 227
momentum integral equation for natural convection, 234
Dimensionless velocity distribution
internal forced convection, 199
for laminar free convection, 226
profile
quartic polynomial, 164
Direction of propagation, 277
Dirichlet boundary conditions, 17

Discretization error
central-difference form, 85
Drag coefficient
by fluid flowing, 152
DuFort-Frankel explicit
numerical algorithms in convection, 248–249

Eckert number, 141, 142, 143
internal forced convection, 195
Eigenfunctions
boundary conditions, 62
definition, 61
problem, 74
Eigenvalue
boundary conditions, 62
problem
definition, 61
Electromagnetic wave spectrum, 275–276
Elemental view factor, 307
Emissivity
curves for, 363
radiation, 281–284
Energy conservation equation
incompressible fluid, 369–370
Energy equations
convection, 131, 246
external forced convection, 184–186
integral method analysis, 154–156
Newtonian fluid
coordinate systems, 138–139
Energy integral equation
constant wall heat flux, 212–215
Euler equation
numerical analysis in convection, 247
Explicit method, 102–107
boundary layer
numerical analysis in convection, 256
example, 106–107

External forced convection, 149–188
Extrusion of thin wire, 377

FDE. *See* Finite difference equation (FDE)
Fiberglass
thermophysical properties, 399
Finite difference
of derivatives, 83–85
approximation, 84
2-D unsteady-state problems
example, 112
rectangular coordinates, 109–113
formulation
differential equation, 92–101
grid
for cylindrical coordinates, 114
method
applied in cylindrical coordinates, 113–115
truncation errors and round-off errors, 115–117
numerical method
convergence, 117–119
representation
of boundary conditions, 88–92
Finite difference equation (FDE)
Crank-Nicholson method, 108–109
for 2-D rectangular steady-state conduction, 86–87
for 1-D unsteady-state conduction
rectangular coordinates, 101–109
implicit methods, 108
numerical algorithms in convection, 248
solution, 92–101
Gaussian, 93–96
matrix, 96–98
relaxation, 99–100
Finite difference formulation
Gaussian elimination method, 93–96

Finite differencing
  of energy equation, 258–259
Fins, 59
First kind (Dirichlet) boundary
    conditions, 17
First law of thermodynamics, 131
First-order backward difference
    approximation
  BTCS, 251
Flow through rectangular porous
    medium, 262
Fluid flow patterns, 4
Forced and free (or natural)
    convection, 5
Forward-difference approximation,
    102–107
  example, 106–107
Forward time backward central space
    (FTBCS) explicit scheme
  numerical algorithms in convection,
    248
Forward time central space (FTCS)
    explicit scheme
  numerical algorithms in convection,
    248
Fourier's conduction law, 5
Fourier's equation, 15
Fourier's heat conduction law, 228
  for isotropic and homogeneous
    media, 13
Fourier's law, 1
  application
    to plane wall, 3
Fourier's law and Newton's law of
    cooling
  relationship, 5
Fourier's number, 22–23, 25
Free convection
  inertial forces, 224
  Nusselt number, 224
Free (or natural) convection, 5, 221–
    242
  definition, 221

Frictional work
  by surface forces, 132
FTBCS explicit scheme
  numerical algorithms in convection,
    248
FTCS explicit scheme
  numerical algorithms in convection,
    248
Fully implicit and Crank-Nicholson
    methods
  incompressible laminar boundary
    layer equations, 257–259
  FORTRAN code, 271–273

Gaussian elimination method, 124
  finite difference formulation, 93–96
  example, 95–96
General heat conduction equation, 13–
    25
Generating function
  Bessel functions, 393
Glass
  thermophysical properties, 399
Gold
  thermophysical properties, 398
Granite
  thermophysical properties, 399
Grashof number
  free convection, 224
Gravitational acceleration at sea level, 397
Graybody, 286
  assumption, 284
Grid points
  convection boundary, 88, 90
  insulated boundary, 90
  irregular boundary, 91
  rectangular grid system, 110
  for rectangular grid system, 86

Heat addition by conduction, 131–137
Heat conduction equation
  with variable thermal conductivity
    in coordinate system, 16

Heat dissipation
  by radiation, 362
Heat energy, 1
Heat flux (heat flow), 67–71
  calculation, 27–28
  energy integral equation, 212–215
  example, 68–69
  incident
    radiation, 281–284
  from surface, 228
Heat transfer
  fundamentals, 1–12, 12
  and Nusselt number, 228–229
  and velocity distribution
    in hydrodynamically and
      thermally developed laminar
      flow, 197–205
Hemispherical reflectivity, 281–282
Higher order terms (H.O.T.), 115–116
Hollow cylinder, 29
Hollow sphere, 30
Hottel's crossed-string theorem
  practical examples, 321–340
  stated, 321
Hottel's theorem, 355
Howarth solution
  for laminar flow, 183
Hydrodynamic and thermal boundary
    layers
  on flat plate, 157–179
    example, 161–179
Hydrodynamic boundary layer, 143

Ice
  thermophysical properties, 399
Implicit methods
  finite difference equation, 108
Incident radiation
  blackbody, 281
  translucent body, 283
Incompressible boundary layer
    equations
  numerical analysis, 243

Incompressible flow
  with constant properties, 243
  convection with, 259–261
Incompressible fluid
  continuity, 369
Inertial forces
  free convection, 224
Infinite-series function
  for circular tube, 208
Infinite-series-solution function
  for circular tube, 210
Initial and boundary conditions, 16–17
Insulated boundary, 90–94
Insulation
  critical radius of, 43–44
Integral energy analysis
  control volume
    of laminar boundary layer, 155
Integral forced convection, 191–220
Integral method analysis, 189
  for energy equation, 154–156
Integral method of solution
  free convection, 234–238
    example, 238
Integrodifferential equation, 360
Internal flows, 220
Iron
  thermophysical properties, 398
Irregular boundaries, 91–92

Kirchhoff's law of radiation, 284–
    285

Laminar boundary layer
  control volume
    for integral energy analysis, 155
  flow
    along flat plane, 368
Laminar flow
  in conduits, 197
  flat plate, 179–184
Laminar fluid flow
  constant wall temperature, 215–216

Laminar steady two-dimensional free
      convection
   no viscous dissipation of energy,
      222
Laplace equation, 15
Laplacian of temperature
   in three coordinate systems, 16
Lead
   thermophysical properties, 398
Linear Burgers equation, 244
   BTCS, 251–252
Longitudinal plate fin, 359
Long two-dimensional fin
   with convective boundary
      conditions, 72

MacCormack explicit
   numerical algorithms in convection,
      249–250
MacCormack implicit
   numerical algorithms in convection,
      250–251
Mass of differential volume element, 127
Matrix inversion
   finite difference formulation, 96–98
   example, 97
Mean liquid temperature
   as function of distance, 380–381
Mercury
   thermophysical properties, 398
Metals
   thermophysical properties, 398
Momentum equations
   conservation of mass, 126–131
   Newtonian fluid
      coordinate systems, 138
Momentum integral method of
   analysis, 149–154
   example, 152–153
Multimode heat transfer, 387

Natural convection, 5, 221–242, 241
   definition, 221

Navier-Stokes equation, 243, 247
Net radiative heat flux, 362
Net rate of heat dissipation by
      radiation, 362
Neumann boundary conditions, 17–18
Neumann's function, 393
Newton's law of cooling, 4, 5, 142
Newton's second law of motion, 126
Nickel
   thermophysical properties, 398
Nonhomogeneous boundary condition,
      70–71, 75
Nonlinear Burgers equation, 244
Nonmetallic material
   thermophysical properties, 399
Numerical algorithms
   numerical analysis in convection,
      247–253
Numerical analysis
   conduction, 83–124
   convection, 243–273
Nusselt number, 143
   distribution of temperature in
      boundary layer, 373–374
   free convection, 224, 237
      vertical surface, 229
   uniform wall heat flux
      constant wall temperature, 233
   values, 210

One-dimensional Cartesian
      coordinates, 27–28
One-dimensional extended surfaces,
      45–56
   fin efficiency, 54–55
      example, 55
   fin equation, 46–47
   heat transfer from finned surface, 56
   temperature distribution and heat
      flow, 47–54
One-dimensional fin equations, 46–47
   for fins with variable cross section,
      47

One-dimensional spherical
    coordinates, 29–30
One-dimensional steady-state heat
    conduction, 27
Orthogonal expansion technique, 61
Ostrach's computations
    natural convection
        vertical surface, 229

Paper
    thermophysical properties, 399
Parabolic-type equation
    two-dimensional porous medium, 263
Partial differential equation (PDE)
    numerical algorithms in convection,
        248
Partially blocked view
    between parallel strips, 323
PDE
    numerical algorithms in convection,
        248
Peclet number, 142
    two-dimensional porous medium,
        263
Perpendicular rectangles
    view factors, 305
Physical constants and thermophysical
    properties, 397
Planck and Boltzmann constants
    blackbody, 278
Platinum
    thermophysical properties, 398
Poisson's equation, 15
Porous media, 243
Prandtl number, 141
    free convection, 224
    integral method of solution for
        natural convection, 237
    internal forced convection, 195
    natural convection, 227–230

Quartic polynomial
    dimensionless velocity profile, 164

Radiant energy per unit time, 277
Radiant heat flux incident
    radiation, 281–284
Radiating longitudinal plate fins, 358
Radiation, 1, 6–8, 9–11
    absorptivity, 281–284
    basic relations, 275–286
    boundary condition
        flow along flat plate, 368
    and conduction, 357–368
        example, 365–368
        in nonparticipating media, 364
    conduction and convection, 377–
        383
        example, 377–383
    and convection, 368–376
        example, 375–376
    effects, 44–45
    emissivity, 281–284
    example, 8, 10
    exchange
        long enclosures, 319–355
    intensity, 277
        and blackbody, 276–281
        notation, 276
    modes of heat transfer, 357–387
Radiative effectiveness, 363
Radiative equilibrium, 284
Radiative exchange
    in nonparticipating medium, 287–
        318
    between two differential area
        elements, 287–291
        example, 288
Radiative heat flux, 360
Radiative heat gain, 360
Radiative heat transfer, 9–10
Radiator
    for space vehicle, 358
Radiosity equation, 360–361
Rate of change of momentum
    momentum integral equation for
        natural convection, 234

Rayleigh number
  free convection, 224
    similarity solution, 230–233
Real surface
  radiation, 281–284
Reciprocity
  between diagonally opposing
    rectangles, 294
Reciprocity relations
  view factor, 304–305
Reciprocity theorem
  diffuse view factor, 292
Rectangular coordinates
  continuity equations, 137
    for two-dimensional steady
      incompressible flow, 126
  energy equations, 138
  momentum equations, 138
Rectangular duct thermal-entry length
  with hydrodynamically fully
    developed laminar flow, 211–
    216
Reflectivity
  radiation, 281–284
  spectral hemispherical, 281–282
Relaxation method
  finite difference formulation, 99–
    101
    example, 100
Reynolds number, 142, 144
Ring element
  annular disk, 366
Robin of mixed boundary conditions, 18
Round-off errors
  finite difference methods, 115–118

Second kind (Neumann) boundary
    conditions, 17–18
Second-order central difference
    approximation
  BTCS, 251
Separation into simpler problems, 70–
  71

Separation of variables, 61, 81
  steps, 71–72
Shear stress
  momentum integral equation for
    natural convection, 234
Silver
  thermophysical properties, 398
Similarity solution
  external forced convection, 179–
    186
  natural convection, 225–230
Slab (one-dimensional Cartesian
    coordinates), 27–28
Slug flow assumption, 243
Solid cylinder, 28
Solid sphere, 29
Spatial boundary conditions
  classification, 17
Spatial derivatives
  BTCS, 251
Spectral blackbody emissive flux, 279
Spectral blackbody emissive power, 280
Spectral blackbody radiation intensity,
  278
Spectral hemispherical reflectivity,
  281–282
Spectral radiant heat flux incident
  radiation, 281–284
Spectral radiation intensity, 277
Speed of light in vacuum, 397
Sphere (one-dimensional spherical
  coordinates), 29–30
Stability
  finite difference numerical method,
    117–119
Standard atmospheric pressure, 397
Steady-state heat conduction
  in rectangular region, 64–67
  in two-dimensional fin, 72–75
Steady-state heat conduction problem
  example, 19
Steel
  thermophysical properties, 398

Stefan-Boltzmann constant, 6, 397
    blackbody, 279
Stefan-Boltzmann law, 6
Stokes theorem
    determine diffuse view factor, 296
    direction of circulation, 295
    stated, 295
Stream function
    natural convection, 225
Sturm-Liouville problem, 74
Superinsulation, 357
Surface integrals, 299
Surface stresses
    on volume element, 129

Taylor series expansion, 115–116
Temperature-dependent thermal
        conductivity, 19–21, 37–42
    one-dimensional with no heat
        generation, 37–38
        example, 38
    Poisson's equation with heat
        generation, 39–42
        example, 42
Temperature difference, 1
    Nusselt numbers
        free convection, 232–233
Temperature distribution
    for different Prandtl numbers, 186
    incompressible fluid, 369
    internal forced convection, 193–194
    for moderate velocities, 200–205
Temperature distribution and heat flow
    fins with convection at tip, 52–54
        example, 54
    fins with negligible heat flow, 50–51
        example, 51
    fins of uniform cross section, 47–54
    long fin, 47–49
Temperature profiles
    for constant heat rate and constant
        surface temperature, 205
    thermal-entry region, 209

Temperature variations
    thermal-entry region, 211
Thermal boundary conditions
    momentum integral equation for
        natural convection, 234
Thermal boundary layer
    development, 212
Thermal conductivity, 1
    calculations
        in solid, 36–37
Thermal contact resistance, 35–37
Thermal diffusivity, 20
Thermal-entry length, 205
    infinite-service solution function,
        208, 210
Thermal radiation, 6
    definition, 275
Thermal resistance, 30–32
    hollow cylinder, 30–31
    hollow sphere, 31–32
    slab, 30
Thin wire
    extruded, 377
Third kind (Robin of mixed) boundary
        conditions, 18
Thomas algorithm
    FORTRAN code, 268
    for tridiagonal systems, 267–269
Time derivative
    BTCS, 251
Time-independent equations
    steady flow over flat plate, 254
Tin
    thermophysical properties, 398
Transient heat conduction
    in slab, 76–78
        example, 78
Translucent body
    incident radiation, 283
Transmissivity
    radiation, 281–284
Tridiagonal matrix
    inversion, 261

Tridiagonal system
  BTCS, 251
Truncation errors
  central-difference form, 85
  finite difference methods, 115–117,
    118
Tungsten
  thermophysical properties, 398
Two-body radiation problem
  example, 7
Two-dimensional computational
    module
  relaxation method, 100
Two-dimensional convection with
    incompressible flow, 261–262
Two-dimensional steady flow over a
    flat plate
  incompressible constant-property
    fluid
    equations, 253

Uniform heat flux
  boundary condition, 208–210
Universal gas constant, 397

Variable cross-section fin, 46
Vectorial coordinate system
  continuity equations, 137
Velocity
  computes both the x- and y-
    components, 256–257
  dimensionless
    distribution, 199, 226
    profile, 164
Velocity boundary conditions
  momentum integral equation for
    natural convection, 234

Velocity distribution
  in hydrodynamically and thermally
    developed laminar flow, 197–
    205
  incompressible flow, 192–193
  incompressible fluid, 369
Velocity profile
  cubic representation, 157
View factors, 318
  concept, 291–293
  heat exchanger, 336–340
  infinitely long semicylindrical
    surfaces, 322
  parallel pipes with opaque plate
    between, 342–346
  of perpendicular rectangles, 305
  relations between, 304–307
    example, 306
  solar field, 331–336
  between surface, 329–330
  between surfaces on sign, 327–328
Viscous Burgers equation, 243, 244

Wood
  thermophysical properties, 399

X-momentum equation
  convection equations, 245
  incompressible fluid, 369

Y-momentum equation
  convection equations, 245

Zinc
  thermophysical properties, 398
Z-momentum equation
  convection equations, 245

9 780367 395568